Dynamic Loading and Design of Structures

Dynamic Loading and Design of Structures

Edited by A. J. Kappos

Routledge
Taylor & Francis Group

LONDON AND NEW YORK

First published 2002
by Spon Press

2 Park Square, Milton Park, Abingdon, Oxfordshire OX14 4RN
52 Vanderbilt Avenue, New York, NY 10017

Routledge is an imprint of the Taylor & Francis Group, an informa business

First issued in paperback 2019

Copyright © 2002 Taylor & Francis

Typeset in 10/12 Garamond by Steven Gardiner Ltd

Library of Congress Cataloging in Publication Data
Dynamic loading and design of structures/edited by A. J. Kappos.
    p. cm.
  Includes bibliographical references.
  ISBN 0-419-22930-2 (alk. paper)
  1. Structural dynamics. 2. Structural design. I. Kappos, Andreas J.

TA654.D94 2001
624.1'7–dc21

                         2001020724

ISBN 978-0-419-22930-8 (hbk)
ISBN 978-0-367-39669-5 (pbk)

Contents

Contributors

M. K. Chryssanthopoulos University of Surrey, UK

G. D. Manolis Aristotle University of Thessaloniki, Greece

T. A. Wyatt Imperial College, London, UK

A. J. Kappos Aristotle University of Thessaloniki, Greece

T. Moan Norwegian Institute of Technology, Trondheim, Norway

A. Watson University of Bristol, UK

J. W. Smith University of Bristol, UK

D. Cooper Flint & Neal Partnership, London

Preface

The dynamical behaviour of civil engineering structures has traditionally been tackled, for design purposes, in an 'equivalent static' way, essentially by introducing magnification factors for vertically applied loads and/or by specifying equivalent horizontal loads. Today the availability of software able to deal explicitly with dynamic analysis of realistic structures with many (dynamic) degrees of freedom, as well as the outcome of the valuable research carried out in the various fields included under 'Dynamics', make this type of analysis a part of everyday life in the design office.

There are also a number of good reasons why dynamical behaviour of buildings, bridges and other structures is now more of a concern for the designer than it used to be 20 or 30 years ago. One reason is that the aforementioned structures currently consist of structural members that are more slender than before, and lighter cladding made of metal and glass or composites rather than of brick walls. This offers a number of architectural advantages, but also makes these structures more sensitive to vibration, due to their reduced stiffness. From another perspective, the risk to environmental dynamic loads, like those from earthquakes, has increased due to the tremendous increase in urbanization witnessed in many countries subject to such hazards. Furthermore, the increased need for building robust and efficient structures inside the sea has also placed more emphasis on properly designing such structures against dynamic loading resulting from waves and currents.

Dealing with all, or even some of the aforementioned dynamic loads in an explicit way is clearly a challenge for the practising engineer, since academic curricula can hardly accommodate a proper treatment of all these loads. Furthermore, the lack of a book dealing with all types of dynamic loading falling within the scope of current codes of practice, makes the problem even more acute.

The main purpose of this book is to present in a single volume material on dynamic loading and design of structures that is currently spread among several publications (books, journals, conference proceedings). The book provides the background for each type of loading (making also reference to recent research results), and then focuses on the way each loading is taken into account in the design process.

An introductory chapter (Chapter 1) gives the probabilistic background, which is more or less common for all types of loads, and particularly important in the case of dynamic loads. This is followed by a chapter (Chapter 2) on analysis of structures for dynamic loading, making clear the common concepts underlying the treatment of all dynamic loads, and the corresponding analytical techniques.

The main part of the book includes Chapters 3–9, describing the most common types of dynamic loads, i.e. those due to wind (Chapter 3), earthquake (Chapter 4), waves (Chapter 5), explosion and impact (Chapter 6), human movement (Chapter 7), traffic (Chapter 8), and machinery (Chapter 9). In each chapter the origin of the corresponding dynamic loading is first explained, followed by a description of its effect on structures, and the way it is introduced in their design. The latter is supplemented by reference to the most pertinent code provisions and an explanation of the conceptual framework of these codes. All these chapters include long lists of references, to which the reader can make recourse for obtaining more specific information that cannot be accommodated in this book that encompasses all types of dynamic loading.

A final chapter (Chapter 10) deals with the more advanced topic of random vibration analysis, which nevertheless is indispensable in understanding the analytical formulations presented in some other chapters, in particular Chapters 3 and 5.

The book is aimed primarily at practising engineers, working in consultancy firms and construction companies, both in the UK and overseas, and involved in the design of civil engineering structures for various types of dynamic loads. Depending on the type of loading addressed, an attempt was made to present code provisions both from the European perspective (Eurocodes, British Standards) and the North American one (UBC, NBC), so the book should be of interest to most people involved in design for dynamic loading worldwide.

The book also aims at research students (MSc and PhD programmes) working on various aspects of dynamic loading and analysis. With regard to MSc courses, it has to be clarified that Loading is typically a part of several, quite different, courses, rather than a course on its own (although courses like 'Loading and Safety' and 'Earthquake Loading', do currently exist in the UK and abroad). This explains to a certain extent the fact that, to the best of the editor's knowledge, no comprehensive book dealing with all important types of dynamic loading has appeared so far. The present book is meant as a recommended textbook for several existing courses given by both Structural Sections and Hydraulics Sections of Civil Engineering Departments.

The contributors to the book are all distinguished scientists, rated among the top few in the corresponding fields at an international level. They come mostly from the European academic community but also include people from leading design firms and/or with long experience in the design of structures against dynamic loads.

Putting together and working with the international team of authors that was indispensable for writing a book of such a wide scope, was a major challenge and experience for the editor, who would like to thank all of them for their most valuable contributions. Some of the contributors, as well as some former (at

Imperial College, London) and present (at the University of Thessaloniki) colleagues of the editor have assisted with suggestions for prospective authors and with critical review of various chapters or sections of the book. A warm acknowledgement goes to all and each of them.

Chapter 1

Probabilistic basis and code format for loading

Marios K. Chryssanthopoulos

1.1 INTRODUCTION

In the last 30 years, practical probabilistic and reliability methods have been developed to help engineers tackle the analysis, quantification, monitoring and assessment of structural risks, undertake sensitivity analysis of inherent uncertainties and make rational decisions about the performance of structures over their working life. These tasks may be related to a specific structure, a group of similar structures or a larger population of structures built to a code of practice. Within a time framework, the structures may be at the design stage, under construction or in actual use. Hence, the methods may be required to back calculate performance and compare with earlier perceptions and observations, or to predict future performance in order to plan a suitable course of action for continued safety and functionality. Clearly, uncertainty is present through various sources and can propagate through the decision making process, thus rendering probabilistic methods a particularly useful tool.

The purpose of this chapter is to summarize the principles and procedures used in reliability-based design and assessment of structures, placing emphasis on the requirements relevant to loading. Starting from limit state concepts and their application to codified design, the link is made between unacceptable performance and probability of failure. This is then developed further in terms of a general code format, in order to identify the key parameters and how they can be specified through probabilistic methods and reliability analysis. The important distinction between time invariant and time variant (or time dependent) formulations is discussed, and key relationships allowing the treatment of time varying loads and load combinations are presented. In subsequent sections, an introduction to the theories of extreme statistics and stochastic load combinations is presented in order to elucidate the specification of characteristic, representative and design values for different types of actions.

This chapter is neither as broad nor as detailed as a number of textbooks on probabilistic and reliability methods relevant to structural engineering. A list of such books is given at the end of the chapter. The reader should also be aware of recent documents produced by ISO (International Organization for Standardization) and CEN(European Committee for Standardization)/TC250 code drafting committees, which provide an excellent up-to-date overview of reliability methods and their potential application in developing modern codes of practice (ISO, 1998; Eurocode 1.1 Project Team, 1996; European Standard, 2001). Finally, it is worth mentioning that many of the topics presented in this chapter have been discussed and clarified within the Working Party of the Joint Committee on Structural Safety (JCSS), of which the author is privileged to be a member. The present chapter draws from the JCSS document on Existing Structures (JCSS, 2001) and in particular the *Annex on Reliability Analysis Principles*, which was drafted by the author and improved by the comments of the working party members.

1.2 PRINCIPLES OF RELIABILITY-BASED DESIGN

1.2.1 Limit states

The structural performance of a whole structure or part of it may be described with reference to a set of limit states which separate acceptable states of the structure from unacceptable states. The limit states are generally divided into the following two categories (ISO, 1998):

- ultimate limit states, which relate to the maximum load carrying capacity;
- serviceability limit states, which relate to normal use.

The boundary between acceptable (safe) and unacceptable (failure) states may be distinct or diffuse but, at present, deterministic codes of practice assume the former.

Thus, verification of a structure with respect to a particular limit state may be carried out via a model describing the limit state in terms of a function (called the limit state function) whose value depends on all design parameters. In general terms, attainment of the limit state can be expressed as

$$g(\mathbf{X}) = 0 \tag{1.1}$$

where $\mathbf{X}$ represents the vector of design parameters (also called the basic variable vector) that are relevant to the problem, and $g(\mathbf{X})$ is the limit state function. Conventionally, $g(\mathbf{X}) \leq 0$ represents failure (i.e. an adverse state).

Basic variables comprise actions and influences, material properties, geometrical data and factors related to the models used for constructing the limit state function. In many cases, important variations exist over time (and sometimes space), which have to be taken into account in specifying basic variables. It will be seen in Section 1.4.1 that, in probabilistic terms, this may lead to a random process rather than random variable models for some of the basic variables. However, simplifications might be acceptable, thus allowing the use of random variables whose parameters are derived for a specified reference period (or spatial domain).

For many structural engineering problems, the limit state function, $g(\mathbf{X})$, can be separated into one resistance function, $g_R(\cdot)$, and one loading (or action effect) function, $g_S(\cdot)$, in which case equation (1.1) can be expressed as

$$g_R(\mathbf{r}) - g_S(\mathbf{s}) = 0 \qquad (1.2)$$

where $\mathbf{s}$ and $\mathbf{r}$ represent subsets of the basic variable vector, usually called loading and resistance variables respectively.

1.2.2 Partial factors and code formats

Within present limit state codes, loading and resistance variables are treated as deterministic. The particular values substituted into eqns (1.1) or (1.2) – the design values – are based on past experience and, in some cases, on probabilistic modelling and reliability calibration.

In general terms, the design value x_{di} of any particular variable is given by

$$x_{di} = \gamma_i x_{ki} \qquad (1.3a)$$

$$x_{di} = x_{ki}/\gamma_i \qquad (1.3b)$$

where x_{ki} is a characteristic (or representative) value and γ_i is a partial factor. Equation (1.3a) is appropriate for loading variables whereas eqn (1.3b) applies to resistance variables, hence in both cases γ_i has a value greater than unity. For variables representing geometric quantities, the design value is normally defined through a sum (rather than a ratio) (i.e. $x_{di} = x_{ki} \pm \Delta x$, where Δx represents a small quantity).

A characteristic value is strictly defined as the value of a random variable which has a prescribed probability of not being exceeded (on the unfavourable side) during a reference period. The specification of a reference period must take into account the design working life and the duration of the design situation.

The former (design working life) is the assumed period for which the structure is to be used for its intended purpose with maintenance but without major repair. Although in many cases it is difficult to predict with sufficient accuracy the life of a structure, the concept of a design working life is useful for the specification of design actions (wind, earthquake, etc.), the modelling of time-dependent material properties (fatigue, creep) and the rational comparison of whole life costs associated with different design options. In Eurocode 1 (European Standard, 2000), indicative design working lives range between 10 to 100 years, the two limiting values associated with temporary and monumental structures respectively.

The latter (design situation) represents the time interval for which the design will demonstrate that relevant limit states are not exceeded. The classification of design situations mirrors, to a large extent, the classification of actions according to their time variation (see Section 1.5). Thus, design situations may be classified as persistent, transient or accidental (ISO, 1998). The first two are considered to act with certainty over the design working life. On the other hand, accidental situations occur with relatively low probability over the design working life. Clearly, whether certain categories of actions (snow, flood, earthquake) are deemed to give rise to transient or accidental situations, will depend on local conditions. Typically, the load combination rules are not the same for transient and accidental situations, and also a degree of local damage at ultimate limit state is more widely accepted for accidental situations. Hence, the appropriate load classification is a very important issue in structural design.

In treating time varying loads, values other than the characteristic may be introduced. These so-called representative values are particularly useful when more than a single time varying load acts on the structure. For material properties a specified or nominal value is often used as a characteristic value, and since most material properties are assumed to be time independent, the above comments are not relevant. For geometrical data, the characteristic values usually correspond to the dimensions specified in design.

Partial factors account for the possibility of unfavourable deviations from the characteristic value, inaccuracies and simplifications in the assessment of the resistance or the load effect, uncertainties introduced due to the measurement of actual properties by limited testing, etc. The partial factors are an important element in controlling the safety of a structure designed to the code but there are other considerations to help achieve this objective. Note that a particular design value x_{di} may be obtained by different combinations of x_{ki} and γ_i.

The process of selecting the set of partial factors to be used in a particular code could be seen as a process of optimization such that the outcome of all designs undertaken to the code is in some sense optimal. Such a formal optimization process is not usually carried out in practice; even in cases where it has been undertaken, the values of the partial factors finally adopted have been adjusted to account for simplicity and ease of use. More often, partial factor values are based on a long experience of building tradition. However, it is nowadays generally accepted that a code should not be developed in a way that contradicts the principles of probabilistic design and its associated rules.

Equation (1.2), lends itself to the following deterministic safety checking code format

$$\gamma_{Sd} s(F_d, a_d, \dots) \leq \frac{1}{\gamma_{Rd}} r(f_d, a_d, \dots) \tag{1.4}$$

where F_d, f_d and a_d are design values of basic variables representing loading, resistance and geometrical variables respectively, which can be obtained from characteristic/representative values and associated partial factors, and γ_{Sd}, γ_{Rd} are partial factors related to modelling uncertainties (loading and resistance functions, respectively).

As can be seen, the safety checking equation controls the way in which the various clauses of the code lead to the desirable level of safety of structures designed to the code. It relates to the number of design checks required, the rules for load combinations, the number of partial factors and their position in design equations, as well as whether they are single or multiple valued, and the definition of characteristic or representative values for all design variables.

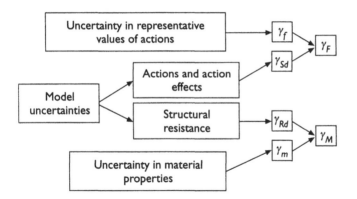

Figure 1.1 Partial factors and their significance in Eurocode 1 (European Standard, 2000).

In principle, there is a partial factor associated with each variable. Furthermore, the number of load combinations can become large for structures subjected to a number of permanent and variable loads. In practice, it is desirable to reduce the number of partial factors and load combinations while, at the same time, ensuring an acceptable range of safety level and an acceptable economy of construction. Hence, it is often useful to make the distinction between primary basic variables and other basic variables. The former group includes those variables whose values are of primary importance for design and assessment of structures. The above concepts of characteristic and design values, and associated partial factors, are principally relevant to this group. Even within this group, some partial factors might be combined in order to reduce the number of factors. Clearly, these simplifications should be appropriate for the particular type of structure and limit state considered. Figure 1.1 shows schematically the system of partial factors adopted in the Structural Eurocodes.

1.2.3 Structural reliability

Load, material and geometric parameters are subject to uncertainties, which can be classified according to their nature. They can, thus, be represented by random variables (this being the simplest possible probabilistic representation; as noted above, more advanced models might be appropriate in certain situations, such as random fields).

In this context, the probability of occurrence of the failure event P_f is given by

$$P_f = \text{Prob} \{g(\mathbf{X}) \leq 0\} = \text{Prob} \{M \leq 0\} \tag{1.5a}$$

where, $M = g(\mathbf{X})$ and $\mathbf{X}$ now represents a vector of basic random variables. Note that M is also a random variable, usually called the safety margin.

If the limit state function can be expressed in the form of (1.2), eqn (1.5a) may be written as

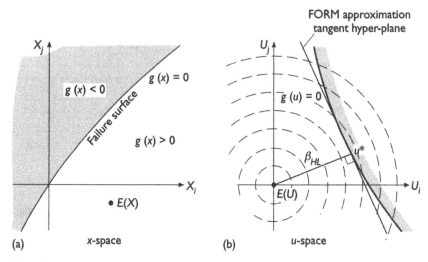

Figure 1.2 Limit state surface in basic variable and standard normal space.

$$P_f = \text{Prob}\left\{r(\mathbf{R}) \leq s(\mathbf{S})\right\} = \text{Prob}\left\{R \leq S\right\} \tag{1.5b}$$

where $R = r(\mathbf{R})$ and $S = s(\mathbf{S})$ are random variables associated with resistance and loading respectively.

Using the joint probability density function of $\mathbf{X}, f_{\mathbf{X}}(\mathbf{x})$, the failure probability defined in equation (1.5a) can now be determined from

$$P_f = \int_{g(\mathbf{X}) \leq 0} f_{\mathbf{X}}(\mathbf{x}) \, d\mathbf{x} \tag{1.6}$$

Schematically, the function $g(\mathbf{X}) = 0$ which represents the boundary between safety and failure is shown in Figure 1.2(a), where the integration domain of eqn (1.6) is shown shaded.

The reliability P_s associated wit\hbox h the particular limit state considered is the complementary event, i.e.

$$P_s = 1 - P_f \tag{1.7}$$

In recent years, a standard reliability measure, the reliability index β, has been adopted which has the following relationship with the failure probability

$$\beta = -\Phi^{-1}(P_f) = \Phi^{-1}(P_s) \tag{1.8}$$

where $\Phi^{-1}(\cdot)$ is the inverse of the standard normal distribution function, see Table 1.1.

The basis for this relationship is outlined in the following section dealing with reliability computation.

In most engineering applications, complete statistical information about the basic random variables $\mathbf{X}$ is not available and, furthermore, the function $g(\cdot)$ is a

Table 1.1 Relationship between β and P_f.

P_f	10^{-1}	10^{-2}	10^{-3}	10^{-4}	10^{-5}	10^{-6}	10^{-7}
β	1.3	2.3	3.1	3.7	4.3	4.7	5.2

mathematical model which idealizes the limit state. In this respect, the probability of failure evaluated from eqn (1.5a) or (1.6) is a point estimate given a particular set of assumptions regarding probabilistic modelling and a particular mathematical model for $g(\cdot)$.

The uncertainties associated with these models can be represented in terms of a vector of random parameters Θ, and hence the limit state function may be rewritten as $g(\mathbf{X}, \Theta)$. It is important to note that the nature of uncertainties represented by the basic random variables $\mathbf{X}$ and the parameters Θ is different. Whereas uncertainties in $\mathbf{X}$ cannot be influenced without changing the physical characteristics of the problem, uncertainties in Θ can be influenced by the use of alternative methods and collection of additional data.

In this context, eqn (1.6) may be recast as follows

$$P_f(\theta) = \int_{g(\mathbf{x},\theta) \leq 0} f_{\mathbf{X}|\Theta}(\mathbf{x}|\theta)\, d\mathbf{x} \tag{1.9}$$

where $P_f(\theta)$ is the conditional probability of failure for a given set of values of the parameters θ and $f_{\mathbf{X}|\theta}(\mathbf{x}|\theta)$ is the conditional probability density function of $\mathbf{X}$ for given θ.

In order to account for the influence of parameter uncertainty on failure probability, one may evaluate the expected value of the conditional probability of failure, i.e.

$$\bar{P}_f = E[P_f(\theta)] = \int_\theta P_f(\theta) f_\Theta(\theta)\, d\theta \tag{1.10a}$$

where $f_\Theta(\theta)$ is the joint probability density function of Θ. The corresponding reliability index is given by

$$\bar{\beta} = -\Phi^{-1}(\bar{P}_f) \tag{1.10b}$$

The main objective of reliability analysis is to estimate the failure probability (or, the reliability index). Hence, it replaces the deterministic safety checking format (e.g. eqn (1.4)), with a probabilistic assessment of the safety of the structure, typically eqn (1.6) but also in a few cases eqn (1.9). Depending on the nature of the limit state considered, the uncertainty sources and their implications for probabilistic modelling, the characteristics of the calculation model and the degree of accuracy required, an appropriate methodology has to be developed. In many respects, this is similar to the considerations made in formulating a methodology for deterministic structural analysis but the problem is now set in a probabilistic framework.

1.2.4 Computation of structural reliability

An important class of limit states are those for which all the variables are treated as time independent, either by neglecting time variations in cases where this is considered acceptable or by transforming time dependent processes into time invariant variables (e.g. by using extreme value distributions). For these problems so-called asymptotic or simulation methods may be used, described in a number of reliability textbooks (e.g. Ang and Tang, 1984; Ditlevsen and Madsen, 1996; Madsen *et al.*, 1986; Melchers, 1999; Thoft-Christensen and Baker, 1982).

Asymptotic approximate methods

Although these methods first emerged with basic random variables described through 'second-moment' information (i.e. with their mean value and standard deviation, but without assigning any probability distributions), it is nowadays possible in many cases to have a full description of the random vector $\mathbf{X}$ (as a result of data collection and probabilistic modelling studies). In such cases, the probability of failure could be calculated via first or second order reliability methods (FORM and SORM respectively). Their implementation relies on:

(1) Transformation techniques

$$\mathbf{T}: \quad \mathbf{X} = (X_1, X_2, \ldots, X_n) \qquad \mathbf{U} = (U_1, U_2, \ldots, U_n) \tag{1.11}$$

where $U_1, U_2, \ldots, U_n$ are independent standard normal variables (i.e. with zero mean value and unit standard deviation). Hence, the basic variable space (including the limit state function) is transformed into a standard normal space, see Figures 1.2(a) and 1.2(b). The special properties of the standard normal space lead to several important results, as discussed below.

(2) Search techniques

In standard normal space, see Figure 1.2(b), the objective is to determine a suitable checking point: this is shown to be the point on the limit–state surface which is closest to the origin, the so-called 'design point'. In this rotationally symmetric space, it is the most likely failure point, in other words its co-ordinates define the combination of variables that are most likely to cause failure. This is because the joint standard normal density function, whose bell-shaped peak lies directly above the origin, decreases exponentially as the distance from the origin increases. To determine this point, a search procedure is generally required.

Denoting the co-ordinates of this point by

$$\mathbf{u}^* = (u_1^*, u_2^*, \ldots, u_n^*)$$

its distance from the origin is clearly equal to

$$\left(\sum_{i=1}^{n} u_i^{*2} \right)^{1/2}$$

This scalar quantity is known as the Hasofer–Lind reliability index β_{HL}, i.e.

$$\beta_{HL} = \left(\sum_{i=1}^{n} u_i^{*2} \right)^{1/2} \tag{1.12}$$

Note that $\mathbf{u}^*$ can also be written as

$$\mathbf{u}^* = \beta_{HL} \alpha \tag{1.13}$$

where $\alpha = (\alpha_1, \alpha_2, \dots, \alpha_n)$ is the unit normal vector to the limit state surface at $\mathbf{u}^*$, and, hence, $\alpha_i (i = 1, \dots, n)$ represent the direction cosines at the design point. These are also known as the sensitivity factors, as they provide an indication of the relative importance of the uncertainty in basic random variables on the computed reliability. Their absolute value ranges between zero and unity and the closer this is to the upper limit, the more significant the influence of the respective random variable is to the reliability. In terms of sign, and following the convention adopted by ISO (1998), resistance variables are associated with positive sensitivity factors, whereas leading variables have negative factors.

(3) Approximation techniques

Once the checking point is determined, the failure probability can be approximated using results applicable to the standard normal space. In a first order (linear) approximation, the limit state surface is approximated by its tangent hyperplane at the design point. The probability content of the failure set is then given by

$$P_{fFORM} = \Phi(-\beta_{HL}) \tag{1.14}$$

In some cases, a higher order (quadratic) approximation of the limit state surface at the design point is desired but experience has shown that the FORM result is sufficient for many structural engineering problems. Equation (1.14) shows that, when using the so-called asymptotic approximate methods, the computation of reliability (or equivalently of the probability of failure) is transformed into a geometric problem, that of finding the shortest distance from the origin to the limit state surface in standard normal space.

Simulation methods

In this approach, random sampling is employed to simulate a large number of (usually numerical) experiments and to observe the result. In the context of structural reliability, this means, in the simplest approach, sampling the random vector $\mathbf{X}$ to obtain a set of sample values. The limit state function is then evaluated to ascertain whether, for this set, failure (i.e. $g(\mathbf{x}) \leq 0$) has occurred. The experiment is repeated many times and the probability of failure, P_f, is estimated from the fraction of trials leading to failure divided by the total number of trials. This so-called Direct or Crude Monte Carlo method is not likely to be of use in practical problems because of the large number of trials required in order to estimate with a certain degree of confidence the failure probability. Note that the number of trials increases as the failure probability decreases. Simple rules may be found, of the form $N > C/P_f$, where N is the required sample size and C is a constant related to the confidence level and the type of function being evaluated.

Thus, the objective of more advanced simulation methods, currently used for reliability evaluation, is to reduce the variance of the estimate of P_f. Such methods can be divided into two categories, namely indicator function methods (such as Importance Sampling) and conditional expectation methods (such as Directional Simulation). Simulation methods are also described in a number of textbooks (e.g. Ang and Tang, 1984; Augusti *et al.*, 1984; Melchers, 1999).

1.3 FRAMEWORK FOR RELIABILITY ANALYSIS

The main steps in a reliability analysis of a structural component are the following:

(1) define limit state function for the particular design situation considered;
(2) specify appropriate time reference period;
(3) identify basic variables and develop appropriate probabilistic models;
(4) compute reliability index and failure probability;
(5) perform sensitivity studies.

Step (1) is essentially the same as for deterministic analysis. Step (2) should be considered carefully, since it affects the probabilistic modelling of many variables, particularly live and accidental loading. Step (3) is perhaps the most important because the considerations made in developing the probabilistic models have a major effect on the results obtained. Step (4) should be undertaken with one of the methods summarized above, depending on the application. Step (5) is necessary insofar as the sensitivity of any results (deterministic or probabilistic) should be assessed.

1.3.1 Probabilistic modelling

For the particular limit state under consideration, uncertainty modelling must be undertaken with respect to those variables in the corresponding limit state function whose variability is judged to be important (basic random variables). Most engineering structures are affected by the following types of uncertainty:

●   Intrinsic physical or mechanical uncertainty; when considered at a fundamental level, this uncertainty source is often best described by stochastic processes in time and space, although it is often modelled more simply in engineering applications through random variables.

- Measurement uncertainty; this may arise from random and systematic errors in the measurement of these physical quantities.
- Statistical uncertainty; due to reliance on limited information and finite samples.
- Model uncertainty; related to the predictive accuracy of calculation models used.

The physical uncertainty in a basic random variable is represented by adopting a suitable probability distribution, described in terms of its type and relevant distribution parameters. The results of the reliability analysis can be very sensitive to the tail of the probability distribution, which depends primarily on the type of distribution adopted. An appropriate choice of distribution type is therefore important.

For most commonly encountered basic random variables, many studies (of varying detail) have been undertaken that contain information and guidance on the choice of distribution and its parameters. If direct measurements of a particular quantity are available, then existing, so-called a priori, information (e.g. probabilistic models found in published studies) should be used as prior statistics with a relatively large equivalent sample size.

The other three types of uncertainty mentioned above (measurement, statistical, model) also play an important role in the evaluation of reliability. As mentioned above, these uncertainties are influenced by the particular method used in, for example, strength analysis and by the collection of additional (possibly, directly obtained) data. These uncertainties could be rigorously analysed by adopting the approach outlined by eqns (1.8) and (1.9). However, in many practical applications a simpler approach has been adopted insofar as model (and measurement) uncertainty is concerned based on the differences between results predicted by the mathematical model adopted for $g(\mathbf{x})$ and a more elaborate model deemed to be a closer representation of reality. In such cases, a model uncertainty basic random variable X_m is introduced where

$$X_m = \frac{\text{Actual value}}{\text{Predicted value}}$$

Uncertainty modelling lies at the heart of any reliability analysis and probability based design and assessment. Any results obtained through the use of these techniques are sensitive to the assumptions made in probabilistic modelling of random variables and processes and the interpretation of any available data. All good textbooks in this field will make this clear to the reader. Schneider (1997) may be consulted for a concise introductory exposition, whereas Benjamin and Cornell (1970) and Ditlevsen (1981) give authoritative treatments of the subject.

1.3.2 Interpretation of results

As mentioned in Section 1.2.4, under certain conditions the design point in standard normal space, and its corresponding point in the basic variable space, is the most likely failure point. Since the objective of a deterministic code of practice is to ascertain attainment of a limit state, it is clear that any check should be performed at a critical combination of loading and resistance variables and, in this respect, the design point values from a reliability analysis are a good choice. Hence, in the deterministic safety checking format, eqn (1.4), the design values can be directly linked to the results of a reliability analysis (i.e. P_f or β and $\alpha_i s$). Thus, the partial factor associated with a basic random variable X_i, is determined as

$$\gamma_{Xi} = \frac{x_{di}}{x_{ki}} = \frac{F_{Xi}^{-1}(\Phi(u_i^*))}{x_{ki}} = \frac{F_{Xi}^{-1}(\Phi(\alpha_i\beta))}{x_{ki}} \tag{1.15}$$

where x_{di} is the design point value and x_{ki} is a characteristic value of X_i. As can be seen, the design point value can be written in terms of the original distribution function $F_x(\cdot)$, the reliability analysis results (i.e. β and α_i), and the standard normal distribution function $\Phi(\cdot)$.

If X_i is normally distributed, eqn (1.15) can be written as (after non-dimensionalizing both x_{di} and x_{ki} with respect to the mean value)

$$\gamma_{Xi} = \frac{1 - \alpha_i\beta v_{Xi}}{1 + k v_{Xi}} \tag{1.16}$$

where v_{Xi} is the coefficient of variation and k is a constant related to the fractile of the distribution selected to represent the characteristic value of the random variable X_i. As shown, eqns (1.15) and (1.16) are used for determining partial factors of loading variables, whereas their inverse is used for determining partial factors of resistance variables. Similar expressions are available for variables described by other distributions (e.g. log-normal, Gumbel type I) and are given in, for example, Eurocode 1 (European Standard, 2000). Thus, partial factors could be derived or modified using FORM/SORM analysis results. The classic text by Borges and Castanheta (1985) contains a large number of partial factor values assuming different probability distributions for load and resistance variables (i.e. solutions pertinent to the problem described by eqn (1.6b)). If the reliability assessment is carried out using simulation, sensitivity factors are not directly obtained, though, in principle, they could be through some additional calculations.

1.3.3 Reliability differentiation

It is evident from eqns (1.15) and (1.16) that the reliability index β can be linked directly to the values of partial factors adopted in a deterministic code. The appropriate degree of reliability should be judged with due regard to the possible consequences of failure and the expense, level of effort and procedures necessary to reduce the risk of failure (ISO, 1998). In other words, it is now generally accepted that 'the appropriate degree of reliability' should take into account the cause and mode of failure, the possible consequences of failure, the social and environmental conditions, and the cost associated with various risk mitigation procedures (ISO,

Table 1.2 Recommended target reliability indices according to Eurocode 1.

Reliability class	Minimum target value for β	
	1 year reference period	50 years reference period
RC3	≥ 5.2	≥ 4.3
RC2	≥ 4.7	≥ 3.8
RC1	≥ 4.2	≥ 3.3

1998; JCSS, 2000). For example, Eurocode 1 (European Standard, 2000) contains an informative annex in which target reliability indices are given for three different reliability classes, each linked to a corresponding consequence class. Table 1.2 reproduces the recommended target reliability values from this document. ISO 2394 (ISO, 1998) contains a similar table, in which target relibility is linked explicitly to consequences of failure and the relative cost of safety measures. Other recently developed codes of practice have made explicit allowances for 'system' effects (i.e. failure of a redundant vs. non-redundant structural element) and inspection levels (primarily as related to fatigue failure) but these effects are, for the time being, primarily related to the target reliability of existing structures.

1.4 TIME-DEPENDENT RELIABILITY

1.4.1 General remarks

Even in considering a relatively simple safety margin for component reliability analysis such as $M = R - S$, where R is the resistance at a critical section in a structural member and S is the corresponding load effect at the same section, it is generally the case that both S and resistance R are functions of time. Changes in both mean values and standard deviations could occur for either $R(t)$ or $S(t)$. For example, the mean value of $R(t)$ may change as a result of deterioration (e.g. corrosion of reinforcement in concrete bridge implies loss of area, hence a reduction in the mean resistance) and its standard deviation may also change (e.g. uncertainty in predicting the effect of corrosion on loss of area may increase as the periods considered become longer). On the other hand, the mean value of $S(t)$ may increase over time (e.g. in highway bridges due to increasing traffic flow and/ or higher vehicle/axle weights) and, equally, the estimate of its standard deviation may increase due to lower confidence in predicting the correct mix of traffic for longer periods. A time-dependent reliability problem could thus be schematically represented as in Figure 1.3, the diagram implying that, on average, the reliability decreases with time. Of course, changes in load and resistance do not always occur in an unfavourable manner as shown in the diagram. Strengthening may result in

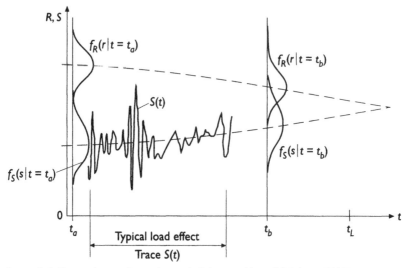

Figure 1.3 General time-dependent reliability problem (Melchers, 1999).

an improvement of the resistance or change in use might be such that the loading decreases after a certain point in time but, more often than not, the unfavourable situation depicted in the diagram is likely to occur.

Thus, the elementary reliability problem described through eqns (1.5) and (1.6) may now be formulated as

$$P_f(t) = \text{Prob}\{\mathbf{R}(t) \leq \mathbf{S}(t)\} = \text{Prob}\{g(\mathbf{X}(t)) \leq 0\} \tag{1.17}$$

where $g(\mathbf{X}(t)) = M(t)$ is a time-dependent safety margin, and

$$P_f(t) = \int_{g(\mathbf{x}(t)) \leq 0} f_{\mathbf{X}(t)}(\mathbf{x}(t)) \, d\mathbf{x}(t) \tag{1.18}$$

is the instantaneous failure probability at time t, assuming that the structure was safe at time less than t.

In time-dependent reliability problems, interest often lies in estimating the probability of failure over a time interval, say from 0 to t_L. This could be obtained by integrating $P_f(t)$ over the interval $[0, t_L]$, bearing in mind the correlation characteristics in time of the process $\mathbf{X}(t)$ – or, sometimes more conveniently, the process $\mathbf{R}(t)$, the process $\mathbf{S}(t)$, as well as any cross correlation between $\mathbf{R}(t)$ and $\mathbf{S}(t)$. Note that the load effect process $\mathbf{S}(t)$ is often composed of additive components, $S_1(t), S_2(t), \ldots$, for each of which the time fluctuations may have different features (e.g. continuous variation, pulse-type variation, spikes).

Interest may also lie in predicting when $S(t)$ crosses $R(t)$ for the first time, see Figure 1.4, or the probability that such an event would occur within a specified time interval. These considerations give rise to so-called 'crossing' problems, which are treated using stochastic process theory. A key concept for such

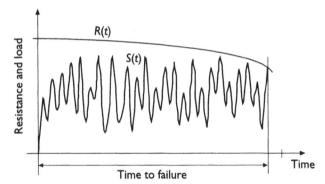

Figure 1.4 Schematic representation of crossing problems.

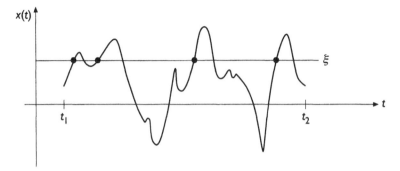

Figure 1.5 Fundamental barrier crossing problem.

problems is the rate at which a random process $X(t)$ upcrosses (or crosses with a positive slope) a barrier or level ξ, as shown in Figure 1.5. This upcrossing rate is a function of the joint probability density function of the process and its derivative, and is given by Rice's formula

$$\nu_\xi^+ = \int_0^\infty \dot{x} f_{X\dot{X}}(\xi, \dot{x}) \, d\dot{x} \tag{1.19}$$

where the rate in general represents an ensemble average at time t. For a number of common stochastic processes, useful results have been obtained starting from eqn (1.19). An important simplification can be introduced if individual crossings can be treated as independent events and the occurrences may be approximated by a Poisson distribution, which might be a reasonable assumption for certain rare load events. Note that random processes are covered in much greater depth and detail in Chapter 10.

Another class of problems calling for a time dependent reliability analysis are those related to damage accumulation, such as fatigue and fracture. This case is depicted in Figure 1.6 showing a threshold (e.g. critical crack size) and a monotonically increasing time dependent load effect or damage function (e.g. actual crack size at any given time).

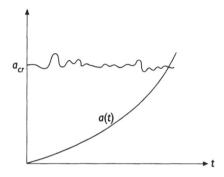

Figure 1.6 Schematic representation of damage accumulation problem.

It is evident from the above remarks that the best approach for solving a time-dependent reliability problem would depend on a number of considerations, including the time frame of interest, the nature of the load and resistance processes involved, their correlation properties in time, and the confidence required in the probability estimates. All these issues may be important in determining the appropriate idealizations and approximations.

1.4.2 Transformation to time independent formulations

Although time variations are likely to be present in most structural reliability problems, the methods outlined in Section 1.2 have gained wide acceptance, partly due to the fact that, in many cases, it is possible to transform a time-dependent failure mode into a corresponding time independent mode. This is especially so in the case of overload failure, where individual time-varying actions, which are essentially random processes, $p(t)$, can be modelled by the distribution of the maximum value within a given reference period T (i.e. $X = \max_T\{p(t)\}$) rather than the point in time distribution. For continuous processes, the probability distribution of the maximum value (i.e. the largest extreme) is often approximated by one of the asymptotic extreme value distributions. Hence, for structures subjected to a single time-varying action, a random process model is replaced by a random variable model and the principles and methods given previously may be applied.

The theory of stochastic load combination is used in situations where a structure is subjected to two or more time-varying actions acting simultaneously. When these actions are independent, perhaps the most important observation is that it is highly unlikely that each action will reach its peak lifetime value at the same moment in time. Thus, considering two time varying load processes $p_1(t), p_2(t), 0 \leq t \leq T$, acting simultaneously, for which their combined effect may be expressed as a linear combination $p_1(t) + p_2(t)$, the random variable of interest is

$$X = \max_T\{p_1(t) + p_2(t)\} \tag{1.20}$$

If the loads are independent, replacing X by $\max_T\{p_1(t)\} + \max_T\{p_2(t)\}$ leads to very conservative results. However, the distribution of X can be derived in few cases only. One possible way of dealing with this problem, which also leads to a relatively simple deterministic code format, is to replace X with the following

$$X' = \max_T \begin{cases} \max_T\{p_1(t)\} + p_2(t) \\ p_1(t) + \max_T\{p_2(t)\} \end{cases} \tag{1.21}$$

This rule (Turkstra's rule) suggests that the maximum value of the sum of two independent load processes occurs when one of the processes attains its maximum value. This result may be generalized for several independent time varying loads. The conditions which render this rule adequate for failure probability estimation are discussed in standard texts. From a theoretical point, the rule leads to an underestimation of the probability of failure, since it is assumed that failure must be associated with the maximum of at least one load process, whereas in reality failure can also occur in other instances.

The failure probability associated with the sum of a special type of independent identically distributed processes (so-called Ferry Borges-Castanheta (FBC) process) can be calculated in a more accurate way, as will be outlined below. Other results have been obtained for combinations of a number of other processes, starting from Rice's barrier crossing formula.

The FBC process is generated by a sequence of independent identically distributed random variables, each acting over a given (deterministic) time interval. This is shown in Figure 1.7 where the total reference period T is made up of n_i repetitions where $n_i = T/\tau_i$. Hence, the FBC process is a rectangular pulse process with changes in amplitude occurring at equal intervals. Because of independence, the maximum value in the reference period T is given by

$$F_{\max_T} X_i(x_i) = [F_{X_i}(x_i)]^{n_i} \tag{1.22}$$

When a number of FBC processes act in combination and the ratios of their repetition numbers within a given reference period are given by positive integers it is, in principle, possible to obtain the extreme value distribution of the combination through a recursive formula. More importantly, it is possible to deal with the sum of FBC processes by implementing the Rackwitz–Fiessler algorithm in a FORM/ SORM analysis.

A deterministic code format, compatible with the above rules, leads to the introduction of combination factors for each time varying load. In principle, these factors express ratios between fractiles in the extreme value and point in time distributions so that the probability of exceeding the design value arising from a combination of loads is of the same order as the probability of exceeding the design value caused by one load. For time varying loads, they depend on distribution parameters, target reliability, FORM/SORM sensitivity factors and on the

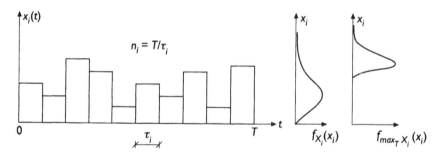

Figure 1.7 Realization of an FBC process.

frequency characteristics (i.e. the base period assumed for stationary events) of loads considered within any particular combination. This is further discussed in Section 1.5.

1.4.3 Introduction to crossing theory

In considering a time dependent safety margin (i.e. $M(t) = g(\mathbf{X}(t))$, the problem is to establish the probability that $M(t)$ becomes zero or less in a reference time period, t_L. As mentioned previously, this constitutes a so-called 'crossing' problem. The time at which $M(t)$ becomes less than zero for the first time is called the 'time to failure' and is a random variable, see Figure 1.8. The probability that $M(t) \leq 0$ occurs during t_L is called the 'first-passage' probability. Clearly, it is identical to the probability of failure during time t_L.

The determination of the first passage probability requires an understanding of the theory of random processes. Herein, only some basic concepts are briefly introduced in order to see how the methods described above have to be modified in dealing with crossing problems. Melchers (1999) provides a detailed treatment of time-dependent reliability aspects.

The first-passage probability $P_f(t)$ during a period $[0, t_L]$ is

$$P_f(t_L) = 1 - P[N(t_L) = 0 \mid \mathbf{X}(0) \in D]P[\mathbf{X}(0) \in D] \tag{1.23}$$

where $\mathbf{X}(0) \in D$ signifies that the process $\mathbf{X}(t)$ starts in the safe domain and $N(t_L)$ is the number of outcrossings in the interval $[0, t_L]$. The second probability term is equivalent to $1 - P_f(0)$, where $P_f(0)$ is the probability of failure at $t = 0$. Equation (1.23) can be rewritten as

$$P_f(t_L) = P_f(0) + (1 - P_f(0)) \, (1 - P[N(t_L) = 0]) \tag{1.24}$$

from which different approximations may be derived depending on the relative magnitude of the terms. A useful bound is

$$P_f(t_L) \leq P_f(0) + E[N(t_L)] \tag{1.25}$$

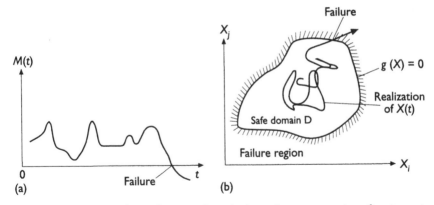

Figure 1.8 Time-dependent safety margin and schematic representation of vector out-crossing (Melchers, 1999): (a) in a safety margin domain, (b) in basic variable space.

where the first term may be calculated by FORM/SORM and the expected number of outcrossings, $E[N(t_L)]$, is calculated by Rice's formula or one of its generalizations. Alternatively, parallel system concepts can be employed.

1.5 ACTIONS AND ACTION EFFECTS ON STRUCTURES

1.5.1 Classification of actions

According to the definition given in ISO 2394 (ISO, 1998),

'an action is

- an assembly of concentrated or distributed mechanical forces on the structure (direct actions), or
- the cause of deformations imposed on the structure or constrained in it (indirect actions).'

Clearly, the above definition is derived bearing in mind the origin of actions. For example, direct actions may be caused by gravity, or can be forces caused by acceleration/deceleration of masses, or by impact. Indirect actions, on the other hand, are the cause of imposed deformations such as temperature, ground settlement, etc.

Actions can also be classified according to their variation in time or space, their limiting characteristics and their nature, which also influences the induced structural response. Table 1.3 summarizes the classification systems which are important in devising an appropriate treatment of actions for design purposes.

The effect of any particular action on structural members or on structural systems is called action effect. Examples of action effects on members include stress resultants (force, moment on any particular beam or column) or stresses, whereas

Table 1.3 Classification of actions.

Origin	Variation in time	Variation in space	Limiting value	Nature/Structural response
Direct	Permanent	Fixed	Bounded	Static
Indirect	Variable	Free	Unbounded	Dynamic
	Accidental			Quasi-static

base shear and top storey lateral deflection may represent action effects on whole structures.

An action should be described by a model, comprising one or more basic variables. For example, the magnitude and direction of an action can both be defined as basic variables. Sometimes an action may be introduced as a function of basic variables, in which case the function is called an action model.

From a probabilistic point of view, the classification of actions according to their variation in time plays an important role, and is examined in detail in the following section dealing with the specification of characteristic and other representative values. Table 1.4 presents, in qualitative terms, the criteria for classifying actions according to time characteristics (Eurocode 1.1 Project Team, 1996). The variability is usually represented by the coefficient of variation (CoV), i.e. the ratio of the standard deviation to the mean value, of the point-in-time distribution of the action. Figure 1.9 shows schematically the three different types of action.

The distinction between static and dynamic actions is made according to the way in which a structure responds to the action, the former being actions not causing significant acceleration of the structure or structural elements, whereas the opposite is valid for the latter. In many cases of codified design, the dynamic actions can be treated as static actions by taking into account the dynamic effects by an appropriate increase in the magnitude of the quasi-static component or by the choice of an equivalent static force. When this is not the case, corresponding dynamic models are used to assess the response of the structure; inertia forces are then not included in the action model but are determined by analysis (ISO, 1998).

Table 1.4 Action classification according to time characteristics.

Action	Permanent	Variable	Accidental
Probability of occurrence during 1 year	Certain	Substantial	Small
Variability in time	Small	Large	Usually large

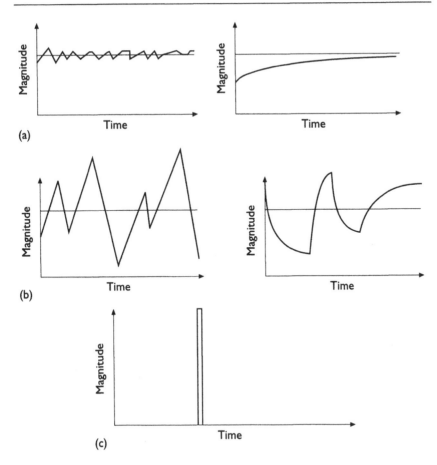

Figure 1.9 Schematic representation of time-varying actions (a) permanent, (b) variable, (c) accidental.

1.5.2 Specification of characteristic values

Permanent actions

The most common action in this category is the self-weight of the structure. With modern construction methods, the coefficient of variation of self-weight is normally small (typically does not exceed 0.05). Other permanent actions include the weight of non-structural elements, which often consists of the sum of many individual elements; hence, it is well represented by the normal distribution (on account of the central limit theorem). For this type of permanent action, the coefficient of variation can be larger than 0.05. An important type of action in this group with high variability is foundation settlement.

According to ISO 2394 (ISO, 1998) and Eurocode 1 (European Standard, 2000), the characteristic value(s) of a permanent action G may be obtained as:

- one single value G_k typically the mean value, if the variability of G is small (CoV ≤ 0.05);
- two values $G_{k,\text{inf}}$ and $G_{k,\text{sup}}$ typically representing the 5 per cent and 95 per cent fractiles, if the CoV cannot be considered small.

In both cases it may be assumed that the distribution of G is Gaussian.

Variable actions

For single variable loads, the form of the point in time distribution is seldom of immediate use in design; often the important variable is the magnitude of the largest extreme load that occurs during a specified reference period for which the probability of failure is calculated (e.g. annual, lifetime). In some cases, the probability distribution of the lowest extreme might also be of interest (water level in rivers/lakes).

Consider a random variable X with distribution function $F_X(x)$. If samples of size n are taken from the population of X : $(x_1, x_2, \ldots, x_n)$, each observation may itself be considered as a random variable (since it is unpredictable prior to observation). Hence, the extreme values of a sample of size n are random variables, which may be written as

$$Y_n = \max(X_1, X_2, \ldots, X_n)$$
$$Y_1 = \min(X_1, X_2, \ldots, X_n)$$

The probability distributions of Y_n and Y_1 may be derived from the probability of the initial variate X. Assuming random sampling, the variables $X_1, X_2, \ldots, X_n$ are statistically independent and identically distributed as X, hence

$$F_{X_1}(x_1) = F_{X_2}(x_2) = \cdots = F_{X_n}(x_n) = F_X(x)$$

The distribution of $F_{Y_n}(y)$ is thus given by

$$F_{Y_n}(y) = P(Y_n \leq y) = P(X_1 \leq y \cap X_2 \leq y \cap \cdots \cap X_n \leq y) \qquad (1.26)$$

which can be written as

$$F_{Y_n}(y) = [F_X(y)]^n \qquad (1.27)$$

Similar principles may be used to derive the distribution of the lowest extreme.

For a time varying load Q the distribution on the left-hand side of equation (1.27) can be interpreted as the maximum load in a specified reference period t_r whereas the distribution on the right-hand side represents the maximum load occurring during a much shorter period, sometimes called the unit observation time τ. In this case, the exponent is equal to the ratio between the two (i.e. $n = t_r/\tau$ and $n > 1$). Equation (1.27) may thus be written as

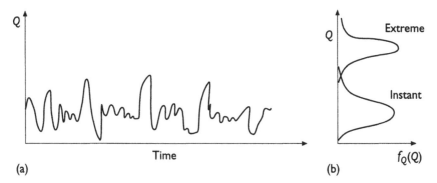

Figure 1.10 Schematic representation of variable action: (a) realization in time, (b) probability distributions.

$$F_{Q[t_r]}(q) = \{F_{Q[\tau]}(q)\}^n \qquad (1.28)$$

where the symbol in square brackets indicates the time period to which the probability distribution is related. As mentioned earlier in this section, the probability of intersection of events can be expressed as a product only if the events are independent; for time varying loads this means that the unit observation time must be chosen so that the maximum value of the load recorded within any such period is independent of the others. Note the similarity of eqn (1.28) with eqn (1.22), which is derived under similar assumptions.

Figure 1.10 illustrates the above concepts and shows schematically the probability distributions associated with the maximum load in different time periods; customarily the lower of the two is called the 'instantaneous' or 'point in time' distribution, whereas the upper one is an 'extreme' distribution. It is clear from eqn (1.28) that the parameters and moments (e.g. mean value) of the extreme distribution are a function of the specified reference period. The longer this is, the greater becomes the gap between the two distributions shown in Figure 1.10.

In principle, for actions of natural origin (e.g. wind, snow, temperature) the 'instantaneous' distribution is determined through observations (i.e. the creation of a homogeneous sample of sufficient size) and classical methods of distribution fitting. However, judgement also plays an important role in refining and improving the statistical model. This is because the direct number of observations may be fairly small. Considering, for example, the snow load the unit observation period may be chosen equal to 1 year, which means that it is unlikely that the number of data points for any particular site will be more than 40 or 50. The distribution of annual maxima could nevertheless be compared with that obtained for different but similar sites, and the final estimates of the distribution may in fact be made on the basis of a larger sample in which the data points from similar sites are combined. Note that when, through eqn (1.28), the distribution of the annual maximum load is transformed to the distribution of, say, the maximum load in 50 years, further uncertainties are introduced, and should be taken into account as far as possible through appropriate judgement. In the case of time-varying loads, these uncertainties may have both systematic and random components. The former can be particularly important for some man-made loads, such as traffic loads on bridges, whereas the latter may include poorly understood environmental influences as well as purely random effects.

The characteristic value of a time varying load Q_k is normally chosen so that events during which the observations exceed the characteristic value are fairly rare. Typically, characteristic values in Eurocode 1 are prescribed for an exceedance probability $p = 0.02$ and a reference period $t_r = 1$ year (European Standard, 2000; Eurocode 1.1 Project Team, 1996). Thus, the characteristic value Q_k may be estimated from

$$F_{Q[1 \text{ year}]}(Q_k) = 1 - p = 0.98$$

In the above the distribution for the annual maximum is used and the reference period is also 1 year. If, for example, the distribution for the monthly maximum was available instead, then for the same criteria (i.e. 2 per cent exceedance probability during one year) and providing monthly maxima were mutually independent, the characteristic value could be estimated from

$$F_{Q[1 \text{ month}]}(Q_k) = (1 - p)^{1/12} \approx 0.99832$$

Note that using a distribution based on observations from a shorter time unit results in a much higher fractile required for estimating a characteristic value based on the same criteria as before. Clearly, many more observations would be needed for the monthly distribution to be sufficiently well described at a 99.8 per cent fractile, than the annual distribution at a 98 per cent fractile. On the other hand, much longer (unit and total) observation periods are associated with the estimation of the distribution of annual maxima. The message is that predictions cannot be improved simply by changing the basis of the distribution used. The most appropriate observation period should be determined on the basis of the characteristics of the action being modelled and the capabilities of the devices/methods used for measurement.

A useful concept in the treatment of time varying loads is the return period T defined as the average time between consecutive occurrences of an event. Again assuming independence between events, and denoting with p the probability of occurrence of the particular event considered, the return period may be determined from

$$T = \frac{1}{p} \tag{1.29}$$

that is, the return period is equal to the reciprocal of the probability of occurrence of the event in any one time interval. In many cases, the chosen time interval is 1 year and p is determined as the probability of occurrence during a year, so that the return period is the average number of years between events.

Thus, for the above example, the return period of the characteristic value of the load Q_k which represents the average time between events $Q > Q_k$ is given by

$$T = \frac{1}{0.02} = 50 \text{ years} \qquad \text{or} \qquad T = \frac{1}{0.00168} = 595.2 \text{ months} \simeq 50 \text{ years}$$

Return periods of 50 to 100 years are reasonable for characteristic values of variable actions used in the design of ordinary permanent buildings. For accidental actions, a longer return period might be appropriate, especially if ultimate or collapse limit states are considered.

Bearing in mind the notation introduced above for the reference (t_r) and unit observation (τ) periods, the return period may be written as

$$T = \frac{\tau}{1 - F_{Q[\tau]}(Q_k)} = \frac{1/n}{1 - (1 - p)^{1/n}} t_r \simeq \frac{1}{\ln(1/(1 - p))} t_r \tag{1.30}$$

where p is defined for a reference period t_r and $n > 1$. The last expression is asymptotically correct as $(1 - p)$ tends to unity, which is compatible with the notion of specifying characteristic values on the basis of fairly rare events; note that the return period becomes independent of the unit observation period τ.

The probability distribution of extreme values is often closely approximated by one of the asymptotic extreme value distributions (Types I, II and III). The characteristics of extreme distributions depend on the initial, or parent, distribution and on the number of repetitions, n. In general, distributions shift to the right with increasing n. Which of the three types is relevant depends on the shape of the upper tail of the parent distribution. Of particular importance in the context of time-varying loads is the Gumbel or Type I extreme distribution for maxima, which is obtained if the initial distribution has an exponentially decreasing upper tail. It has the following probability distribution function

$$F_{Q_n}(q) = \exp\{-\exp[-\alpha_n(q - u_n)]\} \tag{1.31}$$

where u_n and α_n are the distribution parameters. The mean and variance of Q_n are related to the distribution parameters through the following expressions

$$E[Q_n] = u_n + \frac{0.5772}{\alpha_n}, \qquad \text{Var}[Q_n] = \frac{\pi^2}{6\alpha_n^2}$$

An interesting property of this distribution is that the variance is independent of the number of repetitions (i.e. it remains constant). On the other hand, the mean value increases with the number of repetitions. Ang and Tang (1984) present an exposition of extreme value theory as applied to a variety of civil engineering problems.

Accidental actions

In principle, the characteristic values of accidental actions could be determined by extending the procedures presented above for variable actions. Accidental actions are characterized by a (usually) random magnitude and an occurrence rate. For many accidental actions, statistical information is scarce. Hence, in practice, nominal values are often used and sometimes values are agreed for individual projects.

Insofar as seismic actions are concerned, the design values are determined on the basis of a return period of approximately 475 years for use in ultimate limit states and a return period of about 50 years for serviceability limit states (see also Chapter 4).

1.5.3 Other representative values

For variable and accidental actions (i.e. for those actions whose time variation is significant), there is a need to specify a few more representative values, in addition to the characteristic value, for use in codified limit state design. These are briefly reviewed in the following and are schematically shown in Figure 1.11.

Combination value ($\psi_0 Q_k$)

This value is chosen so that the probability that the action effects caused by any particular load combination will be exceeded is approximately the same as by the characteristic value of an individual action. In other words, the combination value is introduced to take account of the reduced probability of the simultaneous occurrence of the most unfavourable values of two or more independent variable actions. The combination value may be expressed as a fraction of the characteristic value through a combination factor $\psi_0 (< 1)$. The combination value is used in load combinations pertaining to the ultimate limit state or to irreversible serviceability limit states.

Using structural reliability theory, values for the combination factor ψ_0 have been derived for load combinations comprising two independent variable actions starting from either FBC load processes or using Turkstra's rule. Expressions for ψ_0 for different probability distributions can be found in code documents (ISO,

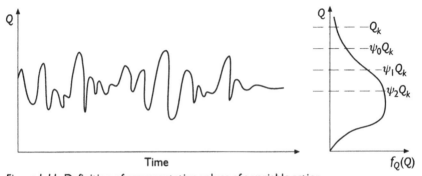

Figure 1.11 Definition of representative values of a variable action.

1998; European Standard, 2000). In operational codes, the values adopted are usually based on historical values linked to successful experience, and are often simplified in order to limit the different values that a designer needs to consider for all the different load cases.

Frequent value ($\psi_1 Q_k$)

The frequent value is used for the dominating variable action in combinations at the ultimate limit state involving accidental actions. It is also used in reversible serviceability limit states. As in the previous case, it can be expressed as a fraction of the principal characteristic value through a factor, $\psi_1 (< 1)$. Typically, the frequent value may be estimated from the point-in-time (or instantaneous) distribution of the action, i.e.

$$F_Q(\psi_1 Q_k) = 1 - q$$

where $q = 0.01$ is suggested for buildings (European Standard, 2000).

The criterion may also be expressed as a return period; for example, for road bridges the frequent value of the traffic load is determined as having a return period of 1 week.

Quasi-permanent value ($\psi_2 Q_k$)

The quasi-permanent value is used for the non-dominating variable action in combinations at the ultimate limit state involving accidental actions. It is also used in reversible serviceability limit states and in the calculation of long term effects in serviceability limit states.

The quasi-permanent value may be regarded as a special case of the frequent value with $q = 0.5$. Thus,

$$F_Q(\psi_2 Q_k) = 1 - q = 0.5$$

It can also be defined as the mean value of the instantaneous probability distribution. In certain cases (e.g. wind or road traffic) the value of ψ_2 is so low that it is set equal to zero (European Standard, 2000).

Table 1.5 summarizes some of the ψ factors as given in the Eurocode (European Standard, 2000). More cases are covered therein, and each code has its own system of factors, broadly based on the principles outlined above. However, as already

Table 1.5 Typical ψ factors for buildings according to Eurocode 1.

Action	ψ_0	ψ_1	ψ_2
Imposed load (domestic, residential, office)	0.7	0.5	0.3
Snow load (Scandinavia and Rest of Europe for altitude >1,000 m)	0.7	0.5	0.2
Snow load (rest of Europe for altitude <1,000 m)	0.5	0.2	0.0
Wind load	0.6	0.2	0.0

mentioned, a number of pragmatic reasons will also influence the final selection of values and all these are normally considered by the code drafting committees.

1.5.4 Duration of actions

The knowledge, and specification, of a maximum action effect individually or in combination is essential for safety checking. In some cases, especially where the sustained live load is high, the duration characteristics, and in particular any inter-mittencies, may also be of interest. In such a case, the components of the stochastic model would increase and may, for example, include an interarrival duration density in addition to a variable describing the number of magnitude changes (e.g. a jump rate which quantifies the number of amplitude changes in a specified period).

In Eurocode 1 (European Standard, 2000; Eurocode 1.1 Project Team, 1996), the frequent and quasi-permanent values of a variable action may also be defined in terms of duration. For example, the frequent value may be specified as that which is exceeded for 5 per cent of the reference period considered; the corresponding per-centage for the quasi-permanent value may be 50 per cent.

1.6 CONCLUDING REMARKS

Structural reliability theory provides a rational basis for the description and quanti-fication of loads and resistances in structural engineering. It enables consistent com-parisons to be made between alternative hazards to which structures are exposed during their service life, and is an indispensable tool for rational decision making in the presence of uncertainty. Whether this uncertainty stems from objective (e.g. future realizations of natural events) or subjective (e.g. limited knowledge of actual material properties in an existing structure) sources, the use of structural relia-bility theory and the allied battery of probabilistic methods has led to very signifi-cant contributions towards an improved design philosophy for structures, both large and small, ordinary or extraordinary.

In the ensuing chapters of this book, the load effects arising from different natural or man-made actions will be described in some detail, with regard to their nature and their treatment in codes. Clearly, the best models for any particualr action and its effect will have to take into account the principal characteristics of the generating phenomenon, as well as the detailed features which come into play as the action interacts with different structural types and forms.

Nonetheless, there are generic features associated with actions and their effects, as well as their consequences. This chapter has attempted to present, within a reason-able length, these generic features, how they might be modelled using probabilistic concepts, what is their significance in terms of the way in which the reliability of the structure may be estimated, and finally how these issues are dealt with in modern codes of practice. The presentation herein has been brief and, hopefully, concise. The bibliography provided is intended to help the reader explore the issues at a much greater depth, should this be required.

1.7 REFERENCES

Ang, A. H. S. and Tang, W. H. (1975/1984) *Probability Concepts in Engineering Planning and Design*, Vols. I and II, John Wiley, New York.

Augusti, G., Baratta, A. and Casciati, F. (1984) *Probabilistic Methods in Structural Engineering*, Chapman and Hall, London.

Benjamin, J. R. and Cornell, C. A. (1970) *Probability, Statistics and Decision for Civil Engineers*, McGraw Hill, New York.

Borges, J. F. and Castanheta, M. (1985) *Structural Safety*, Laboratorio Nacional de Engenharia Civil, Lisbon.

Ditlevsen, O. (1981) *Uncertainty Modelling*, McGraw Hill, New York.

Ditlevsen, O. and Madsen, H. O. (1996) *Structural Reliability Methods*, John Wiley, Chichester.

Eurocode 1.1 Project Team (1996) *Background Documentation for Part 1 of EC1: Basis of Design*, Second Draft, January 1996.

European Standard, Draft prEN 1990(2000) *Eurocode: Basis of Design*, Final Draft, European Committee of Standardization CEN/TC250, Brussels, February.

International Organization for Standardization (ISO) (1998) *General Principles on Reliability for Structures*, ISO/FDIS 2394 (Final Draft).

Joint Committee on Structural Safety (JCSS) (2001) *Assessment of Existing Structures*, RILEM, Publications, Cachan, France.

Madsen, H. O., Krenk, S. and Lind, N. C. (1986) *Methods of Structural Safety*, Prentice-Hall.

Melchers, R. E. (1999) *Structural Reliability: Analysis and Prediction*, 2nd edn, John Wiley, Chichester.

Schneider, J. (1997) *Introduction to Safety and Reliability of Structures*, International Association for Bridge and Structural Engineering (IABSE), Structural Engineering Documents 5.

Thoft-Christensen, P. and Baker, M. J. (1982) *Structural Reliability Theory and its Applications*, Springer-Verlag, Berlin.

Chapter 2

Analysis for dynamic loading

George D. Manolis

2.1 INTRODUCTION

The purpose of this chapter is analysis of structures that are subjected to time varying loads. Despite the fact that the majority of civil engineering structures are built on the assumption that all applied loads are static, there are exceptions which require a distinction between static and dynamic loads to be made, as in earthquake engineering. All loads in nature are time dependent. In many cases, however, loads will be applied to a structure in slowly varying ways, which implies that static conditions can be assumed. The term slow here is quantified through comparison with an intrinsic time of the structure, which is none other than its natural period. Thus, a load varies slowly or is fast only in relation to the time required for the structure to complete a full cycle of oscillation.

There is growing interest nowadays in the process of designing civil engineering structures to withstand dynamic loads (Biggs, 1965; Craig, 1981; Bathe, 1982). As examples, we mention (i) structures which house moving or vibrating equipment, (ii) bridges under traffic, (iii) multistory structures subject to wind and (iv) the case of earthquake induced loads (Clough and Penzien, 1993; Newmark and Rosenblueth, 1971). Essentially, dynamic analyses focus on evaluation of time dependent displacements, from which the stress state of the structure in question can be computed (Paz, 1997; Argyris and Mlejnek, 1991; Chopra, 1995). The most basic pieces of information needed for this are the natural period, which is a function of the structure's mass and stiffness, and the amount of available damping (or, equivalently, the amount of energy that can be absorbed by the structure).

2.2 THE SINGLE DEGREE-OF-FREEDOM OSCILLATOR

The simplest dynamic model is the Single Degree-of-Freedom (SDOF) oscillator shown in Figure 2.1(a). It is an exact model for the simple orthogonal frame with slender columns and a strong inflexible girder, where all the mass can be lumped. Three basic types of vibrations can be considered, namely horizontal, vertical and

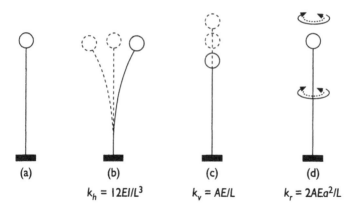

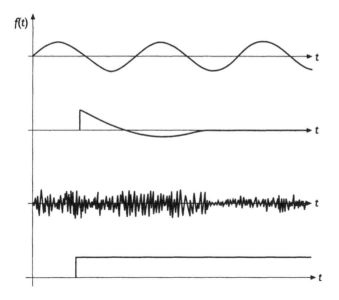

Figure 2.1 (a) SDOF modelling of a single story frame for (b) horizontal, (c) vertical and (d) rotational oscillations.

Figure 2.2 Various types of dynamic loads: harmonic, aperiodic, earthquake and long duration.

rotational, as shown in Figure 2.1(b)–(d). As expected, the SDOF oscillator is used extensively for modelling structural systems, but it should be remembered that it is an approximate model for anything else but the simple frame previously mentioned. Next, some typical dynamic loads are shown in Figure 2.2, where we distinguish between periodic (both harmonic and non-harmonic) and aperiodic (both short and long duration) loads.

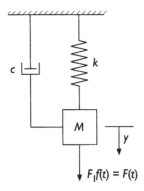

Figure 2.3 SDOF oscillator.

With reference to Figure 2.3, the equation of motion of the SDOF oscillator is

$$M\ddot{y} + c\dot{y} + ky = F_1[f(t)] \tag{2.1}$$

implying that the inertia, damping and restoring forces balance the applied force. Specifically, M is the mass (kg), k is the stiffness (N/m), and c is the damping coefficient (N-sec/m). Furthermore, $y(t)$ is the displacement (m), $\dot{y}(t)$ the velocity (m/sec), $\ddot{y}(t)$ the acceleration (m/sec^2), $F(t) = F_1 f(t)$ the externally applied force (N) with $f(t)$ its dimensionless time variation. Finally, dots denote time derivatives d/dt. Obviously, eqn (2.1) is a second order differential equation that needs to be solved for the displacement $y(t)$.

2.2.1 Motion without damping

2.2.1.1 Free vibrations

The equation of dynamic equilibrium of an SDOF system in the absence of both damping and external force is given below as

$$\ddot{y} + \frac{k}{M}y = 0 \tag{2.2}$$

Thus, the oscillator undergoes free vibrations under the influence of an initial displacement $y(0) = y_0$ and/or initial velocity $\dot{y}(0) = \dot{y}_0$. The solution is simply

$$y = \frac{\dot{y}_0}{\omega}\sin\omega t + y_0\cos\omega t \tag{2.3}$$

and implies a periodic, harmonic motion as shown in Figure 2.4. At this point, we respectively define the circular frequency, the natural period and the frequency as

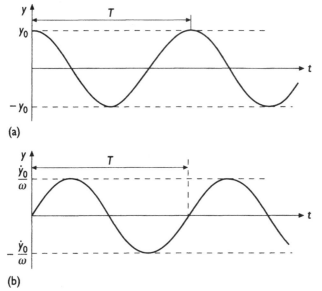

Figure 2.4 Free vibration due to (a) initial displacement and (b) initial velocity.

follows:

$$w \, (\text{rad/sec}) = \sqrt{\frac{k}{M}} \tag{2.4}$$

$$T \, (\text{sec}) = \frac{2\pi}{w} \tag{2.5}$$

$$f \, (\text{cycles/sec or Hz}) = \frac{1}{T} \tag{2.6}$$

2.2.1.2 Forced vibrations

We first look at the case where an external force $F(t)$ is accompanied by zero initial conditions. Specifically, we have a constant load $F(t) = F_1$ applied at time $t = 0$ and subsequently maintained. Equation (2.1) can be written as

$$\ddot{y} + \frac{k}{M}y = \frac{F_1}{M} \tag{2.7}$$

and its solution is

$$y = \frac{F_1}{k}(1 - \cos wt) \tag{2.8}$$

We define as Dynamic Load Factor (DLF) the ratio of the dynamic displacement at

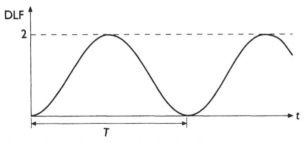

Figure 2.5 DLF for constant load $F(t) = F_1$.

any time instant to the displacement produced by static application of the load $F(t) = F_1$ as

$$\text{DLF} = \frac{y}{y_{st}} = \frac{y}{F_1/k} = \frac{ky}{F_1} \tag{2.9}$$

A simple substitution of eqn (2.8) in eqn (2.9) gives

$$\text{DLF} = 1 - \cos \omega t \tag{2.10}$$

The DLF is dimensionless and measures the amount by which the dynamic displacement in the SDOF system exceeds its equivalent static one. Figure 2.5 plots the DLF for the suddenly applied and maintained load case, where doubling of the response is observed at certain time instances.

2.2.1.3 Forced vibrations for various forcing functions

(a) General solution by superposition of impulses

The general closed form solution can be obtained by synthesis of the SDOF system response to a series of impulses. Assume that the system is at rest and then acted upon by a constant force F with instantaneous time duration t_d. The mass of the oscillator will experience an instantaneous acceleration

$$\ddot{y} = F/M \tag{2.11}$$

which in turn produces an instantaneous velocity

$$\dot{y} = \ddot{y}t_d = \frac{Ft_d}{M} = \frac{I}{M} \tag{2.12}$$

where I is the impulse defined as force times duration.

All dynamic loads can be considered as a sequence of impulses of varying magnitude. Thus, force $F(\tau)$ at time τ and for the ensuing time instant t_d imparts an initial velocity to the SDOF oscillator of the following type:

$$F_1 f(\tau)\, dt/M \tag{2.13}$$

Thus, from eqn (2.3) the system experiences an instantaneous displacement $y(t)$ equal to

$$\frac{F_1 f(\tau)\, d\tau}{M\omega} \sin \omega(t - \tau) \tag{2.14}$$

Finally, the complete displacement history is evaluated by integrating from time $t = 0$ to the present time t as

$$y(t) = \int_0^t \frac{F_1 f(\tau)}{M\omega} \sin \omega(t - \tau)\, d\tau \tag{2.15}$$

If the static displacement due to the load magnitude F_1 is

$$y_{st} = \frac{F_1}{k} = \frac{F_1}{\omega^2 M} \tag{2.16}$$

then

$$y(t) = y_{st}\omega \int_0^t f(\tau) \sin \omega(t - \tau)\, d\tau \tag{2.17}$$

If we finally add the effect of initial conditions at $t = 0$, then we have a generel, closed form expression for the dynamic displacement of the SDOF system in the form of Duhamel's integral as

$$y = y_0 \cos \omega t + \frac{\dot{y}}{\omega} \sin \omega t + y_{st}\omega \int_0^t f(\tau) \sin \omega(t - \tau)\, d\tau \tag{2.18}$$

(b) Suddenly applied load of duration t_d

Here we have a combination of constant load $f(t) = 1$ until time $t = t_d$ and free vibrations past $t > t_d$ with initial conditions $y(t = t_d)\hbox{ and } \dot{y}(t = t_d)$. The resulting DLF factors are:

$$\text{DLF} = 1 - \cos \omega t = 1 - \cos 2\pi \frac{t}{T} \quad t \leq t_d \tag{2.19}$$

$$\text{DLF} = \cos \omega(t - t_d) - \cos \omega t = \cos 2\pi\left(\frac{t}{T} - \frac{t_d}{T}\right) - \cos 2\pi \frac{t}{T} \quad t \geq t_d \tag{2.20}$$

where $y_{st} = F_1/k$. Figure 2.6 plots the above results for two cases, where we observe an intense response when the duration of the load on the oscillator is large ($t_d = 1.2T$). If the load is on the oscillator for a short time ($t_d = 0.1T$), the dynamic response is less than the static one.

(c) Constant load with rise time t_r

The time variation of this load is given by $f(\tau) = \tau/t_r, \tau \leq t_r$ and $f(\tau) = 1, \tau \geq t_r$. The DLF is evaluated as

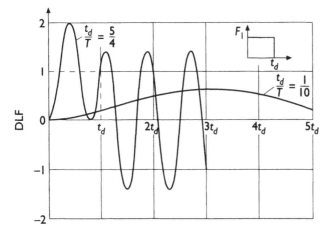

Figure 2.6 DLF for load with duration time t_d.

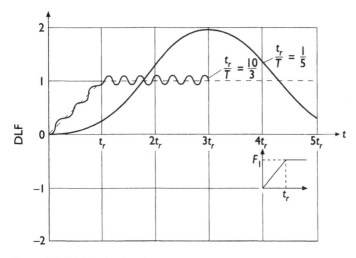

Figure 2.7 DLF for load with rise time t_r.

$$\left. \begin{array}{ll} \text{DLF} = \dfrac{1}{t_r}\left(t - \dfrac{\sin \omega t}{\omega}\right) & t \leq t_r \\[2ex] \text{DLF} = 1 + \dfrac{1}{\omega t_r}\left[\sin \omega(t - t_r) - \sin \omega t\right] & t \geq t_r \end{array} \right\} \qquad (2.21)$$

and Figure 2.7 plots two cases, one with a rapid ($t_r = 0.2T$) and the other with a slow ($t_r = 3.33T$) application. We note that the latter case produces a quasi-static response in the SDOF oscillator. Finally, in Figure 2.8 we have the maximum value of the DLF_{max} as function of the time ratio t_r/T.

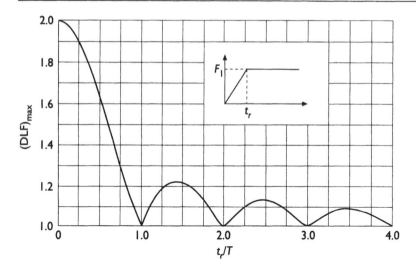

Figure 2.8 Maximum value of the DLF as a function of rise time t_r.

2.2.1.4 Harmonic vibrations

Harmonic loads assume the form $F(t) = F_1 \sin \Omega t$, although the cosine function or the exponential function with an imaginary argument can be used as well. With harmonic loads, there is always danger of resonance (i.e. the structure may experience high or even unbounded vibrations when its natural frequency coincides with that of the load). The equation of motion is

$$M\ddot{y} + ky = F_1 \sin \Omega t \qquad (2.22)$$

and its solution in terms of the DLF (with $y_{st} = F_1/k$) and for $y_0 = \dot{y}_0 = 0$ has the following form:

$$\text{DLF} = \frac{1}{(1 - \Omega^2/\omega^2)} \left[\sin \Omega t - \frac{\Omega}{\omega} \sin \omega t \right] \qquad (2.23)$$

We observe that the oscillations comprise two parts, the free part with frequency ω and the forced part with frequency Ω. Also, an approximate maximum value of the DLF is obtained when $\sin \Omega t = 1$ and $\sin \omega t = -1$, i.e.

$$(\text{DLF})_{\text{max}} = \pm \frac{1}{(1 - \Omega/\omega)} \qquad (2.24)$$

If we ignore the free vibration part, the maximum DLF is

$$(\text{DLF})_{\text{max}} = \frac{1}{(1 - \Omega^2/\omega^2)} \qquad (2.25)$$

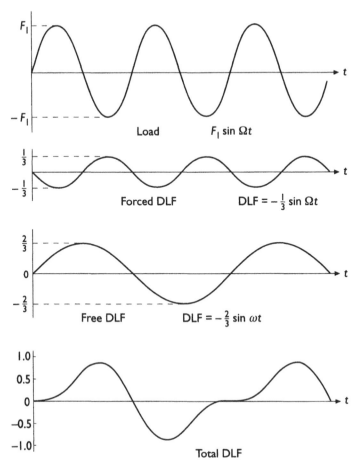

Figure 2.9 DLF for harmonic vibrations when $\Omega = 2\omega$.

When $\Omega \rightarrow \omega$, we have resonance effects and $(DLF)_{max} \rightarrow \infty$. Using L'Hospital's rule in the limit, we obtain that when $\Omega = \omega$,

$$(DLF)_{\Omega=\omega} = \tfrac{1}{2}(\sin \omega t - (\cos \omega t)\omega t) \tag{2.26}$$

Thus, the dynamic displacement diverges, but only after a finite number of oscillations. Also, Figure 2.9 plots the DLF for the case $\Omega = 2\omega$ where we see that the total factor, despite being the superposition of two harmonic functions, is no longer harmonic but only a periodic function of time.

2.2.2 Motion with damping

Damping produces forces which counteract the motions of the SDOF oscillator by absorbing energy. All dynamic systems in practice exhibit a certain amount of damping.

2.2.2.1 Free vibrations

The equation of motion for an SDOF system in the presence of damping is

$$M\ddot{y} + c\dot{y} + ky = F(t) \tag{2.27}$$

and its solution without external loading $(F(t) = 0)$ is given below as

$$y = e^{-\beta t}\left(\frac{\dot{y}_0 + \beta y_0}{\omega_d}\sin\omega_d t + y_0 \cos\omega_d t\right) \tag{2.28}$$

We also define the coefficient of damping and the damped natural frequency as follows:

$$\beta = c/2M \tag{2.29}$$

$$\omega_d = \sqrt{\omega^2 - \beta^2} \tag{2.30}$$

There are three possibilities for β, namely

$$\beta < \omega, \qquad \beta = \omega \qquad \text{and} \qquad \beta > \omega \tag{2.31}$$

which correspond to underdamped, critically damped and overdamped conditions. If $\beta = \omega$, $\sin(\omega_d t) \rightarrow (\omega t)$, and eqn (2.28) becomes

$$y = e^{-\omega t}[\dot{y}_0 t + (1 + \omega t)y_0] \tag{2.32}$$

The displacement is no longer a periodic function of time and the oscillator simply returns to its original position without executing any vibrations. From the condition

$$\omega = \beta = \frac{c_{cr}}{2M} \tag{2.33}$$

we may compute the coefficient of critical damping as

$$c_{cr} = 2M\omega = 2\sqrt{kM} \tag{2.34}$$

Following that, the damping ratio is defined

$$\zeta = c/c_{cr} \tag{2.35}$$

It should be noted here that the coefficient of damping β is seldom used nowadays, with preference given to damping ratio ζ. Obviously, the two coefficients are related as $\zeta = \beta/\omega$. The effect of damping on the natural frequency is minimal; for instance, a 10 per cent of critical damping ratio yields $\omega_d = 0.995\omega$. It is rare to find civil engineering structures exhibiting anything close to critical damping, although many mechanical components (such as shock absorbers) do.

Damping can be experimentally measured by tracing the logarithmic decrement (i.e. the log of the difference between two consecutive peaks in a displacement versus time plot for free vibrations). Referring to Figure 2.10, we have the logarithmic decrement as

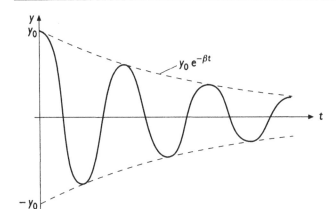

Figure 2.10 Free vibration with damping due to an initial displacement.

$$d = \ln\frac{y(t = t)}{y(t = t + T_d)} = \ln\frac{e^{-\beta t}}{e^{-\beta(t+T_d)}} = \ln e^{\beta T_d} = \beta T_d = \beta\frac{2\pi}{\omega} \qquad (2.36)$$

For instance, when $\zeta = \beta/\omega = 10$ per cent, d is equal to 0.2π and the ratio of two consecutive peaks is $\exp(0.2\pi) = 1.87$. Thus, a damping ratio of 10 per cent reduces the dynamic displacements by a factor of 0.534 during each vibration cycle.

2.2.2.2 Forced vibrations

By analogy to the case of forced vibrations in the absence of damping, we now have that a damped impulse element is

$$\frac{F_1 f(\tau)\, d\tau}{M\omega_d} e^{-\beta(t-\tau)} \sin \omega_d(t - \tau) \qquad (2.37)$$

The complete expression in terms of a Duhamel integral can be found through time integration of the above impulse, to which the effect of initial conditions is subsequently superimposed. Thus,

$$y = e^{-\beta t}\left(\frac{\dot{y}_0 + \beta y_0}{\omega_d}\sin \omega_d t + y_0 \cos \omega_d t\right)$$

$$+ y_{st}\frac{\omega^2}{\omega_d}\int_0^t f(\tau)e^{-\beta(t-\tau)} \sin \omega_d(t - \tau)\, d\tau \qquad (2.38)$$

As a special case consider $F(t) = F_1$; substitution of the time function $f(t) = 1$ in eqn (2.38) gives the solution for a suddenly applied and maintained load as

$$y = \frac{F_1}{k}\left[1 - e^{-\beta t}\left(\cos \omega t + \frac{\beta}{\omega}\sin \omega t\right)\right] \qquad (2.39)$$

Comparing Figures 2.5 and 2.11 clearly shows the effect of damping in reducing the

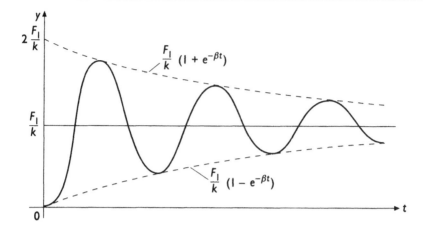

Figure 2.11 Forced vibration with damping due to suddenly applied and maintained load.

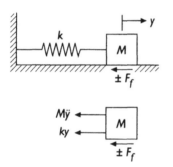

Figure 2.12 Dynamic equilibrium of SDOF system with Coulomb friction.

amplitude of the dynamic displacements and in bringing about, after some time, quasi-static conditions.

2.2.2.3 Coulomb damping

This type of damping is due to friction; as the SDOF oscillator moves on a rough surface, a horizontal force $Ff = \mu_{d}gM$ develops, where μ_d is the dynamic friction coefficient and g is the acceleration of gravity, and acts in direction opposite to the velocity as shown in Figure 2.12. The resulting equation of motion for free vibrations is given below as

$$M\ddot{y} + ky \pm F_f = 0 \tag{2.40}$$

and the solution for an initial displacement $y(t = 0) = y_0$ is depicted in Figure 2.13.

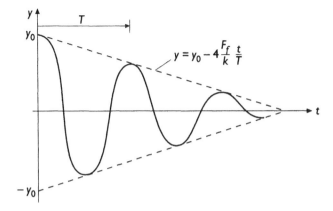

Figure 2.13 Free vibration in the presence of Coulomb damping.

We observe that for every complete cycle of oscillation ($t = T$), the total dynamic displacement $y(t)$ reduces by an amount equal to $4F_f/k$ until all motion ceases.

2.2.2.4 Damped harmonic vibrations

The equation of motion for this case is

$$M\ddot{y} + c\dot{y} + ky = F_1 \sin \Omega t \tag{2.41}$$

and the part of the solution which corresponds to forced vibration with frequency Ω is

$$y(t) = \frac{(F_1/k)[(1 - \Omega^2/\omega^2)^2 + 4(\beta\Omega/\omega^2)^2]^{1/2}\sin(\Omega t + \gamma)}{(1 - \Omega^2/\omega^2)^2 + 4(\beta\Omega/\omega^2)^2} \tag{2.42}$$

where γ is a phase angle. We mention here that the free vibration part with frequency ω dampens out rather quickly, hence it can be ignored. Since the maximum value of the sine is unity and the static displacement is $y_{st} = F_1/k$, the maximum value the DLF attains is

$$(DLF)_{max} = \frac{1}{\sqrt{(1 - \Omega^2/\omega^2)^2 + 4(\beta\Omega/\omega^2)^2}} \tag{2.43}$$

We observe that the amplitude of the vibrations is no longer infinite during resonance $\Omega = \omega$ when there was no damping. Specifically, we have that

$$(DLF)_{max,\Omega=\omega} = \frac{\omega}{2\beta} = \frac{1}{2\zeta} \tag{2.44}$$

Figure 2.14 plots the maximum value of the DLF as a function of the ratio Ω/ω. We observe that when $\Omega \to 0$, the DLF approaches the static value, while as $\Omega \to \infty$,

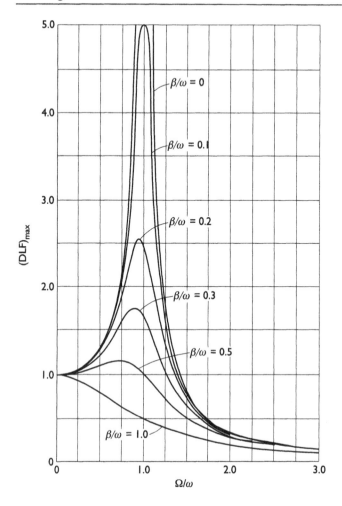

$\beta/\omega = 0$

$\beta/\omega = 0.1$

$\beta/\omega = 0.2$

$\beta/\omega = 0.3$

$\beta/\omega = 0.5$

$\beta/\omega = 1.0$

Figure 2.14 Maximum values for the DLF in the case of harmonic oscillations.

the harmonic load oscillates too rapidly for the SDOF system to respond. As previously mentioned, the dynamic response is most intense at resonance.

2.2.3 Elastoplastic systems

When dynamic loads are intense, the restoring force in the SDOF oscillator is no longer linear, but must be instead written as a generalized function $R(y)$ of the displacement so that non-linear effects can be described.

We will examine the simple case where the restoring force is linear up until the elastic limit y_{el} is reached, past which it assumes a constant value R_m. As shown in Figure 2.15, we consider a suddenly applied and maintained load $F(t) = F_1$,

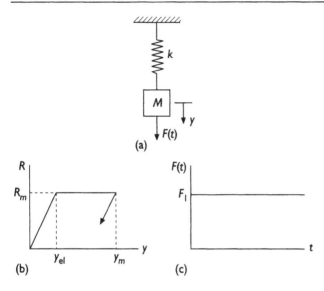

Figure 2.15 (a) Elastoplastic SDOF system, (b) restoring force and (c) loading function.

along with zero initial conditions. The SDOF system response will be divided into three stages. In the first stage, $y \leq y_{el}$ and thus $R(y) = ky$. The response is therefore

$$y(t) = y_{st}(1 - \cos \omega t) \tag{2.45}$$

where $y_{st} = F_1/k$ and $\omega = \sqrt{k/M}$. In the second stage we redefine the time variable as $t_1 = t - t_{el}$, where the elastic response time t_{el} is given by the equation

$$\cos \omega t_{el} = 1 - \frac{y_{el}}{y_{st}} \tag{2.46}$$

The new form for the equation of motion is

$$M\ddot{y} + R_m = F_1 \tag{2.47}$$

under initial conditions

$$y_0 = y_{el}, \qquad \dot{y}_0 = y_{st}\omega \sin \omega t_{el} \tag{2.48}$$

resulting from the first stage. Integrating eqn (2.47) yields

$$y(t) = \frac{1}{2M}(F_1 - R_m)t_1^2 + y_{st}\omega t_1 \sin \omega t_{el} + y_{el} \tag{2.49}$$

By taking the time derivative of the above equation and setting it equal to zero, time t_m during which displacement $y(t)$ attains its maximum value y_m can be evaluated, i.e.

$$t_m = \frac{M\omega y_{st}}{R_m - F_1} \sin \omega t_{el} \tag{2.50}$$

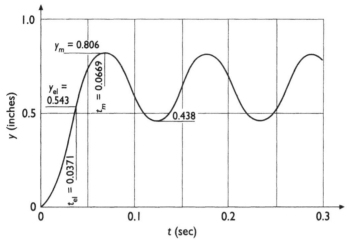

Figure 2.16 Time response of an elastoplastic SDOF system under maintained load $F(t) = F_1$.

In the final stage, we observe that we have harmonic vibration about a neutral position which is given by $y_m - (R_m - F_1)/k$. Redefining a new time $t_2 = t - t_m - t_{el}$, we have the response as

$$y(t) = \left(y_m - \frac{R_m - F_1}{k}\right) + \frac{R_m - F_1}{k}\cos \omega t_2 \tag{2.51}$$

Figure 2.16 plots the dynamic displacement $y(t)$ for the case described above, while Figure 2.17 is a nomograph for the ductility ratio μ of the SDOF elastoplastic oscillator which is defined as the ratio y_m/y_{el} for a load of magnitude F_1 and duration t_d. We finally observe that in order for the elastoplastic SDOF system to behave elastically (i.e. $\mu \leq 1$), the maximum spring resistance R_m must have at least twice the value of the magnitude of the applied load F_1.

2.3 MULTIPLE DEGREE-OF-FREEDOM SYSTEMS

The definition of a Multiple Degree-of-Freedom (MDOF) system is one which requires a second order, ordinary differential equation to describe the motion of each independent DOF. A DOF is an active translation or rotation component of motion at a given nodal point of the structure in question. In three dimensions, we have a total of six DOF per node, namely three displacements and three rotations, while on the x-y plane there is a total of three DOF, namely two displacements and one rotation. As a simple example, we have the two DOF system of Figure 2.18(a) with the following coupled equations of motion:

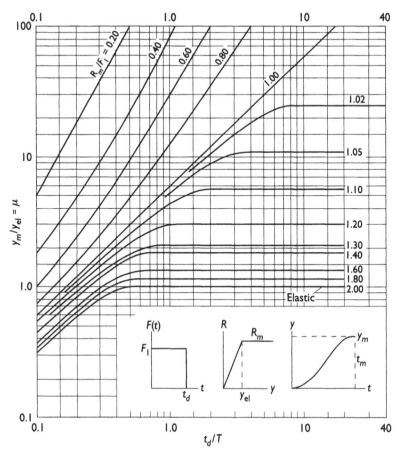

Figure 2.17 Ductility factor μ for an elastoplastic SDOF system as a function of ratio t_d/T.

$$M_1\ddot{y}_1 + k_1 y_1 - k_2(y_2 - y_1) = F_1(t)$$
$$M_2\ddot{y}_2 + k_2(y_2 - y_1) = F_2(t)$$
(2.52)

Next, Figure 2.18(b) depicts a simple rigid foundation on the ground, which is modelled by two springs. The equations of motion, assuming small foundation rotation angle θ, are

$$M\ddot{y} + 2ky = F(t)$$
$$I\ddot{\theta} + (2kd^2)\theta = M_t(t)$$
(2.53)

where M_t is the torque on the foundation and I is its mass moment of inertia. We observe that the above equations are not coupled, which implies that we do not have a true two DOF system; rather, we have a system which can execute two independent motions, namely a translation y and a rotation θ about its centre of mass.

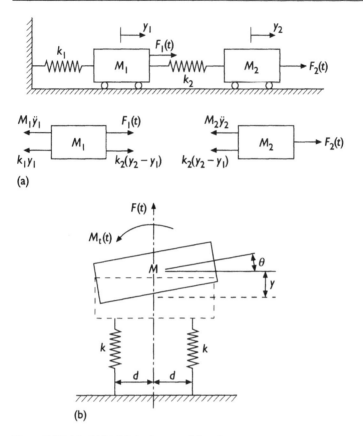

(a)

(b)

Figure 2.18 (a), (b) Multiple degree-of-freedom systems.

2.3.1 Eigenvalues and eigenvectors

The most general form of the equations of dynamic equilibrium of an MDOF system is as follows:

$$M\ddot{y}_1 + k_{11}y_1 + k_{12}y_2 + \cdots + k_{1N}y_N = F_1(t)$$
$$M_2\ddot{y}_2 + k_{21}y_1 + k_{22}y_2 + \cdots + k_{2N}y_N = F_2(t)$$

$$\cdots\cdots\cdots\cdots\cdots\cdots\cdots\cdots\cdots\cdots\cdots\cdots\cdots\cdots\cdots\cdots\cdots$$

$$M_N\ddot{y}_N + k_{N1}y_1 + k_{N2}y_2 + \cdots + k_{NN}y_N = F_N(t)$$

(2.54)

In the free vibration case where $F_1(t) = F_2(t) = \cdots = F_N(t) = 0$, all DOF have the same harmonic time variation $f(t)$. Specifically, we can express all displacement components as

$$y_1 = \alpha_1 f(t), \qquad y_2 = \alpha_2 f(t), \qquad\qquad ,y_N = \alpha_N f(t)$$

(2.55)

where $\alpha_1, \alpha_2, \ldots$ is the amplitude of vibration of each DOF. Substituting this result in the equations of motion yields

$$M_1\alpha_1\ddot{f}(t) + k_{11}\alpha_1 f(t) + k_{12}\alpha_2 f(t) + \cdots + k_{1N}\alpha_N f(t) = 0$$

$$M_2\alpha_2\ddot{f}(t) + k_{21}\alpha_1 f(t) + k_{22}\alpha_2 f(t) + \cdots + k_{2N}\alpha_N f(t) = 0$$

$$\ldots\ldots\ldots\ldots\ldots\ldots\ldots\ldots\ldots\ldots\ldots\ldots\ldots\ldots\ldots\ldots\ldots\ldots\ldots\ldots\ldots\ldots$$

$$(2.56)$$

$$M_N\alpha_N\ddot{f}(t) + k_{N1}\alpha_1 f(t) + k_{N2}\alpha_2 f(t) + \cdots + k_{NN}\alpha_N f(t) = 0$$

From all the above equations we recover the relation $\ddot{f}(t)/f(t) = \text{constant} = -\omega^2$ and thus

$$f(t) = C_1 \sin \omega t + C_2 \cos \omega t = C_3 \sin \omega(t + \gamma) \tag{2.57}$$

where γ is a phase angle. This implies that each and every one of the n DOF undergoes harmonic vibration. Substituting this result in the equations of motion yields the following:

$$(k_{11} - M_1\omega^2)\alpha_1 + k_{12}\alpha_2 + \cdots + k_{1N}\alpha_N = 0$$

$$k_{21}\alpha_1 + (k_{22} - M_2\omega^2)\alpha_2 + \cdots + k_{2N}\alpha_N = 0$$

$$\ldots\ldots\ldots\ldots\ldots\ldots\ldots\ldots\ldots\ldots\ldots\ldots\ldots\ldots\ldots\ldots\ldots$$

$$(2.58)$$

$$k_{N1}\alpha_1 + k_{N2}\alpha_2 + \cdots + (k_{NN} - M_N\omega^2)\alpha_N = 0$$

In order for the above system of equations to have a solution, its determinant must be set equal to zero, i.e.

$$\begin{vmatrix} (k_{11} - M_1\omega^2) & k_{12} & \cdots & k_{1N} \\ k_{21} & (k_{22} - M_2\omega^2) & \cdots & k_{2N} \\ \cdots & \cdots & \cdots & \cdots \\ k_{N1} & k_{N2} & \cdots & (k_{NN} - M_N\omega^2) \end{vmatrix} = 0 \tag{2.59}$$

Upon solution, we recover N values, $\omega_1, \omega_2, \ldots, \omega_N$, for the eigenfrequencies of the system. For each value ω_i which is inserted in eqns (2.58), a vector of coefficients $\{\alpha\}_i^T = [\alpha_{1i}, \alpha_{2i}, \ldots, \alpha_{Ni}]$ results which is the eigenvector corresponding to that particular eigenfrequency. We normalize each eigenvector by setting $\alpha_{1i} = 1$, since they cannot be completely determined from eqn (2.59), and proceed to solve for the remaining components $\alpha_{2i}, \ldots, \alpha_{Ni}$ relative to the first one. In this case, the notation used for the ith normalized eigenvector is $\{\Phi\}_i^T = [\Phi_{1i}, \Phi_{2i}, \ldots, \Phi_{Ni}]$.

2.3.2 Eigenvalue Analysis

A basic property of the eigenvectors is orthogonality with respect to the mass coefficients, i.e.

$$\sum_{r=1}^{N} M_r \Phi_{rn} \Phi_{rm} = 0 \tag{2.60}$$

where m and n correspond to two different eigenvectors. This fundamental property allows for uncoupling the original N coupled equations of motion into N modal equations, each of which is a dynamic equation of equilibrium for an SDOF oscillator whose natural frequency ω_i comes from the discrete spectrum $\omega_1, \omega_2, \ldots, \omega_N$.

Specifically, the nth such equation assumes the following form:

$$\ddot{A}_n(t) \sum_{r=1}^{N} M_r \Phi_{rn}^2 + A_n(t) \sum_{r=1}^{N} k_r \Phi_{\Delta rn}^2 = f(t) \sum_{r=1}^{N} F_r \Phi_{rn} \tag{2.61}$$

The subscript Δ in an eigenvector denotes the difference between two consecutive components (i.e. $\Phi_{\Delta rn} = \Phi_{rn} - \Phi_{(r-1)n}$). A comparison of the above equation with eqn (2.1) for the SDOF system reveals that the equivalent mass, stiffness and loading coefficients for the nth modal equation are

$$\sum_{r=1}^{N} M_r \Phi_{rn}^2, \quad \sum_{r=1}^{N} k_r \Phi_{\Delta rn}^2, \quad \sum_{r=1}^{N} F_r \Phi_{rn} \tag{2.62}$$

respectively. Thus, eqn (2.61) for the nth modal displacement $A_n(t)$ can be re-written as

$$\ddot{A}_n(t) + \omega_n^2 A_n(t) = \frac{f(t) \sum_{r=1}^{N} F_{r1} \Phi_{rn}}{\sum_{r=1}^{N} M_r \Phi_{rn}^2} \tag{2.63}$$

Following the solution procedure outlined for the SDOF system in the previous section, the modal static displacement A_{nst} for the nth equation is given by

$$A_{nst} = \frac{\sum_{r=1}^{N} F_{r1} \Phi_{rn}}{\omega_n^2 \sum_{r=1}^{N} M_r \Phi_{rn}^2} \tag{2.64}$$

since we have that $y_{st} = F_1/k = F_1/\omega^2 M$. For instance, Figure 2.19 plots the eigen-vectors of a two DOF oscillator.

The displacement amplitude given by the nth modal equation can be computed as $A_n(t) = A_{nst}(\text{DLF})_n$ or as $(A_n)_{\max} = A_{nst}(\text{DLF})_{n,\max}$ where the dynamic load factors DLF depend on the particular form of the load's time function $f(t)$ and on natural frequency ω_n. We note that DLFs for various load cases were presented in the previous section on SDOF systems. The final displacement response of the rth DOF of the MDOF system is found by superimposing all the modal displacement

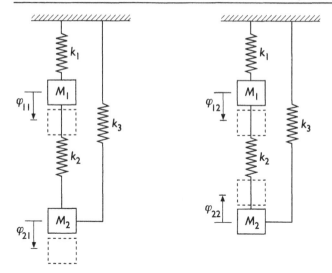

Figure 2.19 Eigenvectors of a simple two DOF system.

amplitudes as

$$y_r(t) = \sum_{n=1}^{N} A_{nst} \Phi_{rn}(\mathrm{DLF})_n \tag{2.65}$$

In sum, there are a number of methods for computing eigenvalues and their associated eigenvectors, which can be grouped into three basic categories as follows: (i) direct methods, which essentially follow the procedure previously described, (ii) iterative methods such as Jacobi's method and (iii) approximate methods (e.g. Rayleigh's method).

Using matrix notation, the equations of motion for an MDOF system assume the form shown below

$$[M]\{\ddot{U}\} + [C]\{\dot{U}\} + [K]\{U\} = \{F\} \tag{2.66}$$

where square and curly brackets respectively denote a matrix and a vector. The orthogonality property of the eigenvectors previously mentioned assumes the following form:

$$[\Phi]^T[M][\Phi] = [\bar{M}]$$
$$[\Phi]^T[K][\Phi] = [\bar{K}] \tag{2.67}$$

where overbars denote a diagonal matrix and superscript T denotes matrix transposition. If eqn (2.66) is premultiplied by $[\Phi]^T$ and if modal co-ordinates are introduced as $\{U\} = [\Phi]\{A\}$, then we recover the following uncoupled form for the equations of motion:

$$[\bar{M}]\{\ddot{A}\} + [\bar{C}]\{\dot{A}\} + [\bar{K}]\{A\} = [\Phi]^T\{F\} = \{\bar{F}\} \tag{2.68}$$

We note that the above uncoupling procedure (i.e. damping matrix $[\bar{C}]$ is also diagonal) will work only in the presence of proportional damping (i.e. if $[C] = a_1[M] + a_2[K]$, where a_1, a_2 are constants. In fact, $[C]$ can be expanded in terms of powers of $[M]$ and $[K]$ and still uncouple eqn (2.66) into N nodal equations (Bathe, 1982).

2.3.3 Damping in MDOF systems

In analogy with the SDOF oscillator, a damping coefficient β or a damping ratio ζ are defined (rather arbitrarily, given the coupling inherent in MDOF systems) for each modal equation as β_n or ζ_n, respectively. From a practical viewpoint, the first modal equation corresponding to the lowest eigenfrequency (or highest modal period) and which approximates the response of the system to quasi-static application of the load, is the dominant one. Thus, it is essential that correct values of damping are prescribed to this mode and also to a few more of the lower ones. Furthermore, it is customary to assign rather large values of damping to the higher modes so as to dampen out unwanted high frequency oscillations in the system.

2.3.4 Time integration methods

As previously mentioned, the equations of motion of an MDOF system need to be solved for the displacement vector $\{U\}$ as a function of time t. An alternative to modal analysis described in the previous section is the use of time marching algorithms, which essentially integrate over time the matrix differential equation (i.e. eqn (2.66)). There are many time marching algorithms in use today, but they all fall into two basic groups: (i) direct methods and (ii) predictor–corrector methods. The accuracy achieved through time integration is a key issue and primarily depends on the size of the time step Δt used, which is obviously judged with respect to the magnitude of the natural periods of the system. Time stepping algorithms can also be subdivided into unconditionally and conditionally stable ones. We note in passing that algorithm stability does not necessarily imply accuracy. Obviously, time marching can be used in conjunction with SDOF systems as well. Also, among the best known algorithms used in structural dynamics are those by Houbolt, Newmark and Wilson (Bathe, 1982). Finally, Table 2.1 presents Newmark's method, while Figure 2.20 compares the results obtained by various algorithms for the second storey displacement of a two-storey frame under lateral loads which vary as sine functions in time. The exact results were obtained through modal analysis in conjunction with the closed form solution given by eqn (2.23) for each of the two modes.

2.3.5 Numerical example

We examine here the three storey plane frame with rigid girders shown in Figure 2.21. In addition to the vertical static loads, the frame is subjected to dynamically

Table 2.1 Newmark's method in algorithmic form.

A. *Initialization*
  1. Formation of stiffness matrix K, mass matrix M and damping matrix C
  2. Initial values ^0U $^0\dot{U}$ and $^0\ddot{U}$
  3. Assign values to time step Δt and to parameters α and δ. Computation of the following integration constants:

$$\delta \geq 0.50, \qquad \alpha \geq 0.25(0.5 + \delta)^2,$$

$$\alpha_0 = \frac{1}{\alpha \Delta t^2}, \quad \alpha_1 = \frac{\delta}{\alpha \Delta t}, \qquad\qquad \alpha_2 = \frac{1}{\alpha \Delta t}, \qquad \alpha_3 = \frac{1}{2\alpha} - 1$$

$$\alpha_4 = \frac{\delta}{\alpha} - 1, \quad \alpha_5 = \frac{\Delta t}{2}\left(\frac{\delta}{\alpha} - 2\right), \quad \alpha_6 = \Delta t(1 - \delta), \quad \alpha_7 = \delta \Delta t$$

  4. Formation of the effective stiffness matrix $\hat{K}$, where $\hat{K} = K + \alpha_0 M + \alpha_1 C$
  5. Triangularization of matrix $\hat{K}$, $\hat{K} = LDL^T$

B. *At each time step level*
  1. Computation of the effective load vector

$$^{t+\Delta t}\hat{R} = {}^{t+\Delta t}F + M(\alpha_0 {}^t U + \alpha_2 {}^t\dot{U} + \alpha_3 {}^t\ddot{U}) + C(\alpha_1 {}^t U + \alpha_4 {}^t\dot{U} + \alpha_5 {}^t\ddot{U})$$

  2. Solution for displacements at time $t + \Delta t$

$$(LDL^T)^{t+\Delta t}U = {}^{t+\Delta t}\hat{R}$$

  3. Computation of accelerations and velocities at time step $t + \Delta t$

$$^{t+\Delta t}\ddot{U} = \alpha_0({}^{t+\Delta t}U - {}^t U) - \alpha_2 {}^t\dot{U} - \alpha_3 {}^t\ddot{U}$$

$$^{t+\Delta t}\dot{U} = {}^t\dot{U} + \alpha_6 {}^t\ddot{U} + \alpha_7 {}^{t+\Delta t}\ddot{U}$$

induced horizontal loads applied at the storey levels. The frame is modelled as a three DOF system and the interstorey stiffness is $k = 2(12EI/h^3)$, where $EI(= 14.67 \text{ kN m}^2)$ is the flexural rigidity of the columns and h is their clear height. The mass lumped at each storey is the total static load pL, where L is the span, divided by the acceleration of gravity g (= 9.81 m/sec²). We thus compute $k_1 = 30.7, k_2 = k_3 = 44.0$ and $M_1 = 141.0, M_2 = 132.0, M_3 = 66.0$ for the stiffnesses and masses, respectively, in units of (kN/m) and (N sec²/m). Finally, we note that the structure s own weight is included in the vertical load.

The equations of dynamic equilibrium are

$$M_1 \ddot{y}_1 + k_1 y_1 - k_2(y_2 - y_1) = F(t)$$

$$M_2 \ddot{y}_2 + k_2(y_2 - y_1) - k_3(y_3 - y_2) = 0.8F(t) \qquad (2.69)$$

$$M_3 \ddot{y}_3 + k_3(y_3 - y_2) = 0.5F(t)$$

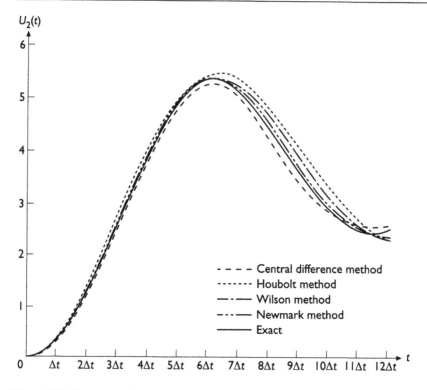

Figure 2.20 Comparison between various commonly used time integration methods for a two DOF system.

We first focus on the free vibration problem, with $f(t) = \sin \omega t$, so that eqns (2.69) assume the form

$$(-M_1\omega_n^2 + k_1 + k_2)\alpha_{1n} + (-k_2)\alpha_{2n} = 0$$

$$(-k_2)\alpha_{1n} + (-M_2\omega_n^2 + k_2 + k_3)\alpha_{2n} + (-k_3)\alpha_{3n} = 0 \qquad (2.70)$$

$$(-k_3)\alpha_{2n} + (-M_3\omega_n^2 + k_3)\alpha_{3n} = 0$$

The eigenfrequencies, natural periods as well as the corresponding eigenvectors which result from solving the above homogeneous system of equations are given in Table 2.2. Also, a sketch of the three eigenvectors appears in Figure 2.22. Next, we continue with modal analysis along the lines developed in the previous section; Table 2.3 presents all intermediate computations plus the final static values of the three modal displacements A_{nst}, $n = 1, 2, 3$ from eqn (2.64). The values for the DLF corresponding to each modal equation depend on the rise time t_r of the applied load (see Figure 2.8) and on the natural periods T_n; they in turn are given in Table 2.4.

Maximum values for the three modal components of the horizontal dynamic dis-

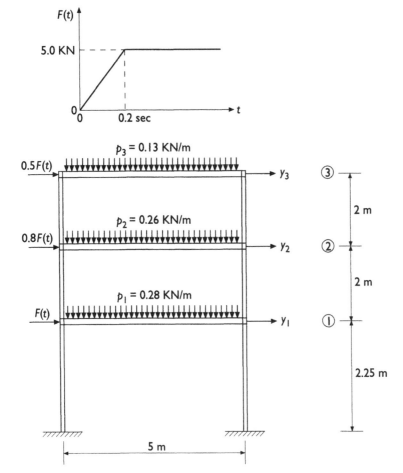

Figure 2.21 Three-storey frame structure with rigid girders.

Table 2.2 Eigenvalues and eigenvectors of the three-storey frame.

Eigenvector	$\omega^2 (rad/sec)^2$	T (sec)	Φ_{1n}	Φ_{2n}	Φ_{3n}
1	69.3	0.755	+1.00	+1.471	+1.639
2	579.0	0.261	+1.00	−0.146	−1.041
3	1231.0	0.179	+1.00	−2.220	+2.680

placement $y_3(t)$ at the third storey level are given below separately (see eqn (2.65)) as

$$y_{31} = (+0358) \times (+1.639) \times (1.89) = +1.11 \text{ cm}$$

$$y_{32} = (0.0146) \times (-1.041) \times (1.28) = -0.02 \text{ cm} \tag{2.71}$$

$$y_{33} = (+0.0018) \times (+2.680) \times (1.11) = +0.005 \text{ cm}$$

Figure 2.22 Schematic view of the eigenvectors of the three-storey frame building.

Table 2.3 Eigenvalue analysis of the three-storey frame structure.

Storey	F_{rl}	M_r	1st eigenvector			2nd eigenvector			3rd eigenvector		
			ϕ_{rl}	$F_{rl}\phi_{rl}$	$M_r\phi_{rl}^2$	ϕ_{r2}	$F_{r2}\phi_{r2}$	$M_r\phi_{r2}^2$	ϕ_{r3}	$F_{r3}\phi_{r3}$	$M_r\phi_{r3}^2$
1	5,000	141	1.000	5,000	141	1.000	5,000	141	1.000	5,000	141
2	4,000	132	1.471	5,884	286	−0.146	−548	3	−2.220	−8,880	650
3	2,500	66	1.639	4,097	177	−1.041	−2,602	72	2.680	6,700	474
$\sum\limits_{r=1}^{3}$				14,981	604		1,814	216		2,820	1265

$$A_{1st} = \frac{14981}{69.3 \times 604} = 0.358 \text{ cm}$$

$$A_{2st} = \frac{1814}{579 \times 216} = 0.0146 \text{ cm}$$

$$A_{3st} = \frac{2820}{1231 \times 1265} = 0.0018 \text{ cm}$$

The total third storey maximum horizontal displacement is approximately the sum of the absolute values of the three modal contributions (i.e. $y_{3max} = 1.13$ cm). The reason for this is that the above maxima do not occur simultaneously in time. As a result, a number of techniques have been devised (Chopra, 1995), for improvement and the value quoted here is obviously a conservative upper bound.

2.4 CONTINUOUS DYNAMIC SYSTEMS

A continuous dynamic system has an infinite number of DOF and eigenvalues, while the associated eigenvectors are continuous functions of the space variables. All structures in reality are continuous dynamic systems and their modelling by SDOF or MDOF systems is approximate and done for practical reasons.

Table 2.4 Maximum values of the DLF for load F(t).

Eigenvector	t_r/T	$(DLF)_{max}$
1	0.26	1.89
2	0.77	1.28
3	1.12	1.11

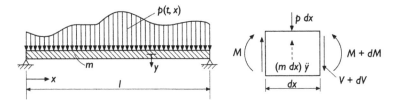

Figure 2.23 The flexural beam as a continuous dynamic system.

2.4.1 Equations of motion for continuous beams

As example, we will examine the flexural beam, which is one of the basic unidimensional structural elements. Referring to Figure 2.23, the equation of dynamic equilibrium of a continuous beam element is

$$EI\frac{\partial^4 y(t, x)}{\partial x^4} + m\ddot{y}(t, x) = p(t, x) \tag{2.72}$$

where EI is the flexural rigidity, m is the mass per unit length, p is the distributed load and $y(t, x)$ is the transverse displacement. For free vibrations, we have that $p(t, x) = 0$ and

$$y(t, x) = \sum_{n=1}^{\infty} f_n(t)\Phi_n(x) \tag{2.73}$$

where $\Phi_n(x)$ is the nth eigenvector. The original equation of motion can be split into two, which respectively govern the temporal and spatial variation of the displacement $y(t, x)$ as

$$\left.\begin{array}{l} \ddot{f}_n(t) + w_n^2 f_n(t) = 0 \\ \dfrac{d^4}{dx^4}\Phi_n(x) - \dfrac{mw_n^2}{EI}\Phi_n(x) = 0 \end{array}\right\} \tag{2.74}$$

The solution for the time function $f_n(t)$ and the eigenvector $\Phi_n(x)$ are given below as

$$f_n(t) = C_1 \sin w_n t + C_2 \cos w_n t \tag{2.75}$$

$$\Phi_n(x) = A_n \sin \alpha_n x + B_n \cos \alpha_n x + C_n \sinh \alpha_n x + D_n \cosh \alpha_n x \tag{2.76}$$

$$\alpha_n = \sqrt[4]{\frac{mw_n^2}{EI}} \tag{2.77}$$

Table 2.5 Eigenvectors of beam of length l under various support conditions

$$\phi(x) = \left(\frac{a}{\ell}\right)_n (\sinh a_n x - \sin a_n x) + \cosh a_n x - \cos a_n x.$$

	Mode	$(a/\ell)_n$	$\int_0^l \phi(x)\, dx$
	1	−0.9825	0.8308l
	2	−1.0007	0
	3	−1.0000	0.3640l
	1	−1.0007	0.8604l
	2	−1.0000	0.0829l
	3	−1.0000	0.3343l
	1	−0.7341	0.7830l
	2	−1.0184	0.4340l
	3	−0.9992	0.2544l

$$\int_0^l [\phi(x)]^2 \, dx = l$$

We obviously have an infinite number of harmonic vibrations with frequency ω_n. Finally, the integration constants appearing in eqn (2.76) depend on the boundary conditions of the beam in question and a few cases are listed in Table 2.5.

As in the case of MDOF systems, a complete eigenvalue analysis is required when non-zero loads are present. In that case, the solution for the transverse dynamic displacement is given by

$$y(t, x) = \sum_{n=1}^{\infty} A_n(t)\Phi_n(x) \tag{2.78}$$

where $A_n(t)$ is the amplitude of vibration of the (uncoupled) nth oscillation component, which is a function of the applied load, while $\Phi_n(x)$ is the corresponding eigenvector.

2.4.2 Examples of various continuous systems

As examples, Figures 2.24–2.27 present the eigenvalues and eigenvectors for four typical types, namely the simply supported beam, the cantilever beam, the fixed end beam and finally the fixed end-simply supported beam.

2.5 BASE EXCITATION AND RESPONSE SPECTRA

The standard method of analysis in earthquake resistant design is through use of response spectra, because in civil engineering practice we are no longer interested

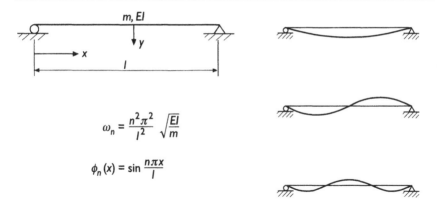

$$\omega_n = \frac{n^2\pi^2}{l^2}\sqrt{\frac{EI}{m}}$$

$$\phi_n(x) = \sin\frac{n\pi x}{l}$$

Figure 2.24 Dynamic properties of the simply supported beam.

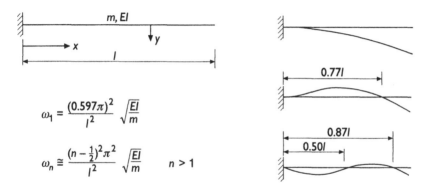

$$\omega_1 = \frac{(0.597\pi)^2}{l^2}\sqrt{\frac{EI}{m}}$$

$$\omega_n \cong \frac{(n-\frac{1}{2})^2\pi^2}{l^2}\sqrt{\frac{EI}{m}} \qquad n > 1$$

Figure 2.25 Dynamic properties of the cantilever beam.

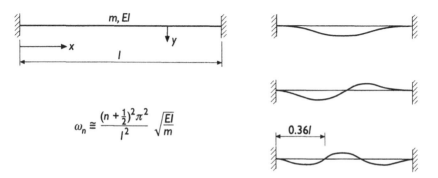

$$\omega_n \cong \frac{(n+\frac{1}{2})^2\pi^2}{l^2}\sqrt{\frac{EI}{m}}$$

Figure 2.26 Dynamic properties of the fixed end beam.

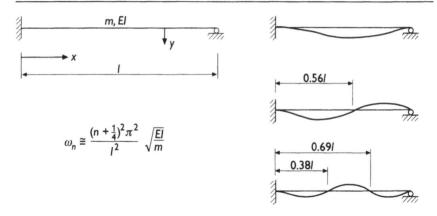

Figure 2.27 Dynamic properties of the fixed end simply supported beam.

in the time evolution of the structural response; instead, we are interested in the maximum values attained by the structure's relative displacements, relative velocities and absolute accelerations since those values control the maximum stresses that ultimately develop.

A response spectrum is defined as the maximum response (be it displacement, velocity or acceleration) of all possible SDOF oscillators, which can be described by their natural frequency and damping coefficient, to a given ground motion. Note that a response spectrum is not the same as the DLF for a SDOF oscillator; both, however, can be used in the analysis of SDOF, MDOF or continuous systems. In Figures 2.28 and 2.29, we respectively present spectra resulting from artificially generated ground accelerations and the true, triple-scale response spectrum for the main shock produced by the Kalamata, Greece 1986 earthquake (Anagnostopoulos *et al.*, 1986).

Response spectra can be classified as either elastic spectra, inelastic spectra, site specific spectra, code prescribed spectra or as design spectra. Here we focus on the first type, as being the most relevant to a first exposure in structural dynamics, and because they form the basis from which the remaining ones can be derived. Specifically, and in order to complete the presentation, we list the equations of motion of the SDOF oscillator subjected to ground displacements $y_s(t)$ and to ground accelerations $\ddot{y}_s(t)$, respectively, as

$$M\ddot{y}(t) + c\dot{y}(t) + ky(t) = ky_s(t) + c\dot{y}_s(t) \tag{2.79}$$

and

$$M\ddot{u}(t) + c\dot{u}(t) + ku(t) = -M\ddot{y}_s(t) \tag{2.80}$$

where

$$u(t) = y(t) - y_s(t) \tag{2.81}$$

is the relative displacement between ground and structure.

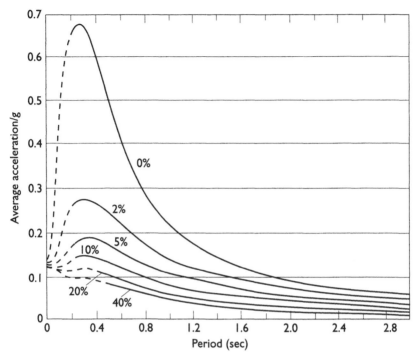

Figure 2.28 Response spectra derived from artificial accelerograms.

The first step in the construction an elastic, relative displacement spectrum S_d (from u) is the solution of eqn (2.80) to a given ground acceleration. The closed form expression for $u(t)$ is Duhamel's integral given by eqn (2.38) for zero initial conditions and for y_{st} defined as equal to $-\ddot{y}_{s0}/\omega^2$, where $\ddot{y}_s(t) = \ddot{y}_{s0}f(t)$ and the difference between ω and ω_d ignored. Given the complexity of ground motion, Duhamel's integral is computed by numerical quadrature, and the maximum value recorded for a given natural frequency and at a given damping level is stored. This process is repeated for a range of frequencies which is considered adequate for design purposes, and for damping ratios up to 20 per cent. The other two spectra (i.e. S_a for the absolute accelerations (from $\ddot{y}$) and S_v for the relative velocities (from $\dot{u}$) are derived from S_d using the following relation given below:

$$S_a = \omega S_v = \omega^2 S_d \tag{2.82}$$

In the above, ω is the natural frequency of the SDOF oscillator. Since eqn (2.82) is exact only in the absence of damping, S_a and S_v are respectively known as spectral pseudo-acceleration and spectral pseudo-velocity. Finally, response spectra are often plotted in terms of the natural period and by using logarithmic scales.

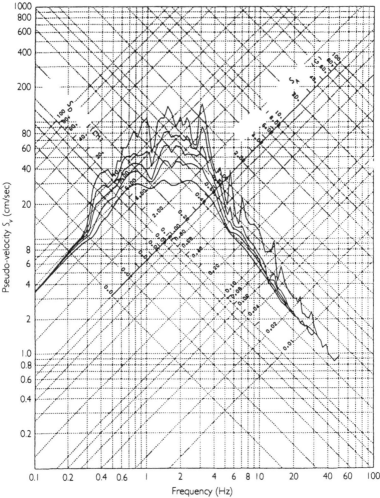

Figure 2.29 Triple spectrum for the Kalamata, Greece 1986 earthquake: velocity (cm/sec) along vertical axis; acceleration (g) along left to right axis; relative displacement (cm) along right to left axis; all versus frequency (Hz). Note: the five curves are for 0%, 2%, 5%, 10% and 20% damping.

2.5.1 Numerical example

In this section, we examine a simple, one-storey warehouse structure which essentially supports a roof loading and is acted upon by the Kalamata, Greece 1986 earthquake (Anagnostopoulos *et al.*, 1986). We employ the response spectrum method and present the solution in algorithmic form. We note here that design of heavy roof slabs with or without strong edge beams is not recommended as good earthquake resistant design. Instead, correct practice is to design for strong columns, weak beams and, if possible, light roof slabs.

(a) Problem description

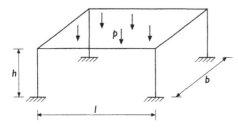

$h = 6$ m

$l = 12$ m

$b = 6$ m

$p = 5$ kN/m^2

$E = 35,000$ MPa

$I = 0.00213$ m^4

Column cross-section: 0.4×0.4 m

Column stiffness computation:

$$Q = 12\ EI/h^3$$

$$k = 4Q = 48EI/h^3 = 16,600 \text{ kN/m}$$

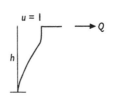

Mass computation:

$$M = (plb)/g = 36.7 \text{ kN sec}^2/\text{m}$$

Damping coefficient:

$$\zeta = c/c_{cr} = 10\% = 0.1$$

(b) SDOF system model

$$\omega = \sqrt{\frac{k}{M}} = \sqrt{\frac{16,600}{36.7}} = 21.3 \text{ rad/sec}$$

$$T = 2\pi/\omega = 0.30 \text{ sec}, \qquad f = 1/T = 3.38 \text{ Hz}$$

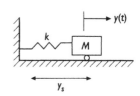

(c) Response spectrum computations

From the triple Kalamata 1986 earthquake response spectrum given in Figure 2.29, we have:

- maximum relative displacement is $u = y - y_s = 1.8$ cm;
- maximum velocity is $y = 35$ cm/sec;
- maximum acceleration is $\ddot{y} = 0.7$ g $= 6.87$ m/sec^2;
- maximum column shear is $V = (ku)/4 = 16,600(0.018)/4 = 74.7$ kN; and
- maximum column shear stress is $\tau = V/A = 74.7/(0.4^2) = 467$ kN/m^2

2.6 SOFTWARE FOR DYNAMIC ANALYSIS

There is much software available today that will, among other things, perform dynamic analyses of typical structural systems such as buildings, bridges, storage tanks, etc. These computer programs are based on the Finite Element Method (FEM) for discretizing the structure so as to produce a mathematical model which can then be used within the context of numerical solution procedures. As one of the earlier programs that was once public domain but in the last 15 years is commercially available, we mention the Structural Analysis Program (SAP) (SAP 2000, 1997). This program is based on the doctoral work of E.L. Wilson at the University of California, Berkeley in the late 1960s when large scale computer implementation of the FEM started (Bathe, 1982). The list of computer programs is quite extensive, and the interested reader is advised to consult current, general information journals in civil engineering (*Civil Engineering* magazine of the ASCE; *New Civil Engineer*, magazine of the Institution of Civil Engineers) where such software is advertised. In this respect, we mention NASTRAN as one of the largest and most complete FEM packages available today, while STAAD/Pro, ANSYS, GT-STRUDL, LUSAS, LARSA, ETABS, IDARC-3D, etc., are some of the better known structural analysis and design software packages in the market. Finally, it is possible to download special purpose, structural dynamics software from the Internet. As example, we mention the numerical integration program NONLIN (NONLIN, 1997) for the SDOF oscillator, which is capable of capturing material nonlinearity and accessible through the MS Windows operating system.

2.7 REFERENCES

Anagnostopoulos, S. A., Theodoulidis, N. P., Lekidis, B. A., Margaris, B. N. (1986) 'The Kalamata 1986 Earthquake', Technical Report No. 86-2 (in Greek), Institute of Technical Seismology and Earthquake Resistant Structures (ITSAK) Publication, Thessaloniki, Greece.

Argyris, I. and Mlejnek, H. P. (1991) *Dynamics of Structures*, North-Holland, Amsterdam.

Bathe, K. J. (1982) *Finite Element Procedures in Engineering Analysis*, Prentice-Hall, Englewood Cliffs, NJ.

Biggs, J. M. (1965) *Structural Dynamics*, McGraw-Hill, New York.

Chopra, A. K. (1995) *Dynamics of Structures: Theory and Applications to Earthquake Engineering*, Prentice-Hall, Englewood Cliffs, NJ.

Civil Engineering, magazine of the American Society of Civil Engineers, 1801 Alexander Bell Dr., Reston, Virginia, twelve monthly issues per year.

Clough, R. and Penzien, J. (1993) *Dynamics of Structures*, 2nd edn, McGraw-Hill, New York.

Craig, R. R. (1981) *Structural Dynamics*, John Wiley, New York.

New Civil Engineer, magazine of the Institution of Civil Engineers, 1 Great George Street, London, twenty-five biweekly issues per year.

Newmark, N. M. and Rosenblueth, E. (1971) *Fundamentals of Earthquake Engineering*, Prentice-Hall, Englewood Cliffs, NJ.

NONLIN (1997) *An Educational Program for Structural Dynamics and Earthquake Engineering, Advanced Structural Concepts*, Golden, CO.

Paz, M. (1997) *Structural Dynamics*, 4th edn, Van Nostrand-Reinhold, New York.

SAP 2000 (1997) *Integrated Finite Element Analysis and Design of Structures*, Version 6.0, Computers and Structures, Berkeley, CA.

Chapter 3

Wind loading

T. A. Wyatt

3.1 WIND GUST LOADING

3.1.1 Basic concepts

The real wind is generally turbulent. Particularly is this so in the extreme storm winds that are the usual focus for conventional wind loads, because the passage of the wind over the irregularities of the ground surface (terrain roughness) will create sufficient disturbance to break down any stable stratification in the wind that may result from a temperature lapse rate that is less than the adiabatic value (an 'inversion'). On the other hand, additional contributions to the turbulence created by thermal instability (convective gusts) are broken down. The presumption is thus a neutral Atmospheric Boundary Layer (ABL) in which the gust structure is dominated by the effect of ground roughness. A standard description is postulated for the profile of mean wind speed with height, with a statistical description of the turbulence superimposed on it, with primary dependence on the ground roughness parameter z_0. In practical terms z_0 is inferred by relating the observed mean speed profile near the ground to a theoretical model. However, a temperature lapse rate lower than the adiabatic value can lead to abnormally smooth flow, which may increase susceptibility to aerodynamic instability in light or moderate winds, discussed later.

In practice, evaluation of gust action is commonly based on classification of z_0 in three steps: $z_0 = 0.003$ m for very smooth surfaces typified by tundra or water (correction for sea surface roughness is possible as a function of storm strength), $z_0 = 0.03$ m for typical UK inland countryside, $z_0 = 0.3$ m for suburban housing and forests. Although higher values of z_0 are possible (e.g. for city centres), the basic presumption of a generalized statistical pattern is unreliable (see buffeting, Section 3.2.6); an estimate of structural response based on the model for suburban roughness will give a conservative estimate of the mean load and probably of the peak gust response, but may underestimate the specifically dynamic action. An *ad hoc* wind tunnel test will give valuable additional information in such cases, but interpretation is beyond the scope of this chapter.

The basic measure of gust strength is the root mean square (r.m.s.) value of the perturbation of the wind speed at a point about the mean; the hourly mean value $\bar{V}$ is the ideal reference for the large storms typical of temperate climates. If a Cartesian co-ordinate system is referred to the mean wind direction, (x alongwind, y horizontal crosswind, z vertical) a corresponding component system (u, v, w, respectively) can be applied to the velocity fluctuation. Notation σ_u, etc. is used to denote the r.m.s. fluctuation, and the intensity of turbulence $I = \sigma_u/\bar{V}$; σ_u as well as $\bar{V}$ is a function of height above ground z. The 'equilibrium' values resulting from a very long fetch of uniform roughness are well established together with reasonable evaluation of the development over changes in roughness (Harris and Deaves, 1981; Cook, 1985).

The UK general code of practice for wind loads, BS6399 Part 2 includes a so-called 'directional procedure' with tabulated coefficients relating the local mean speed $\bar{V}(z)$ and intensity of turbulence $I(z)$ to the basic storm strength (V_b in the notation of BS6399) as a function of the local terrain and topography. Factors S_c and S_t give $\bar{V}$ and I, respectively, for locations in open country, as a function of distance from the coast in each selected wind direction, with the possible addition of allowance for the influence of ground contour (topography, S_h). Further correc-tions (factors T_c and T_t) are given for sites in urban or forest terrain. BS6399 continues with procedures to assess the correlation of gust action over the extent of the structure as a static load process and a simple generalized factor for dynamic augmentation of response.

To proceed further to address the dynamic effects of gusts, in the sense of effects influenced by the inertia of the structure, the methodology is extended by represen-tation in the frequency domain using the Fourier integral transform as outlined in Section 3.1.2. The Fourier integral (spectral) approach can also be used in assessing the spacial correlation of quasi-static pressures, which is dominated by the action of frequency components well below any resonant frequency of the structure, where inertial effects are negligible. Some of the approximations made in the dynamic analysis given in this chapter become questionable at low frequen-cies, and the writer advocates instead a direct approach to this part of the wind-load problem, either by formal static correlation analysis (Wyatt, 1981) or by the postulate of a critical gust duration proportional to the quotient of the size of the loaded surface in question with the mean windspeed (Cook, 1985). This so-called 'TVL' approach (gust averaging time, windspeed, loaded length) is developed in BS6399 Part 2.

The distinctive characteristics of gust response, by contrast especially with earth-quake effects (see Chapter 4), are:

- long duration, the storm persisting near peak intensity for duration of order 1 hour;
- a coexistent mean load, permitting no reversal of inelastic deformations;
- the forces are randomly spatially variable over the structure;
- effects are significant over a frequency range extending down to a few cycles per hour.

Any individual increments of inelastic deformation must therefore be kept very small, and a linear structural analysis is sufficient, but a sophisticated treatment of the correlation of gust actions over the extent of the structure is essential. The duration and random nature of the process makes the power spectrum (see Chapter 10) an attractive analytical tool.

The ensuing presentation concentrates on the basic application of spectral analysis to the gust loading problem pioneered by Davenport (1961, 1962), using the neutral ABL model, although there is increasing recognition of the potential importance of convective effects (Wyatt, 1995). This analysis further presumes simple 'quasi-steady' aerodynamics; the companion problems of dynamic effects caused by flow-pattern instabilities or by feedback of structural motion to the aerodynamic forces are considered under the heading of Aerodynamic Instability, Section 3.2.

3.1.2 Spectral description of wind loading

The turbulent velocities are described by Cartesian components (u, v and w) superimposed on the mean windspeed $\bar{V}$; u is in the mean wind direction, w is commonly used for the vertical component. It is generally presumed that the turbulence components can be treated for analytic approximations as small compared to $\bar{V}$; the instantaneous windspeed $V(t)$ is thus $V(t) = \bar{V} + u(t)$, and v and w can be treated as causing small changes in the instantaneous wind direction. The notation σ_u^2 is used throughout this chapter for the variance of quantity u, and correspondingly for other input and response quantities.

In a severe temperate-climate windstorm, the mean windspeed and the associated statistical description of the gusts carried by it remain constant ('stationary' in the statistical sense) for a sufficient duration that analysis in the frequency domain using power spectra is the preferred approach. Provided the gusts are the result of surface roughness over a long fetch, rather than being substantially influenced by specific discrete obstacles in the immediate vicinity, the input spectra take universal normalized forms. In this way, the spectrum of each turbulence component is fully defined by the r.m.s. value and a timescale parameter (for normalization of frequency), the Harris–von Karman algebraic formulation being widely accepted (see below, especially Figure 3.2a, Section 3.1.3) . Cross-spectra describing the spatial correlations are also crucial in this application (Harris and Deaves, 1981; ESDU, 1986b).

Given standard algebraic descriptions of the input spectra, the subsequent analytic steps are straightforward in application, and *ad hoc* numerical Fourier-transform operations are not normally required except for interpretation of wind tunnel data or full size monitoring studies; such specialist aspects are not considered further here. Full description of spectral procedures can be found in sources such as Newland (1993), in which the mathematical basis is developed, crucially equations 10.20–10.22, and 10.71. Attention is specially drawn to the discussion given with these equations. Generally good guidance can be drawn from the simple concept that the spectrum defines the strength of an infinite number of infinitesimally spaced sinusoid components which are sustained so that the steady state response is attained for each component. The essential randomness results from the infinite complexity of the beat phenomena between such components. The magnitudes are defined in mean-square terms; the spectrum describes the distribution of the variance (mean square deviation from the mean) of the process on a frequency abscissa, and the ordinates are thus values of (process)^2 per unit of frequency. Throughout this chapter the notation $\sigma^2(\cdot)$ is used for the variance of the quantity indicated in the parenthesis (in the case of windspeed, σ_u is used for $\sigma(u)$ to facilitate concise presentation).

Wind engineering is exceedingly fortunate that the choices of spectral definition and notation in the seminal presentations (Davenport, 1961, 1962) have been universally followed. The basic spectrum is used in the single-sided form with frequency (n) expressed in Hz, which is denoted $S(n)$ (although $W(n)$ has become more common as the notation for this form in other fields). The numerical values of ordinates in the wind engineering format are thus 4π times the values for the double-sided circular-frequency form given as $S(\omega)$ in eqn 10.20. It is further general in wind engineering to present spectra in the normalized non-dimensional format of $nS(n)/\sigma^2$ plotted on a logarithmic scale abscissa n. Noting that

$$\int_{-\infty}^{\infty} nS(n)\, d(\log n) = \int_{0}^{\infty} S(n)\, dn = \sigma^2 \qquad (3.1)$$

this preserves the visual interpretation of the area of the spectral plot as the variance of the process. It gives a clear graphical representation despite the considerable frequency range present in the natural wind and has the great convenience that scaling parameters applicable to the frequency abscissa have the effect only of a 'rigid body' shift of the normalized shape.

The Harris–von Karman normalized form (Figure 3.2a, Section 3.1.3) for the alongwind gust component (u) is

$$nS_u(n)/\sigma_u^2 = 0.6\tilde{n}/(2+\tilde{n}^2)^{5/6} \qquad (3.2)$$

The frequency normalization favoured by the present author is $\tilde{n} = 12nT$, in which T is the timescale, the (one-sided) integral of the autocorrelation function of the windspeed. Theory derived for Homogeneous Isotropic Turbulence (HIT), which ignores the distortion of the turbulence field resulting from proximity to the ground (where w must clearly be zero), gives the numerical factor as $2\sqrt{(2\pi)}\Gamma(\frac{1}{3})/\Gamma(5/6) = \sqrt{(2\times 70.78)} = 11.9$. The length scale $^xL_u \equiv T\bar{V}$ may alternatively be used as the input parameter for frequency normalization. Dynamic analysis is focused on the upper tail of the spectrum $\tilde{n} \gg 1$, as demonstrated later; in this range

$$\frac{nS_u(n)}{\sigma_u^2} = \frac{0.6}{(12nT)^{2/3}} \qquad (3.3)$$

Table 3.1 Modified timescale T_s (seconds).

Reference windspeed V_b		25 m/s			32 m/s		
Location (terrain)		Coast	Country	Town	Coast	Country	Town
Roughness length z_0 (m)		0.003	0.03	0.3	0.003	0.03	0.3
Height above	10	4.4	3.5	2.8	3.6	2.9	2.1
ground, z(m)	20	6.3	6.0	5.4	5.6	5.0	4.2
	50	7.3	9.2	10.6	7.0	8.4	8.7
	100	7.9	11.0	15.1	7.3	9.8	12.7
	200	8.5	12.7	19.4	7.9	11.4	16.6

Storm strength is generally best defined by the *hourly mean* windspeed; a 10-minute average may be substituted where this is the basis of local records, or in climates where the storm peak is not stationary for the longer period. Unfortunately, although definition of local values of $\bar{V}$ and σ_u for a defined synoptic storm strength is well established for the case of extreme winds in temperate climatic locations, this is less true for the scale parameter. The preference here for T rather than ^xL_u reflects the postulate that the depth of the surface boundary layer, and thus ^xL_u, increases with storm strength. Extensive further discussion will be found in *ESDU Data Items 85020*, Revision E (ESDU, 1990) and 86010. Item 85020 develops a more complex spectral form, but upper-tail ordinates may still be evaluated from the basic expression given above, by substitution for T of an effective timescale T_s, which is given by $(^xL_u/12\bar{V})(0.6/A)^{3/2}$ in the ESDU notation. Representative values are given in Table 3.1 (Maguire and Wyatt 1999).

Where there is a discrete obstacle in the upstream flow of size comparable to the structure under consideration, the response to turbulence may be significantly enhanced, even at separations exceeding 20 times the width of the obstacle, and the only solution may be specific wind tunnel modelling. For this reason no values are given in Table 3.1 for city centre conditions. An indication of response may be obtained by applying 'town' input parameters, which are based on extensive suburban ground roughness, in the expectation that enhanced turbulence would be compensated by reduced mean windspeed, but this is not necessarily conservative.

Wind loading is based on the concept of a 'kinematic pressure' q, given by $q(t) = \frac{1}{2}\rho V^2(t)$, in which ρ is the density of air (about $1.2\,\text{kg/m}^2$ at normal altitudes). The actual pressure on a structural surface is obtained by multiplying q by an appropriate coefficient; catalogues of such coefficients are given in the various design guides and codes. The alongwind force ('drag' P, say) on a simple structure such as a signboard face on to the wind can thus be written $P = qAC_D$ in which A is the loaded area and C_D is the drag coefficient. For practical manipulation in the frequency domain, the basic drag force formulation is first linearized, i.e.

$$P(t) = \tfrac{1}{2}\rho V^2(t)AC_D = \tfrac{1}{2}\rho(\bar{V}+u)^2 AC_D = \tfrac{1}{2}\rho(\bar{V}^2 + 2u\bar{V})AC_D \qquad (3.4)$$

It is then simplified by writing $\frac{1}{2}\rho A C_D \bar{V}^2 = \bar{P}$, so that

$$P(t) = \bar{P}\{1 + 2u(t)/\bar{V}\} \tag{3.5}$$

The corresponding spectral relationship describing the force fluctuation is then

$$S_P = \bar{P}^2(2/\bar{V})^2 S_u \tag{3.6}$$

or

$$nS_P/\bar{P}^2 = 4I^2 nS_u/\sigma_u^2 \tag{3.7}$$

3.1.3 Structural response: the line-like structure

It has thus far been assumed that the area A is sufficiently small that gust perturbation u can be assumed uniform over the structure. In practice, however, to evaluate dynamic response, it is always necessary to make allowance for imperfect spacial correlation. The basic analysis addresses the case of a structure whose loading can reasonably be defined by reference to a single coordinate, such as height above ground for a slender tower or a chimney, or location along a cable or a bridge deck. It is further assumed that the force acting at any point is fully defined by the windspeed at that location that would have arisen in free stream, so that the spacial correlations of the load are the same as for the incident wind.

Consider a member of length H and cross-section such that the force exerted on an element of length dz can be written $dp = p \, dz = qbC_D \, dz$, in which the drag coefficient reference dimension b may be a function of the location co-ordinate z. The modal generalized force is $P = \int p(z)\mu(z) \, dz$, in which $\mu(z)$ is the shape function. After linearization as introduced at eqn (3.4), the time varying component is

$$P(t) = \int_0^H \rho C_D \mu(z) b(z) \bar{V}(z) u(z, t) \, dz \tag{3.8}$$

To simplify presentation, the variation of $\bar{V}$ with z will initially be ignored. The autocovariance function at time-lag $\tau (C_P(\tau)$ say) is then

$$C_P(\tau) = (\rho \bar{V} C_D)^2 \left\langle \int_0^H \int_0^H \gamma(z)\gamma(z')u(z, t)u(z', t + \tau) \, dz \, dz' \right\rangle \tag{3.9}$$

in which $\gamma(z)$ has been written to comprise the quantities variable with location but invariant with time, in this case $\gamma(z) \equiv \mu(z)b(z)$, and the brackets $\langle \ \rangle$ signify time-average.

In HIT, $\langle u(z, t) u(z', t + \tau) \rangle = \langle u(z, t) u(z', t - \tau) \rangle$, and the cross-spectrum of u for these points, written $S_{uu}(z, z'; n)$, is a real quantity. The HIT presumption clearly also implies that it is a function of the separation distance $\lambda = |z, z'|$, not of z or z' individually, and can conveniently be normalized by division by the single point spectrum. The resulting 'normalized co-spectrum' is commonly written $R_u(\lambda, n)$, i.e.

$$R_u(\lambda, n) = S_{uu}(z, z'; n)/S_u(n) \tag{3.10}$$

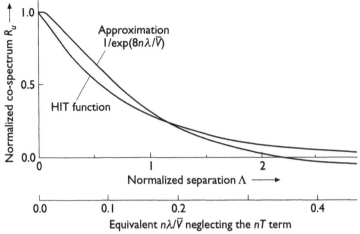

Figure 3.1 The normalized co-spectrum function (HIT).

A solution from turbulence theory is available for this function (Figure 3.1) (Harris and Deaves 1981, Irwin 1979), based on a universal normalized independent variable

$$\Lambda = \frac{2\pi n\lambda}{\bar{V}}\left(1 + \frac{1}{(8.4nT)^2}\right)^{1/2} \tag{3.11}$$

At the resonant frequencies for virtually all practical structures, $nT > 0.5$, so that the second term in eqn (3.11) is negligible, making $R_u(\lambda, n)$ a universal function of $n\lambda/\bar{V}$ only. Significant correlation is then restricted to separations that are commonly sufficiently small by comparison with height above ground to make HIT a credible basis. Values are conveniently presented in ESDU 86010 (ESDU, 1986b). An important derived parameter is the integral scale for frequency component n, which is $L_n = \int R_u(\lambda, n)\, d\lambda\, (0 < \lambda < \infty)$. The HIT formulation gives $L_n = \bar{V}/8.9n$, but this includes negative values of R_u at large separations, which are probably of limited practical reality; limitation to positive ordinates suggests $L_n = \bar{V}/8n$ for practical evaluation.

The exact HIT functional relationships are desirable for interpretation of full-scale or wind tunnel measurements, and for applications involving cross-wind components of turbulence, but for the base case of response of a conventional structure to the alongwind (u) component, the much simpler formulation

$$R_u = \exp(-\lambda/L_n) \tag{3.12}$$

is an acceptable approximation, with $L_n = \bar{V}/8n$ (Figure 3.1). An even smaller value is proposed in Annex B (informative) of the Eurocode ENV 1991–2–4, which should be viewed with caution.

The Fourier transform of the autocovariance (eqn (3.9)) gives the spectrum of the modal generalized force as

$$S_{Pj} = (\rho \bar{V} C_D)^2 S_u(n) \int_0^H \int_0^H R(z, z'; n) \gamma(z) \gamma(z') \, dz \, dz' \tag{3.13}$$

Noting that the mean value is $\bar{P}_j = \frac{1}{2} \rho \bar{V}^2 \int \gamma(z) \, dz$, this is commonly expressed in the form

$$\frac{n S_{Pj}}{\bar{P}_j^2} = 4 I^2 J^2 \frac{n S_u}{\sigma_u^2} \tag{3.14}$$

in which the correlation transfer function, or 'aerodynamic admittance', J^2, is

$$J^2 = \iint \gamma(z) \gamma(z') R(z, z'; n) \, dz \, dz' \bigg/ \left(\int \gamma(z) \, dz \right)^2 \tag{3.15}$$

(Davenport, 1962; Bearman, 1981; Dyrbye and Hansen, 1997, Section 6.4.3). The integrals comprise the whole structure. For a uniform slender horizontal structure, $\gamma(z)$ reduces to the mode shape function.

The foregoing development has presumed homogeneous wind ($\bar{V}$ and gust parameters σ_u and T_s invariant with location on the structure). Variation in the input wind parameters, generally the case with vertical structures, greatly increases algebraic complexity if approached rigorously (Wyatt, 1981), but in practice it is usually sufficient to use constant values, evaluated for a reference height selected by judgement (e.g. three-quarters of the height of a tower or chimney). In this event, $\bar{P}_j$ for insertion in eqn (3.14) should be evaluated consistently.

The full sequence of the spectral analysis is shown by Figure 3.2:

(a) the wind spectrum in the universal form (eqn (3.2)), defined on abscissa $\tilde{n} = 12 n T_s$, is multiplied by
(b) the aerodynamic admittance expressing spatial correlation, reflecting H/L_n, defined here on abscissa $nH/\bar{V}$, where H is the size of the structure (loaded length);
the product (a) × (b) × $(2\sigma_u/\bar{V})^2$ gives the spectrum of the modal generalized force (normalized on the mean value $\bar{P}_j$);
(c) which is divided by the square of the modal generalized stiffness (K_j) and multiplied by the square of the steady state dynamic magnifier (i.e. $|\mathcal{H}^2| = 1/ \{[1 - (n/n_j)^2]^2 + [(\delta/\pi)(n/n_j)]^2\}$) in which δ is the damping (as log dec) and n_j is the natural frequency (abscissa n/n_j);
(d) to obtain the spectrum of modal generalized displacement S_Y.

The lowest natural frequency of any given structure is broadly predictable from its size. Buildings typically have $n_1 = 46/H$, approximately. Towers and bridges are particularly well defined, given definition of the structural form and basic geometric proportions (chapter 3 in Maguire and Wyatt, 1999). Consider a tower of height $H = 100 \, \text{m}$, which is likely to have a frequency of about 0.67 Hz. A

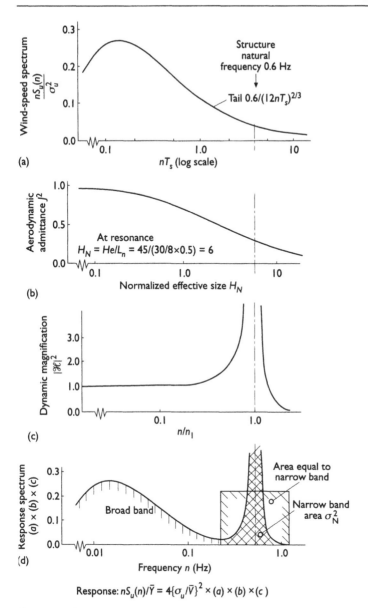

Response: $nS_u(n)/\bar{Y} = 4\{\sigma_u/\bar{V}\}^2 \times (a) \times (b) \times (c)$

Figure 3.2 Spectra and transfer functions for wind gust response analysis.

typical value of spectral timescale, $T_s = 10$ sec, gives $\tilde{n} = 80$, which is in the extreme tail of the spectrum; an unusually low value has been adopted for clarity of illustration in Figure 3.2. In wind $\bar{V} = 25$ m/s at 10 m above ground or 32 m/s at a 'representative' height $0.75H = 75$ m, the scale length of the 'resonant' gust is $L_n = \bar{V}/8n_j = 6$ m. Thus $H/L_n \gg 1$, and significant contribution to the volume

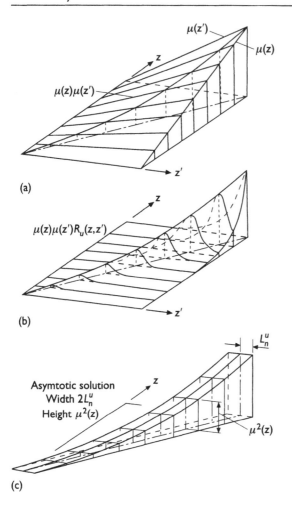

$\mu(z')$

$\mu(z)$

$\mu(z)\mu(z')$

z

z'

(a)

$\mu(z)\mu(z')R_u(z,z')$

z

z'

(b)

Asymtotic solution
Width $2L_n^u$
Height $\mu^2(z)$

$\frac{L_n^u}{}$

z

$\mu^2(z)$

(c)

Figure 3.3 The diagonal approximation to the admittance integral.

integral constituting the numerator of J^2 becomes restricted to close to the diagonal $z = z'$ of the area of integration ($0 < z < H, 0 < z' < H$), so that the numerator double integral is well approximated by $2L_n \int \gamma^2(z)\,dz$ (Davenport, 1962). This approximation is illustrated by Figure 3.3, using for clarity the 'straight line' weighting function $\gamma(z) = z/H$.

An 'effective size' H_e can be defined by the quotient of integrals

$$H_e = \left\{ \int \gamma(z)\,dz \right\}^2 \Big/ \left\{ \int \gamma^2(z)\,dz \right\} \qquad (3.16)$$

leading to a corresponding normalized value, $H_N = H_e/L_n$ (Wyatt, 1981). The solution of eqn (3.15) for uniform weighting (i.e. $\gamma(z) \equiv 1$) in conjunction with the simple exponential form for R_u, which is

$$J^2 = \frac{2}{H_N}\left(1 - \frac{1}{H_N}(1 - 1/\exp H_N)\right) \tag{3.17}$$

(plotted as Figure 3.2b), then constitutes an excellent approximation to J^2 for any weighting in which $\gamma(z)$ is of uniform sign throughout the structure. This covers the fundamental mode of most cantilever structures (towers, chimneys and translational modes of buildings) and other single-span structures; it is not restricted to $H_N \gg 1$. The less sophisticated approximation $J^2 = 2/H_N$ remains applicable for any weighting at large H_N, with error of order $1/H_N$. For the example given, with the first cantilever mode approximated by $\gamma(z) = (z/H)^{1.5}$, $H_e = 0.64H = 64\,\text{m}$, so that with $L_n = 6\,\text{m}$, the normalized effective size $H_N = 10.7$, and $J^2 = 20.17$.

For white noise excitation (i.e. S_P invariant with frequency) there is a closed form solution for the response variance, i.e. for displacement Y_j in mode j,

$$\sigma^2(Y_j) = (\pi^2/2\delta)n_j S_P(n_j)/K_j^2 \tag{3.18}$$

in which n_j is the natural frequency and δ is the damping expressed as logarithmic decrement.

As the peak spectral response ordinate comprises dynamic magnifier $(\pi/\delta)^2$, this implies an effective bandwidth $n_j\delta/2$. For practical values of δ, the magnifier is so large, and the response bandwidth so small compared to the bandwidth of the input spectrum, that this result gives an excellent approximation to the area under the resonance peak, identified as $\sigma_N^2(Y_j)$ on Figure 3.2(d), i.e.

$$\frac{\sigma_N^2(Y_j)}{\bar{Y}_j^2} = 4I^2\left(\frac{\pi^2}{2\delta}\right)J_u^2(n_j)\frac{0.6}{(12n_j T_s)^{2/3}} \tag{3.19}$$

in which $\bar{Y}_j = \bar{P}_j/K_j$ is the mean value of modal displacement. Clearly the r.m.s. value $\sigma_N(Y_j)$ follows

$$\frac{\sigma_N(Y_j)}{\bar{Y}_j} = 2I\frac{\pi}{\sqrt{(2\delta)}}J_u(n_j)\frac{0.78}{(12n_j T_s)^{1/3}} \tag{3.20}$$

The corresponding value of modal narrowband (quasi-resonant) contribution to any structural load effect F (e.g. stresses or stress resultants such as bending moments) can be obtained by multiplying the displacement by the respective modal influence coefficient β_{Fj} (say), i.e. $\sigma_{Nj}(F) = \beta_{Fj}\sigma_N(Y_j)$. Values for the modal influence coefficients can now generally be obtained from the computer modal solution output, but if in doubt concerning accuracy of modelling, or when using hand computation, they should be evaluated as the static effect of the 'inertia loading' given by the product for all masses of mass and acceleration per unit modal displacement (i.e. $Y_j = 1$ m), i.e.

$$\beta_{Fj} = 4\pi^2 n_j^2 \int i_F(z)m(z)\mu_j(z)\, dz \tag{3.21}$$

in which $i_F(z)$ is the conventional static influence function for load effect F and μ_j is the mode shape function.

The response to lower frequency input components is effectively quasi-static, and space does not permit detailed consideration in this dynamics text. The r.m.s. value for load effect F is commonly denoted $\sigma_B(F)$, 'B' signifying spectral broadband of frequencies. It can be evaluated by purely static correlation analysis (Wyatt, 1981), which is particularly well-suited to cases where the load on some part of the structure has the effect of relieving the net load effect. $\sigma_B(F)/\bar{F}$ is equivalent to the 'background factor' in Davenport based design codes and recommendations, and can also be inferred from conventional static results such as the detailed method of BS 6399 Part 2. Denoting the codified peak quasi-static value as F_{QS} (prior to application of dynamic factor C_r) and the *hourly mean* value as $\bar{F}$, and writing $F_{QS} = \bar{F} + g_s\sigma_B(F)$, the crest factor for quasi-static response can be taken as $g_s = 4.1 - 0.25\log_{10}H_e$ (Wyatt, 1981), for H_e expressed in metres. $\sigma_B(F)$ is then readily evaluated.

The static (broadband) and dynamic (narrowband) effects are statistically independent, and can be combined by root sum square $\sigma_T(\sigma_N^2 + \sigma_B^2)^{1/2}$. Design is commonly based on the expected maximum value $F_{des} = \bar{F} + g\sigma_T(F)$, in which g is the 'crest factor' (Davenport, 1964) given by $g = (\ln 2\nu\tau)^{1/2} + 0.577/(\ln 2\nu\tau)^{1/2}$. In the latter, τ is the storm-strength averaging time (e.g. 3,600 sec) and ν is the effective frequency, which can be taken as $\nu = (\sigma_N/\sigma_T)\, n_j$. The upper tail nature of both spectrum and admittance function are such that the first mode dominates dynamic gust response, but if necessary further mode contributions can be added to $\sigma_T(F)$ by root sum square.

3.1.4 Further cases of gust load spectra

In the case of a lattice tower of significant face width compared to the integral scale L_n, the foregoing presumption that the load on any element is fully defined by a single-point windspeed remains acceptable but it is necessary to allow for the correlation in two dimensions (i.e. with reference to location co-ordinates z_1 and z_2). This case is referred to as the 'lattice plate'. The numerator of the admittance function thus becomes a quadruple integral. For the case where L_n is small compared to the structure dimensions H_1 and H_2 in two orthogonal directions, the analogue to the approximation $2L_n \int \gamma(z)^2\, dz$ for a single line becomes $A_n \iint \gamma(z_1, z_2)\, dz_1\, dz_2$, in which $A_n = \int \lambda R_u(\lambda)\, d\lambda$ (limits 0 to ∞). For the exponential approximation to R_u, $A_n = 2\pi L_n^2$. For the HIT form, $A_n = 5.9L_n^2$ (considering only positive ordinates of R_u). The resulting approximation to the admittance (J) is given to a good approximation by $J = J_1 \times J_2$, where J_1 and J_2

are values for the two dimensions evaluated separately. J_1 is evaluated from the normalized effective size $H_{N1} = H_{e1}/L_n$ (eqn (3.16)) for whichever dimension gives the larger value of H/L_n, but the smaller normalized value is reduced for evaluation of J_2, to $H_{N2} = (4/5.9)H_{e2}/L_n$ (Wyatt, 1981).

In the case of a clad structure such as a building, it is empirically established that the correlation of pressure fluctuation over the upwind face is better than that of the free stream velocity over the same distances (Cook, 1985). This effect is partly countered by relative weakening of the effective load fluctuation on the downwind face; the net effect is not addressed explicitly in Davenport based design formulations. The author's personal practice is to apply the lattice plate solution as above, but evaluated taking $L_n = \bar{V}/6n$, an increase of one-third over the free stream value. This procedure cannot be expected to offer precision comparable to the line-like or true lattice plate cases, but may be sufficient for decision whether specialist investigations are necessary, in particular for subjective comfort criteria. Pressure fluctuations near the corners, especially in 'glancing' winds, are likely to be important for lateral or torsional excitation.

The foregoing discussion has considered only the alongwind (u) component of turbulence. Crosswind components may also be important, commonly treated in two independent orthogonal components, v (horizontal) and w (vertical). A number of formulations are available for spectra and net r.m.s. values. The HIT solution gives the upper-tail ordinates of S_v and S_w as 4/3 times S_u at the given frequency. This can be used as a practical approximation to S_v at frequencies such that L_n is less than (say) one-fifth of the height above ground, but becomes increasingly conservative at lower frequencies. S_w can be treated similarly, but with greater conservatism.

For modal analysis of response the forces are generally required in body axis components. In the basic case of a vertical structure for which the drag coefficient has constant value C_D for all directions and the crosswind force coefficient is uniformly zero, which is a reasonable approximation for many lattice towers, the body axis force perpendicular to the mean wind direction is $(\frac{1}{2}\rho\bar{V}^2AC_D) \times (v/\bar{V})$. The analysis then follows the alongwind treatment described above, with $(v/\bar{V})\bar{P}$ replacing $(2u/\bar{V})\bar{P}$, and thus S_v (or S_w) replaces $4S_u$. The HIT solution for L_n for the v component on vertical separation (and likewise the w component on horizontal separation) is, however, twice as large as the value for u, being $\bar{V}/4.43n$ when integration extends over the full range, including negative ordinates of R_v or R_w. The practical validity of this increase remains controversial, and some authorities retain the same values as for u, a nonconservative assumption. The vectorial analysis leading to generalized expressions for excitation of an element at an arbitrary inclination is highly complex (Strømmen and Hjorth-Hansen, 1995). Practical approximations for inclined tower structures are, however, available (Wyatt, 1992).

For bridges, gust dynamic response is generally dominated by vertical motion with excitation based on $dC_L/d\alpha$, in which C_L is the lift force coefficient and α is the angle of inclination of the wind to the deck. The analysis of correlation along the deck ('spanwise') follows the methodology introduced above, but it is usual to include a further admittance factor which takes account of the width of the deck ('chordwise'). The theoretical solution for an aerofoil (Sears' function) serves well in many cases (Walshe and Wyatt, 1983). A comparative survey of published formulations is given by Hay (1992). Much more sophisticated models are available for integration of gust action with the feedback of the effect of structural motion on the forces, as discussed in section 3.2.5.

3.1.5 Aerodynamic damping

The alongwind response of skeletal structures such as lattice towers is commonly significantly reduced by aerodynamic damping. The narrowband response is essentially harmonic (sinusoidal), modulated relatively slowly, and the relative velocity $(\bar{V} + u - y)$ (where $y = dy/dt$ is the velocity of the structure in the downwind direction) thus includes a sinusoidal perturbation. For a tower of natural frequency $n = 1\,\text{Hz}$, comprising members of width $d = 0.3\,\text{m}$ and in wind of mean speed $\bar{V} = 30\,\text{m/s}$, the reduced velocity $V_R = \bar{V}/nd$ is of order 100 (i.e. the fluid advance in the duration of one cycle of oscillation is 100 times the significant reference dimension of the structure). The induced perturbation of the drag force will therefore be closely quasi-static with amplitude $(2y/\bar{V})\bar{P}$, given the usual linearization. Examination of the equation of motion shows this to be equivalent to a viscous damper with coefficient $c = 2\bar{P}/\bar{V}$, thus making a contribution to damping logarithmic decrement

$$\delta_a = 2\pi c/c_{\text{crit}} = \bar{P}/Mn_j\bar{V} \tag{3.22}$$

in which use has been made of the standard result $c_{\text{crit}} = 4\pi Mn_j$. In modal analysis this becomes

$$\delta_a = \int (\bar{p}/\bar{V})\mu_j^2 dz/n_j \int m\mu_j^2 dz \tag{3.23}$$

where $\bar{p}$ and m are the mean load and mass per unit length and μ_j is the mode shape function. Equation (3.22) can also be expressed in terms of a Scruton number (eqn (3.31), Section 3.2.2) and reduced velocity $V_R = \bar{V}/nd$ (where d is the reference dimension for the drag coefficient C_D), i.e.

$$K_{Sa} = 2m\delta_a/\rho d^2 = C_D V_R \tag{3.24}$$

For a crosswind motion, the postulate 'drag coefficient C_D constant, crosswind C_L zero' gives damping one-half of this value (cf. the comparison between v and u gust excitation, above).

To show the likely importance, consider a lattice tower of tubular steel members of wall thickness t, having total mass (M) twice the mass of the members exposed in one face and total drag coefficient $C_D = 1.2$ based on the shadow area (A) of that face. This gives

$$\delta_a \bar{P}/Mn_j \bar{V} = \tfrac{1}{2}\rho \bar{V}^2 A C_D/(2 \times \pi A t \rho_s \times n_j \bar{V})$$

$$= (C_D/4\pi) \times (\rho/\rho_s) \times (\bar{V}/n_j t) \tag{3.25}$$

With material density $\rho_s = 8\,\text{t/m}^3$ and illustrative values $C_D = 1.2$, $\bar{V} = 3.2\,\text{m/s}$, $t = 10\,\text{mm}$ and $n_j = 0.8\,\text{Hz}$, the aerodynamic damping $\delta_a = 0.06$. A comparable tower of angle section members would give about twice this value. For buildings, the conditions for the quasi-steady assumption will be less well satisfied, but aerodynamic damping is also likely to be much less effective due to the lower ratio of drag to weight. For a typical building with drag coefficient 1.0 and mass $4\,\text{t/m}^2$ of face area (mass density $400\,\text{kg/m}^2$ and alongwind plan dimension $20\,\text{m}$, say), $\bar{V} = 32\,\text{m/s}$ and $n_j = 0.8\,\text{Hz}$, the above formulation gives $\delta_a = 0.006$.

Aerodynamic damping is commonly very important for the vertical or torsional gust dynamics of bridges. Taking the deck width B as the reference dimension for coefficients and reduced velocity, for vertical motion

$$\frac{2m\delta_a}{\rho B^2} = \frac{1}{2}\frac{dC_L}{d\alpha}V_R \tag{3.26}$$

The quasi-steady assumption is, however, significantly non-conservative for $dC_L/d\alpha$. If specific wind tunnel data are not available, the theoretical solution for an aerofoil with harmonic perturbation generally gives a useful approximation (Walshe and Wyatt, 1983). The effective value of $dC_L/d\alpha$ at practical frequencies is between 3 and 4. In strong winds δ_a can be significantly larger than δ_s; for example, for a deck of width $B = 25\,\text{m}$, mass $m = 15\,\text{t/m}$ and natural frequency $0.5\,\text{Hz}$ (typical for span $250\,\text{m}$), in a wind of $30\,\text{m/s}$, $\delta_a = 0.21$. The aerodynamic damping of bridge decks can also be expressed using the 'derivatives' as discussed in Section 3.3.4.

3.2 AERODYNAMIC INSTABILITY

3.2.1 Introduction

'Aerodynamic instability' is a very convenient generic term to cover a wide range of dynamic responses to wind, but means little more than a statement that the response in question is not sufficiently described in terms of the gust action considered above. There are intrinsically two distinct mechanisms:

- flow instability excitation, or simple 'vortex shedding';
- aeroelastic excitation.

In the *former*, the flow pattern is unstable, even when the structure is stationary. In the common example of a slender prismatic structure such as a chimney, vortices associated with flow separation from the flanks of the structure grow until they are carried away by the flow. The *latter* results from the changes of the aerodynamic forces consequent upon motion or deformation of the structure. It is instructive first to consider these two mechanisms separately, but unfortunately interaction between them is commonly crucial to the severity of the effects on the structure.

Anyone encountering such problems is well advised to read the seminal survey by Scruton and Flint (1964). Further introduction, particularly relevant to bridges, is available in Wyatt and Scruton (1981). The basic introduction to the stochastic model of vortex shedding given by Vickery and Basu (1984) is also desirable preliminary reading. Blevins' 'Flow induced vibrations' (1994) is widely respected for reference. There is a very wide range of design specifications, which will be introduced later.

3.2.2 Vortex shedding: deterministic representation

The starting point for the dynamic effects of vortex shedding is the von Karman vortex street, illustrated by Figure 3.4. This represents a cross-section through the flow field round a long prismatic structure; the circular section has been selected for illustration partly on the grounds of familiarity, but also because of the freedom from galloping-type aeroelastic behaviour. The quasi-steady behaviour of this section has already been addressed in the context of gust action, and has been shown to give unconditionally positive damping of structural oscillation, increasing in proportion to windspeed. The vortex street as shown indicates that vortex growth occurs alternately on opposite sides of the structure. When there is a large attached vortex on one side, the wake is displaced laterally and there is a lateral component of force on the structure. When this vortex is shed and (in this case) replaced by growth on the other side, and the process repeated, there is clearly a cyclic lateral excitation. Although the actual variation is unlikely to be sinusoidal, it is usual to extract the first Fourier component to give a coefficient of fluctuating lift $\check{C}_L$; that is, describing force $p(t)$ per unit length of prism as

$$p(t) = \tfrac{1}{2}\rho V^2 D \check{C}_L \sin 2\pi n_s t \qquad (3.27)$$

in which n_s is the frequency of the shedding.

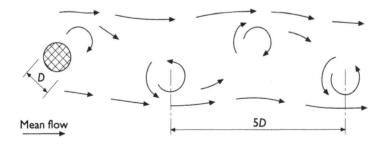

Mean flow

5D

Figure 3.4 The vortex street.

The vortex street proves to have a characteristic geometry, scaled to the diameter D, irrespective of windspeed; for the circular section, the distance between successive vortices on the same side is a little under $5D$. As the vortices are carried downstream at a speed only marginally less than the flow V, this implies that the cycle periodicity, equal to the time between successive vortex shedding, is about $5D/V$. In general, this is usually expressed in terms of the shedding frequency n_s, i.e.

$$n_s = S_t V/D \qquad (3.28)$$

in which S_t is a constant known as the Strouhal number, 0.2 in this case.

As critical conditions are likely to be associated with synchronism between shedding and a natural frequency of the structure, engineering interpretation is more commonly based on the reciprocal of the Strouhal number (i.e. $V/n_s D$). Experimental data, from wind-tunnel tests or full size, are appropriately related to a normalised representation of the windspeed, $V_R = V/nD$, in which n is the relevant natural frequency. The velocity at which resonance ($n = n_s$) occurs may be denoted V_C ('critical' velocity) so that

$$V_{RC} = V_C/nD = 1/S_t \qquad (3.29)$$

The term 'critical', here corresponding to resonance, must be treated with some caution. Although most often indeed the critical condition, significant response will occur at somewhat lower speeds, which may be important in terms of human subjective response, or even of structural fatigue, when account is taken of the increased duration of occurrence of such lower speeds. The maximum response may actually occur at a higher speed, as a result of persisting resonance due to 'lock on', considered later.

The above relationships can be combined to give a prediction of the steady-state response. For simplicity, a 'rigid body' motion is considered first, typified by a 'section model' wind-tunnel test. A rigid model is mounted on springs; if the mass per unit length is m and the prism length L, the spring stiffness must be $k = mL\omega^2 = 4\pi^2 mLn^2$. At resonance, the steady state dynamic magnifier is π/δ, where δ is the damping expressed as logarithmic decrement (or $\frac{1}{2}\zeta$ in terms of proportion of critical damping) so the response amplitude ($\tilde{y}$, say) is given by

$$\tilde{y} = \frac{\pi \frac{1}{2}\rho V^2 DL}{\delta \, 4\pi^2 mLn^2} \tilde{C}_L \qquad (3.30)$$

This is re-expressed in terms of the normalized quantities by writing $\eta = \tilde{y}/D$ and the normalized damping

$$K_s = \frac{2m\delta}{\rho D^2} \qquad (3.31)$$

giving the normalized amplitude

$$\eta = \frac{1}{4\pi} \frac{\check{C}_L V_{RC}^2}{K_s} \tag{3.32}$$

K_s is known as the Scruton number, and expresses the energy dissipation by structural damping by comparison with the potential aerodynamic input. The latter is represented by the work term $qDL \times \tilde{y}$, in which q is the kinematic pressure $\frac{1}{2}\rho V^2$, with the appropriate normalization of V and $\tilde{y}$. K_s is widely used as the basis for interpretation of scale model tests and other empirical comparisons.

For a flexible structure, the above result is easily extended by a modal decomposition, in which only the resonant mode is likely to have significant effect. Denoting the mode shape as $\mu(x)$, where x is the location coordinate, with maximum value of μ_T, the Scruton number should be evaluated using an equivalent value of mass per unit length, $m_e = \int m(x)\mu^2(x)\,dx / \mu_T \int \mu(x)\,dx$. The above expression for η then gives $\tilde{y}/D$ for the point of maximum displacement. The numerator integral is taken over the whole structure, but the denominator is evaluated only over the length of prism subject to the resonant excitation.

If m, V and D are constant ($m = m_0$, say), as is common for a bridge deck, or for individual members in a truss, $m_e < m_0$; for example, for a uniform simply supported beam $m_e = (\pi/4)m_0$, and the peak displacement is $4/\pi$ times the single-degree of freedom estimate for the given values of $\check{C}_L$ and m_0. For chimneys, it is common practice to use the mean value of $m(x)$ taken over the top third of the height. This rolls up in a rough and ready way the increase given by the quotient of modal integrals with the observed decrease of $\check{C}_L$ near the free end of the prism. The degree of reduction by the end effect is believed to be affected by chimney efflux, but this is ill explored. The effects of taper and of the variation of mean windspeed with height are addressed later.

The normalized reduction of the deterministic response equations is noteworthy; there is no independent mention of frequency or mode order. The phenomenon of lock-on (see also Section 3.2.4) causes the vortex street to reverse phase at the nodes, so that each internodal length adds consistently to the excitation (i.e. the integral in the denominator of the equation for m_e should be written $\int |\mu(x)|\,dx$). Thus if resonance in more than one mode is possible within the possible range of windspeed, the predicted displacement amplitudes will be similar, subject only to marginal correction according to the mode shape integrals. The internal stresses developed in the structure will, however, increase. For a long simply supported beam, the shape of mode j is given by $\mu_j = \sin(j\pi x/L)$, so the curvature and thus bending stresses increase as j^2, and a broadly similar pattern applies to most flexural structures. For a cable, the structural criterion is likely to be angular deflection at the attachments, which increases linearly with j.

Fortunately, the values of $\check{C}_L$ and K_s are such that η is usually small; values exceeding 0.2 are uncommon. Furthermore, the aerodynamic input is self-limiting at values of $\tilde{y}/D$ less than unity, although values at the worst point, such as the tip of a cantilever, especially in modes other than the fundamental, may somewhat exceed unity, being driven by excitation at sections where the amplitude is lower.

3.2.3 Vortex shedding: slender elements, cables

In most cases the fundamental mode of vibration is the dominant concern, because this clearly gives the lowest critical speed. Particularly in the case of chimneys, the large (typically sixfold) frequency difference between first and second modes commonly has the effect that the critical windspeed for the second mode is in excess of the maximum that may occur at the location in question. At the other extreme, very slender members, including cables, may reach the resonance condition up to quite a high mode number. The vertical motions of the deck of the first Tacoma Narrows bridge that persisted for a large part of its life (as opposed to the eventual destructive torsional motion), providing the spectacular film footage of vehicles in deck waves in which they almost disappeared from view, were of this kind. The switching between modes with change of windspeed, up to a seven-node case at speed 14 m/s, clearly identified vortex shedding and gave no concern for early structural failure.

The once familiar audible frequency vibration of overhead electric telegraph and telephone lines comes in this category, and this mechanism has been referred to as 'aeolian vibration' on the presumption that this was the Aeolian harp of classical mythology. Consider the suspenders supporting the deck of a suspension bridge. The stretched string natural frequency for a cable with material density ρ_s carrying tensile stress f_T is

$$n = \frac{1}{2\lambda}\left(\frac{f_T}{\rho_s}\right)^{1/2} \tag{3.33}$$

in which λ is the internodal length (half-wavelength). Thus for $f_T = 300\,\text{N/mm}^2$ (say) and $\rho_s = 7.8\,\text{t/m}^3$, $n \approx 100/\lambda$ (Hz, m units). For a cable of diameter $D = 50\,\text{mm}$, the critical windspeed is thus $V_c = 5nD = 25/\lambda$ (m/sec, m units). The curvature associated with a defined displacement amplitude increases with increasing mode order, inversely as the square of λ, so with a spiral laid cable, the mode sequence will be curtailed by increasing damping due to interwire friction. Low modes in this case will give a trivial value of critical speed and structural stressing. The suspender cables on the Severn Bridge, for example, showed oscillation over a range of modes immediately after construction, prior to fitment of dampers. The most serious observed response of a long suspender ($L = 80\,\text{m}$) was considered to be that with fourteen intermediate nodes, $\lambda = 5.3\,\text{m}$, $n = 19\,\text{Hz}$, occurring at $V_c = 4.8\,\text{m/s}$.

It is normal practice to protect such cables (also major electricity transmission lines) by additional damping, commonly by Stockbridge-type dampers. These comprise a substantial mass (18 kg for protection of the above example, equal to about 2 per cent of the cable mass) attached to the cable by a short length of spiral strand. The latter is clamped to the cable so that it acts as a cantilever in bending with substantial damping derived from interwire friction; the device is thus a simple form of inertial damper, albeit not optimized as a Tuned Mass Damper (TMD) for any one specific mode. It is attached near the end of the cable, accepting a reduction of effectiveness overall in order to avoid the circumstance of coincidence with a node in any of the modes for which protection is required. Such augmentation of damping is only necessary for such a heavy prism (equivalent density about 6 t/m^2 for steel spiral strand, 3.5 t/m^2 typical for composite steel and aluminium conductors) because the structural damping at low amplitudes is exceptionally low; values as low as logarithmic decrement $\delta = 0.003$ have been quoted. The Reynolds number is also unfavourable.

3.2.4 Reynolds number, size number, lock-on

Reynolds number $R_e = VD/v$, has a strong influence on vortex shedding from members where flow separation takes place from a curved surface. The kinematic viscosity of air $v = 1.5 \times 10^{-5} \, \text{m}^2/\text{sec}$ under normal ambient conditions, so the above basic definition gives $R_e = 0.7 \times 10^5 \, VD$. For a circular section a change in the mean position of separation tends to occur at about $R_e = 3.5 \times 10^5$. For a limited range above this value, the 'supercritical range', vortex shedding tends to be less well organized, giving much weaker excitation than in the subcritical range. Wootton (1969) pointed out that substituting the critical value of reduced velocity to replace V in Reynolds number gave a 'size number' nD^2/v, such that the Reynolds number at resonance is equal to the size number divided by V_{RC}. With $V_{RC} = V/nD = 5$, the condition $nD^2 \approx 1$ is likely to ensure freedom from serious excitation. It will be noted that the cable example gives a much lower value of size number; for 19 Hz, $nD^2 \approx 0.05$.

Lighting columns and masts give low values of nD^2 in the fundamental mode, and generally also in the second mode. As an indication of the latter, a 15 m mast may have a second mode frequency of 5 Hz. The crucial diameter for lock on is likely to be around the mid-height, typically less than 0.2 m, giving $nD^2 < 0.2$. Chimneys, however, tend to give much higher values. A concrete chimney of 2 : 1 taper and height ten times the base diameter will have a natural frequency about $70/h$ (Hz, given height h in metres). 1.0 Hz would then be associated with height 70 m and the diameter at, for example, 0.8 h above ground would be 4.2 m, giving $nD^2 = 18$. In this 'transcritical' range relatively well organized vortex shedding is again observed. $\check{C}_L = 0.7$–1.0 have been recommended for the subcritical range, 0.25–0.40 for the transcritical range.

The favourable range is of greatest significance for individual tubular members making up lattice structures. The natural frequency for a circular steel tube member of length L can be expressed by

$$n = c_f \frac{\pi}{2} \left(\frac{EI}{mL^4} \right)^{1/2} \approx 2860 c_f \frac{(D-t)}{L^2} \tag{3.34}$$

in which t is the wall thickness and c_f is a fixity factor, equal to unity for simple supports. The dimensions must be expressed in metres. For chord members, c_f is commonly not much greater than unity, whereas for bracing members c_f is typically about 1.9 (corresponding to EI/L values for the bracing around one-quarter of the value for the chord). For full end fixity, $c_f = 2.27$. To achieve $nD^2 \approx 1$, the above equation gives the limiting slenderness as

$$L/D \approx \{2860c_f(D-t)\}^{1/2} \qquad (3.35)$$

For a tubular bracing with $D - t = 0.2$ m and $c_f = 1.9$, the favourable size number range corresponds to $L/D \approx 33$. Significantly higher values of L/D than given by this condition will lead to unfavourable subcritical resonance, which should be avoided unless a high value of K_s is assured, exceeding 20 or 25.

It has long been appreciated that motion of the structure led to locking of the shedding frequency to the structural frequency over a range of reduced velocity extending from marginally below the value given by the reciprocal of the stationary body Strouhal number to a value typically some 30 per cent larger. Very small movements are sufficient, possibly as low as $\tilde{y} = 0.015D$. As the coefficient of alternating lift can be sustained over much of this range, the maximum response may be increased, occurring at a higher speed than given by the stationary body V_{RC}.

It is apparent, however, that this effect has an equally important action in encouraging coherent excitation in the face of contrary factors, which include the variation of mean speed over the height of a vertical or inclined structure, taper of the structure (giving a pro-rata variation of the nominal resonance speed), gusts (continually varying local speeds) and indeed the inherent randomness of the above-critical separated flow. A parametric study of chimneys at Reynolds numbers up to 2×10^6 was carried out in the National Physical Laboratory compressed air-wind tunnel (Wootton, 1969). It was noted that whereas reduction of the Scruton number from 16 to 8 typically caused an increase of r.m.s. response tip displacement from $0.01D$ to $0.015D$ (in line with prediction by the stochastic model described below presuming relatively poor synchronization of shedding over the length of the structure), further reduction to $K_s = 4$ caused the displacement to rise sixfold, to more than $0.1D$.

The lock-on effect is of special importance in promoting a well organized net excitation in turbulent or sheared flow, although a larger amplitude may be required to achieve an equal result. Noting that for the fundamental mode of a chimney, the majority of the modal excitation is derived from (say) the top third of the height, lock-on is commonly presumed to be effective over this length if its taper does not exceed about 20 per cent (± 10 per cent on mean diameter). Shear, as represented by the change in the mean speed, is generally less than this (e.g. a power law with index 0.15 gives a variation of only ± 3 per cent over the top third of the height.

3.2.5 Stochastic modelling of vortex shedding

Ignoring, for the moment, the feedback of structural motion, the degree of randomness inherent in boundary layer and wake effects suggests application to this problem of the procedures developed for gust response analysis. This has proved particularly fruitful for circular sections in the transcritical flow regime and in the presence of turbulence in the incident flow (Vickery and Basu, 1984; ESDU, 1986a).

The crosswind force per unit length $p(z)$ is expressed by the power spectrum

$$S_p(n) = q^2 D^2 S_{CL}(n) \tag{3.36}$$

in which q is the kinematic pressure corresponding to the mean windspeed. Following the gust analysis model, this is normalized to give

$$n S_p(n) = q^2 D^2 \sigma^2(\check{C}_L)[n S_{CL}(n)/\sigma^2(\check{C}_L)] \tag{3.37}$$

and a universal shape is postulated for the term in square brackets. Following Vickery, the algebraic form of the Gaussian probability density function is commonly used, defining a bandwidth parameter B_S such that the peak ordinate (at the central frequency $n = n_s$ determined from the Strouhal number) is $1/B_S\sqrt{\pi}$.

The methodology of the gust analysis is further followed to write the value of the spectrum of the modal generalized force P at the resonant frequency n_j

$$n_j S_P(n_j) = J^2 \left(\int \mu \, dz \right)^2 q^2 D^2 n_j S_{CL}(n_j)$$

$$= J^2 \left(\int \mu \, dz \right)^2 \{\tfrac{1}{2}\rho(V_{RC}n_j D)^2\}^2 D^2 \sigma^2(\check{C}_L)/(B_S\sqrt{\pi}) \tag{3.38}$$

The aerodynamic admittance expresses the correlation of the excitation along the length of the prism, and depends on the normalized co-spectrum and the mode shape functions as before.

In the absence of lock on, in the case of turbulent incident flow, transcritical Reynolds number and low structural damping, the correlation decays rapidly with increasing separation and is sufficiently expressed for all practical purposes by the integral scale L_C of the normalized cospectrum of $\check{C}_L$ (denoted R_{CL}); L_C is usually referred to as the 'correlation length'. For locations z and z', R_{CL} is a function only of the separation $\lambda = |z - z'|$ (i.e. $L_C = \int R_{CL}(\lambda) \, d\lambda \, (0 < \lambda < \infty)$). The Davenport 'diagonal' approximation $J^2 = 2L_C \int \mu^2 \, dz/(\int \mu \, dz)^2$ and its extension by applying eqn 3.17 through the concept of an effective height $H_e = (\int \mu \, dz)^2/\int \mu^2 \, dz$ is equally useful here as in gust analysis. Finally, for the likely low value of structural damping, the bandwidth of the mechanical admittance (frequency response function) is presumed small compared with the bandwidth of the excitation, and the white noise closed form solution for the response variance is a good conservative approximation, i.e.

$$\sigma^2(Y_j) = (\pi^2/2\delta)n_j S_P(n_j)/K_j^2 \tag{3.39}$$

in which Y_j and K_j are the modal generalized displacement and stiffness, respectively. An approximate quasi-static loading for a design check is readily defined from the modal analysis.

It will be noted that three parameters define the effective excitation: $\sigma(\check{C}_L)$, B_S and L_C. All are presumably sensitive to motion of the structure, but in an overall physical visualization of lock-on, the dominant effect may perhaps best be envisaged as a constraint on the phase of shedding. This is incompatible with simple spectral visualization. The familiar action in which phase, relative to the elastic forces, is crucial is damping, and Vickery therefore visualized lock-on as a negative damping action superimposed on the basic excitation, as well as the simpler effect of ensuring a uniform central frequency of shedding over a range of height in the presence of moderate taper and mean speed profile. This negative damping is normalized in the same way as positive structural damping, as a modified Scruton number, which will be negative. For chimneys, presuming the critical event to lie in the transcritical regime and turbulence levels typical of the neutral stability atmospheric boundary layer, Vickery suggests $\sigma(\check{C}_L) = 0.15$, $B_S = 0.3$ and $L_C = 1.0D$ to $1.5D$, together with aerodynamic damping equivalent to $K_{sa} = 2m\delta_a/\rho D^2 = -7.6$. The aerodynamic damping is combined with structural damping (δ_s) in the response variance equation (i.e. $\delta = \delta_s + \delta_a$, with δ_a negative), so for a structure with basic Scruton number $K_s = 15$, lock on would halve the effective damping and increase response by a factor of 2, but if the basic value were only $K_s = 7.6$, the response would increase without limit. A term modelling damping forces proportional to the cube of the response amplitude can be added to the linear damping which leads to simple Scruton number normalization, in order to express the self-limiting nature of vortex excitation at amplitudes of the order of D (Vickery, 1981).

The ESDU approach commences with evaluation of a so-called broadband formulation, by which is denoted an input spectrum broad by comparison with the frequency response function, as in Vickery's model. The structural response will, however, be narrowband dominated by the structural natural frequency, albeit with a broadly modulated amplitude, and the maximum value is taken as four times the r.m.s. Another spectrally based model has been postulated for the locked on condition, in which the force bandwidth is treated as if narrow by comparison with the frequency response function. Although originating from the same school as Vickery, the continuing application has been in these data items (ESDU 85038, etc.) (ESDU, 1986a). In both formulations $\sigma(\check{C}_L)$ and L_C are treated as increasing with amplitude, but in the second model the bandwidth assumption gives reversion to the same functional form as the basic deterministic model, with the addition of a correlation admittance. The response in the second model is deemed to be constant amplitude sinusoidal, giving the peak value as $\sqrt{2}$ times the r.m.s. The outcome is presumed to be whichever model gives the greater peak response.

3.2.6 Other vortex shedding problems: proximity, alongwind and ovalling excitation

Any structure placed in a vortex street originating from another structure nearby is likely to suffer stronger periodic excitation than the originating body. This commonly arises where a power station, for example, is served by two or more chimneys. Clearly this only operates when the wind is within a small range of direction, but extends to separations as large as 15 times the diameter. The most unfavourable effects occur when the chimneys are identical, as resonant motion of the upwind element leading to enhanced regularity of the vortex street will coincide with resonance of the affected downwind element. For this case it has been suggested that response of the downwind element may be twice that predicted for an isolated stack if the separation is $5D$, or 1.5 times the isolated value if the separation is $10D$ (Vickery, 1981; see also informative annex C.3.2.3 of the Eurocode ENV 1991-2-4).

Serious consideration must be given to this problem when slender modern structures are located in proximity to existing structures causing greater disturbance to the flow. The effect may be to present a significantly organized vortex street, changing with increasing separation to greater resemblance of a normal turbulence field but with strongly enhanced strength in the range of frequencies likely to embrace the structural resonant frequency. The term 'buffeting' has been expressly applied to this effect in UK usage, in distinction from its application to turbulence effects in general in the aeronautical field and elsewhere. The seminal example was the decision to build a stiff lattice arch bridge in place of the proposed suspension bridge over the Mersey at Runcorn, in proximity to the nineteenth century through truss railway bridge (Scruton et al., 1955; Grillaud et al., 1992; Bietry et al., 1994).

Two other resonant potential responses to vortex shedding should be borne in mind, although generally out of range or readily circumvented when the excitation derives from the natural wind. There is a weak in-line excitation, with one cycle for each vortex shed, giving resonance at one-half of the cross-flow resonance windspeed. The Scruton number limit to effectively avoid a structural problem is perhaps $K_s = 7$. This has been observed with wind excitation only in exceptional circumstances such as aluminium tubular members in a frame with very low damping, but can be a serious problem in water (e.g. for pile columns supporting a jetty). The second phenomenon possible at this reduced velocity is excitation of the ovalling mode of the structural section, which should be circumvented by ensuring a sufficiently high frequency, by stiffening if necessary, so that resonance is out of the practical windspeed range.

3.2.7 Vortex shedding: design rules for circular sections

The windspeed corresponding to resonance at the Strouhal frequency for a stationary structure can be robustly estimated, and if this exceeds the practical windspeed for the site, no further action is required. If not (as is commonly the case), the stresses caused by resonant response must be checked. Lock on may lead to a slightly greater response at a windspeed perhaps 20 per cent greater than the basic Strouhal resonance; for transcritical Reynolds number (nD^2 significantly exceeding unity) it is arguable whether such refinement is appropriate, given the inherent uncertainty in this problem.

BS 8100 (Lattice towers and masts) uses a deterministic model with a rather complex notation and presentation. In the notation of this chapter, $\check{C}_L = 0.3$ (transcritical Reynolds number is presumed) and primary resonance is assumed at $V_{cr} = 5n_jD$. To allow for lock on at higher speeds, a correction factor k_e is presented graphically; the graphical presentation is poor but it is apparently intended that windspeed $1.2V_{cr}$ gives an effective $\check{C}_L V^2$ about 8 per cent greater than the basic value $0.3V_{cr}^2$.

Stochastic models can be expected to give a smaller response. The negative aerodynamic damping concept was implemented in Commentary B to the Canadian National Building Code (NBC) in 1980. The parameter values suggested by Vickery, as given on page 89, are supplemented by a formulation for admittance which can be written as $J = J_{as}K_{AR}$, in which J_{as} is the Davenport 'diagonal' value for the correlation factor, and K_{AR} combines allowances for the approximation therein and for the aerodynamic effect of the free end. The basic value given is $K_{AR} = (h/16D)^{1/2}$ (but $\not> 1$), in which h is the height of the chimney (or, for moderately tapered chimneys, three times the length deemed to have shedding locked on). A closer approximation (but disregarding the end effect) would be given by $K_{AR}^2 = 1 - 1/H_e$ (cf the gust analysis, page 77); for a typical first mode shape this would agree at $h/D = 12$ and be somewhat smaller than the Code value for more slender chimneys.

The first mode resonance solution was expressed in the NBC Commentary by an equivalent static load (P_L, say, per unit length) acting over the top third of the height. This is set equal to the theoretical inertial load intensity at the top of the chimney, which with the input values for normal turbulence wind conditions, expressed in the notation of this chapter, is

$$F_L = \frac{3K_{AR}qD}{\left\{\frac{h}{D}\left(\frac{\delta}{2\pi} - \frac{0.6\rho D^2}{m}\right)\right\}^{1/2}} \tag{3.40}$$

in which the mass per unit length m is averaged over the top third. As noted earlier (Section 3.2.5), the denominator can be written in terms of the Scruton number, emphasizing the importance of this normalized parameter. The corresponding normalized tip deflection can then be written as

$$\eta = \frac{V_{RC}^2}{4\pi K_s} \frac{3K_{AR}(2\delta D/\pi h)^{1/2}}{(1 - 7.6/K_s)^{1/2}} \tag{3.41}$$

By comparison with the deterministic solution taking $\check{C}_L = 0.3$, using the mode shape approximation $\mu = (z/h)^{3/2}$ (which gives the modal integral quotient $\mu_T \int \mu(z)\,dz / \int \mu^2(z)\,dz = 1.6$), the stochastic result is smaller by a factor

$$\frac{2.4K_{AR}(\delta D/h)^{1/2}}{(1 - 7.6/K_s)^{1/2}} \bigg/ (1.6 \times 0.3) \tag{3.42}$$

For a slender concrete chimney with $h = 12D$ and $\delta = 0.04$ (say), this factor is $0.25/(1 - 7.6/K_s)^{1/2}$. If furthermore $D/t = 40$ (say) at the reference height for evaluation of K_s (height $z = 5h/6$ is suggested), then $K_s = 12$ (taking structural mass only (i.e. unlined)) and the given stochastic estimate is 40 per cent of the deterministic value.

For low turbulence conditions, the Canadian code gives doubled values for both the basic exciting force coefficient and the negative damping factor. K_s must therefore robustly exceed $2 \times 7.6 = 15$; the above example would be unacceptable. The possibility of low turbulence must clearly be approached with severe caution, with regard to the frequency of occurrence (or upper limit of co-existent windspeed) of a stably stratified flow with temperature lapse rate inversion. A deterministic check may particularly be advisable.

The CICIND (1999) recommendations for steel chimneys present a rather complicated algebraic formulation for η. This is based on curve fitting the foregoing stochastic model for small amplitude response, combined with a sharp lock on effect as the tip amplitude exceeds about 0.01D (r.m.s.) but with a self-limiting reduction of excitation for amplitudes exceeding about 0.2D (r.m.s.). Four different parameter sets are given to cover subcritical and transcritical Reynolds number and normal and low levels of turbulence (threshold speeds for the latter being specified). The larger amplitude response predictions are based largely on experience in Denmark and in Poland where chimneys with very high slenderness h/D have allowed survival of such amplitudes (cf eqn (3.45)), but it is questionable how far this should be exploited in design.

The guidance for concrete chimneys produced by the American Concrete Institute, the *ACI Manual of Concrete Practice* part 307 (ACI 307-95) is another elaboration of this format. The analysis remains essentially unchanged, but many of the parameters treated hitherto as simple constants are now dependent on intensity of turbulence and/or aspect ratio. Guidance is given on application to the second cantilever mode, and to combination with the effect of alongwind forces when the critical speed is approaching the design windspeed. The net changes operative in turbulent wind are typically modestly favourable, but the problems posed by low turbulence appear to be viewed very lightly in this Code.

The Eurocode ENV 1991–2–4 informative Annex C presents a compromise procedure developed by Ruscheweyh (1982, see also Ruscheweyh et al., 1988), supplemented in ENV 1993–3–2 for chimneys. The basic response equation (ENV 1991–2–4 equation C.4) has the form of the deterministic model; the excitation is defined in terms of a coefficient for the r.m.s. value of load per unit length (denoted c_{lat}) and a 'correlation length' (denoted L_j) over which the modal force integral is evaluated following the deterministic format. L_j is a function of the response amplitude, but unfortunately this parameter combines the consideration of correlation and the factor to be applied to the r.m.s. to obtain a design value. The given values are thus not readily comparable with other procedures; for response amplitudes less than 0.1D, $L_j = 6D$, but increases to 12D if the response amplitude is more than 0.2D. The high threshold for the effect of motion on L_j

implies the presumption of a substantial intensity of turbulence. c_{lat} is given as 0.7 for Reynolds number less than 3×10^5 ($nD^2 = 0.2$), falling rapidly to 0.2, which is applicable for $1.5 < nD^2 < 12$. The eventual transcritical value $c_{lat} = 0.3$ is only reached at $nD^2 = 30$.

Further review of the Eurocode procedure is given by Dyrbye and Hansen (1997), with extensive comment on recent practical experience and design comparisons with the Canadian recommendations. It is interesting to note that procedures with broad differences in formulation converge to give similar predictions for middle of the range structures.

3.2.8 Vortex shedding: design impact and countermeasures

It was suggested in Section 3.2.4 that *individual structural members* may benefit from the intrinsic weakness of excitation in the range $nD^2 \approx 1$, which for bracing members in trusses with full continuity connections implies limitation of L/D to about 33, corresponding to a structural slenderness ratio of $0.7L/r = 65$. As a more slender member is commonly more economic structurally, exploration of the limit of the favourable range is highly desirable. A sophisticated extension of the simple Reynolds' number has been presented by ESDU (1986a) taking account of small scale components of turbulence and the surface roughness of the structure to give an effective value R_{ee}. This is difficult to interpret and to calibrate against existing experience, especially with regard to surface roughness. The lower limit of the favourable range should therefore not be presumed substantially below $nD^2 = 1$.

A greater L/D implies subcritical resonance, and to ensure freedom from the lock on enhancement of excitation (and of the cumulative time over which sufficient response to cause fatigue damage could be sustained) a high value of K_s is called for; $K_s = 25$ has been suggested in guidance for welded tubular towers for service offshore prepared on behalf of the UK Department of Energy (BRV, 1990). For a steel tube, wall thickness t, K_s can be re-expressed

$$K_s = \frac{2\pi(D - t)\rho_s \delta}{\rho D^2} \approx \frac{38000\delta}{(D/t)} \tag{3.43}$$

Unfortunately the value of δ is difficult to predict. For fully welded structures the paramount source of damping is the attachment of non-structural 'ancillaries', commonly involving bolting and/or frictional grips. If a member has no such attachments, damping may be very low, perhaps as low as 0.15 per cent critical (Doucet and Nordhus, 1987) or logarithmic decrement $\delta = 0.009$. In practice there is often significant dispersion of energy through the structure, giving an enhanced effective value of K_s.

Robust expectation of satisfactory performance with $\delta = 0.009$ would require limitation of D/t to about 14, which is clearly contrary to current practice. For compression members, slenderness ratio considerations commonly encourage much higher values towards the constraint imposed by local buckling at

$D/t = 0.076E/f_y$, which is 44 for yield stress $f_y = 350\,\text{N/mm}^2$. The Department of Energy guidance (BRV, 1990) includes response prediction based on the ESDU (1986a) analysis which appears non-robust as a result of very high sensitivity to the damping estimate. It will be seen that the nD^2 and K_s criteria conflict; for a given member capacity, increasing D to meet an nD^2 criterion will diminish K_s. This question remains controversial.

For *steel chimneys* economic design commonly pushes D/t to 200 or even 250. At $\delta = 0.03$ (cf. ENV, 1991) an unlined chimney thus has $K_s \approx 5$, increased for a lined chimney pro rata to the mass. However, $\delta = 0.03$ is not a robust lower bound for an unlined chimney, and countermeasures should generally be applied. The most common aerodynamic countermeasure is the spiral strake (Walshe and Wootton, 1970). This typically comprises a three-start spiral projecting about $0.1D$. It has the disadvantage of broadly doubling the quasi-static wind load in the transcritical regime, the effective force coefficient related to the basic diameter (D) being about 1.4. Strakes are commonly applied to the top third of the stack to protect the first mode. A smaller drag penalty but at greater structural complexity is offered by the perforated shroud.

An alternative of increasing popularity is the addition of a damping device. A Tuned Mass Damper (TMD) optimized for the control of harmonic excitation can give very high values of K_s. In the ideal case the nominal logarithmic decrement is $\delta = 0.5(m_D/0.2mb)^{1/2}$, in which m_D is the damper mass (e.g. $\delta = 0.15$ if $m_D = 0.02mb$). Practical departure from optimal values of the damper parameters will substantially reduce this, and it is common to 'overdamp' the auxiliary mass to reduce sensitivity to error and to reduce the magnitude of its relative motion. 'Sloshing fluid' dampers are also available in proprietary form. Dampers have also been applied to similar problems with lighting masts (including use of elastomer inserts in the base mounting) and with guyed masts (including hanging chain impact dampers).

It is instructive to consider the cantilever first mode stress influence function. The bending stress at the base can be written

$$f_B = 1.7c_{tr}E\eta\frac{D_T D_B}{h^2} \tag{3.44}$$

in which D_T and D_B are the diameters at top and bottom, respectively, $\eta = y/D_T$ and c_{tr} is a factor taking account of the taper profile. For a uniform cantilever $c_{tr} = 1$; it is only very weakly affected by non-uniform mass, and only modestly by non-uniform second moment of area I. Writing I_B and I_T for the values of I at the bottom and top of the chimney, respectively, the expressions

$$c_{tr} = 1.86 - (\ln I_B/I_T)^{0.2} \qquad \text{for } I_B/I_T > 3 \tag{3.45a}$$

$$c_{tr} = 1 - 0.15\ln(I_B/I_T) \qquad \text{for } I_B/I_T < 3 \tag{3.45b}$$

give a close approximation for cases of linear diametral taper, and a satisfactory working approximation for practical steel chimneys with two or more cylindrical sections and conical transitions. Thus, for $f_B = 25\,\text{N/mm}^2$ (say, giving stress range $50\,\text{N/mm}^2$ to limit fatigue damage) the response limit for a uniform diameter chimney is $\eta = 0.07 \times 10^{-3}(h/D)^2$ (e.g. $\eta = 0.03$ for $h/D = 20$). For an unlined stack at $D/t = 200$, the Canadian formulation then requires $K_s = 16$ (turbulent flow). For low turbulence conditions, the deterministic estimate with $\check{C}_L = 0.3$ (for high nD^2) and the modified Canadian formulation both require $K_s = 32$.

For *concrete chimneys*, if unlined, $K_s = 12000\delta/(D/t)$. The American Concrete Institute code ACI 307 allows presumption of 1 per cent of critical damping, $\delta = 0.063$. Thus with $D/t = 40$, $K_s = 20$; a 50 per cent addition to the mass by a refractory liner would increase this to 30. Thus, η values are comparable with the above steel example. However, H/D values are commonly lower, and for an untapered multi-flue windshield at (say) $h/D = 14$, the associated strains will be higher; $\eta = 0.03$ then gives concrete stress (at $E = 35\,\text{kN/mm}^2$) $f_B = 9\,\text{N/mm}^2$. A $1:0.6$ diametral taper, retaining $h/D_B = 14$ and with $I_B/I_T = 8$ (say, giving $c_{tr} = 0.70$), would reduce the stress to $4\,\text{N/mm}^2$. Greater taper would lead to more than ±10 per cent change of diameter over the top third and thus to reduction of the predicted response according to the above rules.

The final word of this section must be to highlight the importance in all assessments of

- the value of structural damping;
- the possibility of low turbulence wind at the critical speed.

The occurrence of low turbulence conditions varies very greatly according to location, and is generally ill explored in engineering guidance.

3.2.9 Vortex shedding: bridges

The case of Tacoma Bridge has already been mentioned. However, the pursuit of improved deck cross-section profiles to ensure freedom from strong torsional motion generally also has the effect of reducing the strength of excitation by vortex shedding. Nevertheless, it is in practice impossible to eliminate it entirely, and it is necessary to check both fatigue and the subjective reaction of users of the structure. The subjective reaction criterion incorporated in the UK Design Rules for Aerodynamic Effects on Bridges BD49/93 is an acceleration value, centred on an amplitude of $0.8\,\text{m/s}^2$, or 8 per cent of the acceleration due to gravity. The corresponding structural stresses will be about 8 per cent of the stresses due to dead load; although such stresses might pose a significant fatigue risk (depending on fatigue detail classification and on the frequency of the wind condition for resonance) the subjective reaction criterion is commonly more significant.

The amplitudes thus accepted are generally sufficient for lock on ensuring vortex shedding correlated over a substantial part of the span, especially in the relatively low turbulence environment typical of long spans (estuarial and/or high level valley crossings). The deterministic model, giving a response inversely proportional to the Scruton number, is therefore used for prediction and scaling of wind tunnel results. The section depth (d) is generally used both for reduced velocity and Scruton number definition, revealing a somewhat clearer pattern than the option of normalization on the deck width (B). Typically, the reduced velocity $V_{RC} = V_c/nd$ is given by $V_{RC} = 6 + 0.5B/d$ for sections with $B/d < 6$, $V_{RC} = 1.5B/d$ for more slender sections, but with considerable scatter (cf. figure 2 in Wyatt and Scruton, 1981).

Excitation strength is very sensitive to details of the cross-section, especially at the leading edge, and (because of the feedback by lock-on) may be sensitive to damping. Very low scaled wind tunnel speeds may also create problems if the full bridge is modelled. The section model technique, using a large-scale model of a fraction of the span which is spring borne with independently controllable frequencies and damping is most strongly desirable for this purpose. Unfortunately, it is then not possible to model atmospheric turbulence at correct scale. The effects of turbulence should be thought of as comprising two distinct actions: the lower spectral frequencies which reflect large size gusts will be seen as a change in the incident speed, while the high frequencies are associated with localized momentum transfer affecting the boundary layer and tending to promote reattachment of separated flow. The former may be overborne by lock-on in conditions of modest turbulence, while the latter is generally beneficial.

There are much larger numbers of road bridges and viaducts in the span range up to about 70 m, but commonly in locations where much greater turbulence is the norm. In most cases the critical speed is sufficient to be out of range, or at least sufficiently high to make subjective response to motion of little practical concern and to give only a low potential rate of accumulation of fatigue cycles. Care should be taken that the high frequency components of turbulence are not over-represented in testing, suggesting a lower target value for the total intensity, as illustrated by Figure 3.5. Footbridges may give greater concern, although the need to avoid adverse response to pedestrian excitation commonly leads to structural forms giving frequencies over 3 Hz or to special provision for damping. The deck may, however, be thin; for example, a deck $d = 800$ mm, $B = 3{,}200$ mm at 3 Hz gives a critical speed of 19 m/s. At a height of (say) 6 m above ground, this gives the assurance of a high level of turbulence, typically intensity 0.25 or more. Nevertheless, robust design assessment may be difficult.

The UK rules BD49 were based on an extensive parametric study, which suggested that for decks with a slender leading edge detail in conjunction with a substantial deck cantilever beyond the main face of the supporting structure, the excitation factor $\check{C}_L V_R^2$ could be taken as $5.8B/d$ (see the Rules for strict definition of B, d and limits to edge details). Recognizing the greater potential effect of reattachment on wider decks, a reduction factor $K_R = 0.87(d/B)^{1/2}$ was then applied to account for the beneficial effect of turbulence (Smith and Wyatt, 1981). A default estimate of three times this excitation was postulated for decks not satisfying the edge cantilever requirement, but the combination of a solid edge parapet with girder structure below the deck was explicitly excluded as leading to more severe

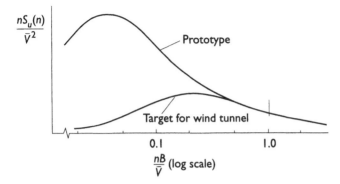

Figure 3.5 Turbulence spectra matching for wind tunnel testing.

excitation. Damping values are suggested of logarithmic decrement 0.03, 0.04, 0.05 for steel, composite and concrete bridges, respectively. Experience with longer spans suggests those values may then be optimistic, especially for cable stayed structures using parallel wire strands (or assemblies of small strands); spiral laid strands may also have low damping below a limiting friction threshold amplitude, perhaps a deck amplitude of span/5,000.

Applied to the footbridge introduced above, the Scruton number might be about 80, and the deterministic prediction of amplitude (allowing $4/\pi$ as the quotient of the mode shape integrals) gives values of

$$\frac{y}{d} = \frac{K_R}{\pi^2} \frac{\check{C}_L V_R^2}{K_s} \tag{3.46}$$

of 0.01 to 0.03 according to the edge arrangement. An amplitude of $0.03d$ with $d = 3$ m and $n = 3$ Hz gives an acceleration of more than 8 m/s^2, clearly intolerable. Even a concrete deck with edge overhanging would raise serious concern. The absence of adverse reports from structures of this type suggests that the much higher turbulence, and indeed more generally disturbed flow, has a greater beneficial effect than can yet be robustly quantified.

The most significant experience with a modern cable stayed bridge is perhaps Kessock Bridge (on the Moray Firth near Inverness) (Cullen-Wallace, 1985). Although the critical speed is as high as 22 m/s ($d = 3.25$ m, $n = 0.52$ Hz, $B = 7d$, $V_{RC} = 13$) the winter condition of cold water in the Moray Firth and relatively warm winds led to low turbulence conditions at resonance and amplitudes exceeding 200 m. Tuned mass dampers were then added. This bridge has no edge cantilevers. A similar profile at Longs Creek in Canada had earlier shown severe oscillation under winter conditions of frozen surface and build-up of ice against the deck-side barriers (Wardlaw, 1981), countered by adding an inclined face cladding. However, the constant depth trapezoidal beam valley crossing at Milford Haven (Cleddau Bridge) was provided with a tuned mass damper *ab initio* on wind tunnel evidence of strong vortex shedding excitation. This bridge crosses a curving, steep sided valley, and the inference from its behaviour in service is that the resulting disturbance might have presented serious response, even in the absence of supplementary damping (Wex and Brown, 1981).

Although solid parapets are strongly adverse, porous windshielding barriers have only moderate adverse effect. The initial feasibility study for the Second Severn Crossing suggested that with full length 3 m/50 per cent solidity windshielding, a full 'streamlined' enclosure of the girder structure would be necessary to reduce excitation. In the event, satisfactory performance has been achieved by painstaking wind tunnel optimization of the edge detail supplemented by two non-structural longitudinal dividers below the deck. The criterion for this design was an acceleration amplitude limit of $0.2n^{-1/2}$ m/s² (actual $n = 0.33$ Hz).

3.3 AEROELASTIC EXCITATION

3.3.1 The quasi-steady model: galloping

The concept of change in aerodynamic forces in response to the vector resultant relative velocity has already been introduced, with respect to aerodynamic damping, in Section 3.1.5. There are, however, circumstances in which aerodynamic damping becomes negative. If the net damping (algebraic sum of structural and aerodynamic components) becomes negative, a harmonic response at the natural frequency will develop. Unless structural failure (or enhanced damping due to inelastic behaviour at large amplitudes) intervenes, the amplitude reached will be limited by non-linearity of the force coefficient relationship to amplitude, but such amplitudes may be very large. The classic example is the overhead line with ice accretion, in which amplitudes of several metres have occurred on cables of a few centimetres diameter, commonly appearing as travelling waves and giving the phenomenon of the generic name 'galloping'.

The basic linear quasi-steady model is shown in Figure 3.6(a), which postulates the form of variation of lift force coefficient with angle of incidence that is shown by rectangular prisms (i.e. $dC_L/d\alpha$ negative over a range of incidence close to in-line with the longer side, as shown). A perpendicular motion downwards as drawn, causing the relative incidence vector to be inclined upwards, will thus result in a change of lift tending to reinforce the motion. For the case shown, the downward velocity is $\dot{y} = \tilde{y} \times 2\pi n \cos 2\pi nt$. The apparent angle of incidence is thus $\alpha = \tan^{-1}(\dot{y}/\bar{V})$. For small values of α, writing $\sin \alpha = \alpha$ and $\cos \alpha = 1$, this becomes $\alpha = \dot{y}/\bar{V}$ and the body axis force per unit length of the prism is

$$Z = \tfrac{1}{2}\rho \bar{V}^2 d \left(\frac{dC_L}{d\alpha} - C_D \right) \alpha = \tfrac{1}{2}\rho \bar{V} d \left(\frac{dC_L}{d\alpha} + C_D \right) \dot{y} \qquad (3.47)$$

in which C_L and C_D are the coefficients for lift and drag forces as shown, referred to reference dimension D. It will be seen that the positive direction for Z opposes

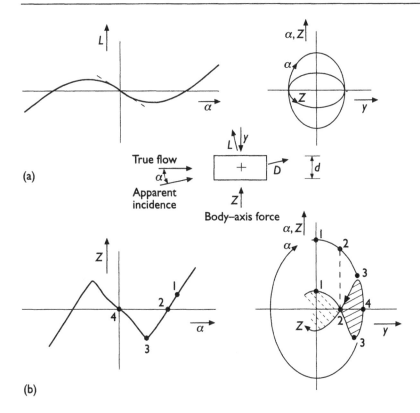

(a)

(b)

Figure 3.6 (a) Galloping: definition diagram; (b) galloping: construction of force/displacement loop.

the motion. The energy input to the system per cycle ΔU_a is thus

$$\Delta U_a = \pi(-\tilde{Z})(\tilde{y}) = -\tfrac{1}{2}\rho \bar{V} d\left(\frac{dC_L}{d\alpha} + C_D\right) 2\pi n \tilde{y}^2 \tag{3.48}$$

The maximum value of the kinetic energy provides a simple estimate of the energy of oscillation per unit length of prism U (say). The energy dissipated by damping by cycle ΔU_δ is thus

$$\Delta U_\delta = 2\delta_s U = 2\delta_s \times \tfrac{1}{2}m(2\pi n\tilde{y})^2 \tag{3.49}$$

The condition for instability $\Delta U_a > \Delta U_\delta$ thus gives the critical velocity $\bar{V}_C$, or the corresponding reduced velocity V_{RC},

$$V_{RC} = \frac{\bar{V}_c}{nd} = \frac{2K_s}{-\left(\dfrac{dC_L}{d\alpha} + C_D\right)} \tag{3.50}$$

in which K_s is the Scruton number $2m\delta_s/\rho d^2$.

The foregoing linearized analysis is generally sufficient for civil engineering cases, but for flexible systems tolerant of very large displacements, it may be necessary to estimate limiting amplitudes taking account of the full pattern of C_L (or C_Z) as a function of incidence. Figure 3.6(b) shows a graphical construction of the force displacement loop. Starting from a postulated amplitude $\tilde{y}$, with corresponding $\tilde{\alpha}$, consideration of successive pairs of values of $\dot{y}$ and y leads to a plot which may comprise both energy input (continuous shading) and dissipation (broken-line shading). The limiting amplitude is found by trial and error such that the net input balances the dissipation by damping. Closed form algebraic procedures have also been presented, commencing with a polynomial curve fit for C_Z (Parkinson, 1965, Novak, 1972). Although the dominant parameters are normalized in the same grouping as for the vortex shedding phenomenon, the resulting behaviour patterns are distinct:

- vortex shedding – critical speed V_{RC} fixed, response amplitudes sensitive to K_s;
- galloping – critical speed V_{RC} proportional to K_s, amplitudes likely to rise to much larger values than typical of vortex shedding when $V_R > V_{RC}$.

Unfortunately interactions between these mechanisms of excitation commonly distort the clarity of interpretation. Figure 3.7 shows three rectangular prisms tested as part of the Department of Transport study (Wyatt and Scruton, 1981) undertaken to support the UK Design Rules (BD49; Smith and Wyatt, 1981). The first case (deck width B equal to the depth d) shows vortex shedding at $V_R = 7$ and galloping fairly distinct at perhaps $V_R = 0.5K_s$. The third case ($B = 3d$) shows very clear vortex shedding at $V_R = 10$, but no evidence of galloping within the range of the tests ($V_R < 1.5K_s$). The intermediate case clearly has some characteristics of both mechanisms, strongly modified. To show these values in perspective, a steel box (e.g. a bridge girder during erection) $B = 2d$, plate thickness $d/150$ (plus allowance 50 per cent to mass to allow for stiffeners, transverse elements, etc.) and damping log dec 0.03, would have K_s approximately 20.

3.3.2 Flutter of bridge decks

An aerofoil, whether a flat plate or a slender smooth outline prism, does not show the 'negative lift slope' which is the key to galloping. However, violent self-excited oscillation of aircraft wings has long been recognized as a potential hazard, under the name 'flutter'. Analysis based on the aerofoil flutter model has proved remarkably useful for slender bridges. This proves to be essentially a *coupling* phenomenon, combining modes of vibration which in still air are quite distinct, and dependent on the departure of flow patterns and resulting forces from the quasi-steady model. This departure is not only a question of magnitude, but also of phase shift between motion and force. A common method of description is by defining coefficients for the force components proportional to instantaneous values of the rate of change of the displacements as well as to the displacements themselves; these 'derivative coefficients' are discussed further in Section 3.3.4.

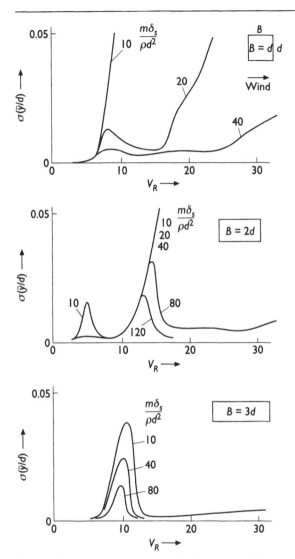

Figure 3.7 Interaction of vortex shedding with galloping excitation.

For an ideal aerofoil there is an analytic solution, conveniently written in complex number notation (Fung, 1955). The quasi-static solution is a force equivalent to a lift coefficient $C_L = 2\pi\alpha_a$ (taking the deck width B as the reference dimension) which acts at the quarter-chord point ($B/4$ from the upwind edge). For this purpose the apparent instantaneous angle of incidence (α_a) is based on the ratio of the net vertical velocity at the three-quarters-chord point to the free stream velocity. All forces in harmonic motion (components in phase and in quadrature of lift and torque associated with both vertical and torsional motions) are then

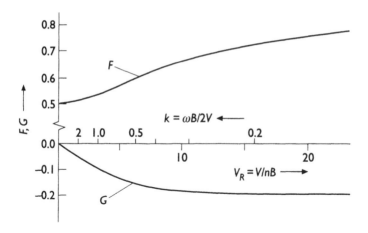

Figure 3.8 Ideal aerofoil behaviour: Theodorsen's function.

given by scaling the quasi-static solution by a single complex factor, generally given as Theodorsen's function, $C = F + jG$, in which $j = \sqrt{-1}$. Tables of F and G are available as a function of reduced velocity (e.g. $V_R = \bar{V}/nB$), or its reciprocal, a reduced frequency commonly written following aeronautical practice as $k = \omega b/\bar{V}$, in which ω is the circular frequency and b is the semichord ($b = \frac{1}{2}B$) (Fung, 1955). Thus $V_R = \pi/k$. Figure 3.8 shows the variation of F and G over the range of V_R of practical interest for bridges. The severity of departure from the quasi-steady solution ($F = 1$, $G = 0$) will be noted.

Because the lift acts at a distance $B/4$ in front of the centre line, it acts to increase twist, analogous to a negative stiffness, and the torsional natural frequency thus falls with increasing windspeed. The torsional natural frequency (n_θ) of practical bridge structures is higher than the vertical (n_y), so the differential is reduced. Classical flutter is the culmination of this process, when the forces resulting from motion combine to sustain an oscillation combining vertical and torsional motions at the same frequency. The critical windspeed is revealed by discovery of a combination of speed, relative vertical and torsional amplitudes, and phase angle, satisfying the equations of motion but in which the actual response magnitude becomes indeterminate.

The ideal aerofoil solution for the case of a deck with exactly matching vertical and torsional mode shapes and $r^2 = 0.1B^2$ (in which r is the mass radius of gyration), undamped, is given in Figure 3.9. The solution is not very sensitive to modal mismatch or the r/B ratio, and is insensitive to structural damping. The response grows very rapidly if the critical speed is exceeded. The importance of resistance to coupling by a high frequency ratio or high inertia is clear, although high inertia alone is not sufficient if the frequency ratio is unfavourable. Selberg (1961) showed that for a wide range of practical circumstances, an excellent

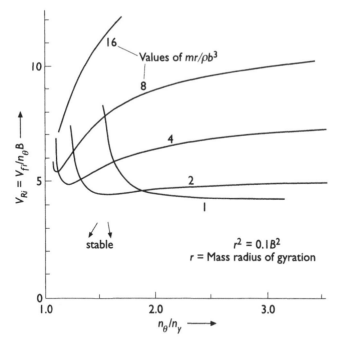

Figure 3.9 Ideal aerofoil behaviour: critical speeds for flutter.

approximation to the ideal aerofoil flutter speed V_{fi} is given by

$$V_{Ri} = V_{fi}/n_\theta B = 3.7\{mr/\rho B^3\}^{1/2}\{1 - (n_\theta/n_y)^2\}^{1/2} \qquad (3.51)$$

Many slender bridges, especially those with inclined web box stiffening structure, can achieve a good approximation to the aerofoil behaviour, suggesting definition of an 'aerofoil efficiency' of the cross-section profile, η (say), defined by

$$V_{Rf} = \eta \times V_{Ri} \qquad (3.52)$$

in which V_{Rf}, V_{Ri} are respectively the 'actual' critical value of reduced velocity, and the 'ideal' value from the chart or approximated by the Selberg formula. For sections such as the Severn Bridge η reaches more than 0.9. Generalized values for simple slender deck shapes have been proposed by Klöppl and Thiele (1967).

3.3.3 Strong torsional excitation: 'Tacoma Syndrome'

If the bridge deck assembly presents considerable vertical faces, especially as the ratio of deck width to depth falls (say, below 15 : 1), it is likely that the flutter efficiency concept will fail to offer a useful representation of the actual sensitivity to frequency ratio, with the eventual development of strong excitation of single degree of freedom torsional motion, such as the destroyed Tacoma Narrows in 1940. This remains poorly understood, and design validation depends entirely on

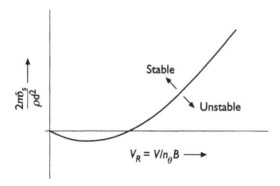

Figure 3.10 Stability envelope for bluff section bridge.

empirical evidence, mostly obtained from section model wind tunnel testing although 'discrete vortex' computational fluid dynamics is now adding to such studies (Larsen, 2000). The critical speed is also likely to show increased sensitivity to structural damping, as shown on Figure 3.10.

3.3.4 Comprehensive description of motion dependent forces: Scanlan's notation

The calculation procedures developed to evaluate flutter speeds from the aerofoil solution for forces can be extended to accept empirical values of the derivatives. A number of notations have been proposed, with the system developed over many years by Scanlan (Simiu and Scanlan, 1986) gaining widest acceptance. The lift (L) and torsional couple (M) resulting from harmonic vertical motion y and rotation α (and their time differentials $\dot{y}$, $\dot{\alpha}$) are commonly written as

$$L/(\tfrac{1}{2}\rho V^2 B) = KH_1^*\dot{y}/V + KH_2^*\dot{\alpha}B/V + K^2 H_3^*\alpha + K^2 H_4^* y/B \qquad (3.53a)$$

$$M/(\tfrac{1}{2}\rho V^2 B^2) = KA_1^*\dot{y}/V + KA_2^*\dot{\alpha}B/V + K^2 A_3^*\alpha + K^2 A_4^* y/B \qquad (3.53b)$$

defining the eight derivatives $H_i^*(K)$ and $A_i^*(K)$. It will be noted that an adaptation of aeronautical notation is used; great care is needed in interpretation of papers on this topic in view of numerical factors arising from the usage of b or B and sign changes according to whether the positive direction of displacement is the same or opposing that of the respective force. If the normalized frequency is taken as $K = B\omega/V$ ($2k$ in the aeronautical usage given above) and forces are taken positive in the same direction as the respective displacements, the values of the derivatives agree with those given in Dyrbye and Hansen (1997).

The forms taken by these derivatives in terms of Theodorsen's function for the case of ideal aerofoil behaviour are set out in full by Dyrbye and Hansen (1997: 151), who also provide a valuable critique of this increasingly prominent approach. In the wind tunnel, the derivatives can be measured directly on a section model which is externally driven in harmonic motion, but estimates can also be made by 'system identification' techniques applied to free vibration responses. The approach can be extended, for example to include alongwind (horizontal) motion, making a potential set of eighteen derivatives.

It will be noted that if the response is restricted to torsion only (or that the practical magnitude of vertical response is too small to have a significant effect through the coupling derivatives A_1^* and A_4^*), A_2^* corresponds to an aerodynamic damping

$$\delta_{a\alpha} = -(\pi\rho B^4/2mr^2)A_2^* \tag{3.54}$$

For an aerofoil A_2^* is unconditionally negative, giving positive damping, albeit fairly small. For sections subject to torsional instability A_2^* replicates Figure 3.10. For simple vertical motion the corresponding aerodynamic damping is

$$\delta_a = -(\pi\rho B^2/2m)H_1^* \tag{3.55}$$

For an aerofoil noting $H_1^* = -FV_R$ (F being the real part of Theodorsen's function, Figure 3.8), the aerodynamic damping commonly substantially reduces the vertical response to gusts. It has also been found that this estimate of vertical motion damping is useful for a relatively wide range of non-aerofoil sections.

3.4 REFERENCES

3.4.1 Books and journals

Bearman, P. W. (1981) 'Aerodynamic loads on buildings and structures', *Wind Engineering in the Eighties*, CIRIA, London (chapter 5).

Bietry, J., Chauvin, A., Redoulez, P. and Augustin, V. (1994) 'Elorn River Bridge, wind effects modelling and structural analysis; paper given at Conference Cable Stayed and Suspension Bridges, Deauville 2, pp. 153–62, APFPC Bagneux, Paris.

Blevins, R. D. (1994) *Flow-induced Vibrations.*

BRV (1990) *A Criterion for Assessing Wind-induced Crossflow Vortex Vibration in Wind Sensitive Structures* (supplement to the Department of Energy *Offshore Installations: Guidance on Design and Construction*), Brown and Root, London.

Cook, N. J. (1985) *The Designer's Guide to Wind Loading of Building Structures (Part 1)*, Butterworths, London.

Cullen-Wallace, A. A. (1985) 'Wind influence on Kessock Bridge', *Engineering Structures* 7: (January).

Davenport, A. G. (1961) 'The application of statistical concepts to the wind loading of structures', *Proc. Instn. Civil Engineers* 19 (August): 447–72.

—— (1962) 'The response of slender, line-like structures to a gusty wind', *Proc. Instn Civil Engineers* 23 (November) 389–407.

—— (1964) 'A note on the distribution of the largest value of a random function', *Proc. Instn. Civil Engineers* 28 (June): 187–196.

Doucet, Y. J. and Nordhus, A. (1987) 'Vibration monitoring of a flare boom', Texas OTC5523, Offshore Technology Conference, Houston.

Dyrbye, C. and Hansen, S. O. (1997) *Wind Loads on Structures*, Wiley, Chichester.

ESDU (1986a) Data Items 85038 Revn A and 85039 Revn A, *Circular-cylindrical Structures, Dynamic Response to Vortex Shedding: Part 1, Calculation Procedures and Derivation, Part 2, Simplified Calculation Procedures and Derivation*, ESDU International, London.

—— (1986b) Data Item 86010, *Characteristics of Atmospheric Turbulence near the Ground: Part 3, Variations in Space and Time for Strong Winds*, ESDU International, London.

—— (1990) Data Item 85020 Revn E, *Characteristics of Atmospheric Turbulence near the Ground; Part 2, Single Point Data for Strong Winds*, ESDU International, London.

—— (1996) Data Item 96030, *Response of Structures to Vortex Shedding: Structures of Circular or Polygonal Cross-section*, ESDU International, London.

Fung, Y. C. (1955) *An Introduction to the Theory of Aeroelasticity*, Wiley, New York.

Grillaud, G., Chauvin, A. and Bietry, J. (1992) 'Comportement dynamique d'un pont à haubans dans une turbulence de sillage', *J. Wind Eng. Industrial Aerodynamics* 41: 1181–9.

Harris, R. I. and Deaves, D. M. (1981) *Wind Engineering in the Eighties*, CIRIA, London (chapter 4).

Hay, J. (1992) *The Response of Bridges to Wind*, HMSO, London.

Irwin, H. P. A. H. (1979) 'Cross spectra of turbulence velocities in isotropic turbulence', *Boundary Layer Meteorology* 16: 237–43.

Klöppl, K. and Thiele, F. (1967) 'Wind tunnel tests for design of bridges against wind-excited oscillation', *Der Stahlbau* 36: 12.

Larsen, A. (2000) Aerodynamics of the Tacoma Narrows bridge – 60 years later, *Structural Engineering International (IABSE)*, 4: 243–8.

Maguire, J. and Wyatt, T. A. (1999) *Dynamics: An Introduction for Civil and Structural Engineers* (ICE design and practice guide), TTL, London.

Newland, D. E. (1993) *An Introduction to Random Vibrations, Wavelet and Spectral Analysis*, 3rd edn, Longmans, London.

Novak, M. (1972) 'Galloping oscillations of prismatic structures', *Proc. ASCE (Engineering Mechanics Divn.)* 98 EMI (February): 27–46.

Parkinson, G. V. (1965) 'Aeroelastic galloping in one degree of freedom', *Wind Effects on Buildings and Structures*, HMSO, London.

Ruscheweyh, H. (1982) *Dynamische Windwirkung an Bauwerken*, Bauverlag, Wiesbaden.

Ruscheweyh, H. and Sedlacek, G. (1988) Crosswind vibrations of steel stacks – critical comparison between some recently proposed codes', *J. Wind Eng. Industrial Aerodynamics*, 30: 173–83.

Scruton, C. and Flint, A. R. (1964) 'Wind-excited oscillation of structures', *Proc. Instn. Civil Engineers* 27 (April): 673–702.

Scruton, C., Woodgate, L. and Alexander, A. J. (1955) Aerodynamic Investigation for the Proposed Runcorn–Widnes Suspension Bridge, Report NPL/Aero 291.

Selberg, A., (1961) 'Oscillation and aerodynamic stability of suspension bridges', *Acta Polytechnica Scandinavica*, Ci: 13.

Simiu, E. and Scanlan, R. H. (1986) *Wind Effects on Structures*, 2nd edn, John Wiley, New York (chapter 13).

Smith, B. W. and Wyatt, T. A. (1981) 'Development of the draft Rules for aerodynamic stability', *Bridge Aerodynamics*, TTL, London (chapter 2).

Strømmen, E. and Hjorth-Hansen, E. (1995) 'The buffeting wind-loading of structural members at an arbitrary attitude in the flow', *J. Wind Eng. Industrial Aerodynamics* 56: 267–90.

Vickery, B. J. (1981) 'Across-wind buffeting in a group of four in-line model chimneys', *J. Wind Eng. Industrial Aerodynamics* 8: 177–93.

Vickery, B. J. and Basu, R. I. (1984) 'Response of reinforced concrete chimneys to vortex shedding', *Engineering Structures* **6**: 324–33.

Walshe, D. E. J. and Wootton, L. R. (1970) 'Preventing wind-induced oscillations of structures of circular section', *Proc. Instn Civil Engineers* **47**: 1–24.

Walshe, D. E. J. and Wyatt, T. A. (1983) Measurement and application of the aerodynamic admittance function for a box girder bridge, *J. Wind Eng. Industrial Aerodynamics* **14**: 211–22.

Wardlaw, R. L. (1981) 'Some observations on the effects of turbulence on the aerodynamic stability of bridge road decks', *Bridge Aerodynamics*, TTL, London (chapter 5).

Wex, B. P. and Brown, C. W. (1981) 'Existing bridges or new rules – which is right?' *Bridge Aerodynamics*, TTL, London (chapter 9).

Wootton, L. R. (1969) 'The oscillations of large circular stacks in wind', *Proc. Instn Civil Engineers* **43**: 573–98.

Wyatt, T. A. (1981) 'Evaluation of gust response in practice', *Wind Engineering in the Eighties*, CIRIA, London (chapter 7).

—— (1984) 'An assessment of the sensitivity of lattice towers to fatigue induced by wind gusts', *Engineering Structures* **6** (October): 262–7.

—— (1992) 'Dynamic gust response of inclined towers', *J. Wind Eng. Industrial Aerodynamics* **41**: 2153–63.

—— (1995) 'Engineering applications and requirements of prediction of extreme wind gust effects', *Proc. Instn Civil Engineers Structures Buildings* **110**(3) (April): 322–5.

Wyatt, T. A. and Scruton, C. (1981) 'A brief survey of the aerodynamic stability problems of bridges', *Bridge Aerodynamics*, TTL, London (chapter 1).

3.4.2 Selected codes of practice

ACI 307-95/ACI 307R-95: *Standard practice for the design and construction of reinforced concrete chimneys*, American Concrete Institute, Detroit, MI.

BD 49: *Design rules for aerodynamic effects on bridges* (part of volume 1 section 3 of the Design Manual for roads and bridges), Highways Agency, London.

BS 6399: *Loading for buildings; part 2, Code of practice for wind loads*, BSI (British Standards Institution), Milton Keynes.

BS 8100: *Lattice towers and masts; part 1, Code of practice for loading; part 2, Guide to background and use; part 4, Code of practice for loading of guyed masts*, BSI (British Standards Institution), Milton Keynes.

CICIND *Model Code for steel chimneys* (revision 1999), CICIND (Comité International des Cheminées Industrielles), Zurich.

ENV 1991–2–4: *Eurocode 1, Basis of design and actions on structures; part 2–4, Wind actions* (prestandard), CEN, Brussels.

ENV 1993–3–2: *Eurocode 3, Design of steel structures; part 3–2, Chimneys* (prestandard), CEN, Brussels.

NBC: *National Building Code of Canada (supplement), Commentary B, Wind loads*, National Research Council of Canada, Ottawa.

Chapter 4

Earthquake loading

Andreas J. Kappos

4.1 INTRODUCTION

Earthquakes give rise to dynamic loads that have a high potential for disastrous consequences for structures, as well as humans. There are different ways in which structures are affected by earthquakes, the vibration of the ground being the most common, but not the only one. Other earthquake effects, not specifically addressed in this chapter, are ground failures such as *liquefaction* (loss of strength in silt or sand layers due to build-up of pore water pressure), landslides and mudflows (usually triggered by liquefaction); further effects include sea waves (*tsunamis*) and lake waves (*seiches*). By far, most of the damage due to earthquakes is caused by the ground motion, but other effects can also be quite devastating, as shown, for instance, by the July 1998 tsunami that hit the coast of Papua–New Guinea, causing over 2,000 deaths and complete destruction of the villages near the coast.

In the remainder of this chapter, following a brief description of the earthquake phenomenon and the methods of assessing seismic hazard, the focus will be on the different ways the seismic actions (loads) can be defined in a design project, which strongly depend on the type of analysis chosen, and range from simple sets of horizontal forces to response spectra (deterministic or probabilistic) or acceleration time histories. The chapter will conclude with a brief discussion of the principles governing the design of structures to resist earthquakes, touching on issues beyond the seismic loading itself (structural configuration, hierarchy of member strength, systems for response control).

4.2 EARTHQUAKES AND SEISMIC HAZARD

4.2.1 Generation of earthquakes

Earthquakes are generated wherever the accumulation of strain at geological *faults* (discontinuities of the rock) leads to their rupture and to slip along the fault, until a new stable state is reached. Fault rupture gives rise to waves propagating in all directions and causing ground movement in the areas around the fault. Given the appropriate geological conditions, earthquake motions can be felt (and even cause losses) in areas located several hundreds of kilometres away from the initial rupture. The point on the fault where rupture initiates is called *focus* or *hypocentre*, while its projection on the earth surface is called *epicentre*; the distance between these two points is called *focal depth*.

In the 1960s the mechanism of strain accumulation at faults was understood and the theory of *plate tectonics* was developed, whereby the *lithosphere* (i.e. the upper part (or shell) of the earth) including the crust as well as part of the mantle, consists of several discrete segments, called *plates*, which move with respect to each other at the rate of a few centimetres a year; this relative movement is caused by convection currents in the mantle of the earth. The six main tectonic plates, as well as other smaller ones, are shown in Figure 4.1; note that some continents are on a single plate, whereas others straddle more than one plate. The plate boundaries can be either *divergent* (sea floor spreading at mid-ocean ridges), or *convergent*; particularly important in the latter case is the phenomenon of *subduction* (i.e. when a plate is pushed below the neighbouring plate). As seen in Figure 4.1, that depicts the distribution of epicentres of recent (1960–2000) earthquakes, the most serious tectonic activity takes place at the boundaries of the plates (different size and colour of circles correspond to different magnitude and focal depth). Earthquakes occurring close to the plate boundaries are called *interplate* events, while earthquakes remote from the boundaries are referred to as *intraplate* events; the latter are far less common and much more difficult to explain than the former (Bolt, 1993; Reiter, 1991).

Although earthquakes can be triggered by other phenomena, such as volcanic eruptions, sudden changes in the stress state of soil layers due to filling of reservoirs behind dams, 'mine-burst' (masses of rock collapsing explosively in mines), or even underground nuclear explosions (Bolt, 1993), the vast majority of them are due to faulting. There are essentially two types of faults, those associated with horizontal movement (*strike-slip*), and those associated with vertical movement (*dip–slip*). Fault orientations have a strong effect on the resulting earthquake motion; for instance, reverse dip–slip faults are usually the ones associated with the most catastrophic ground motions.

As mentioned previously, fault rupture gives rise to seismic waves. These propagate either by compression and dilation (like sound waves), with the ground particle motion in the same direction as the propagation, and are called *longitudinal* or *P-waves*, or by shear (particle motion perpendicular to the direction of the propagation), and are called *transverse* or *S-waves*; these two types of waves are referred to as *body waves*. The velocity of shear waves is given by

$$v_s = \sqrt{\frac{G}{\rho}} \tag{4.1}$$

where G is the shear modulus of the ground and ρ its mass density; v_s is a very useful

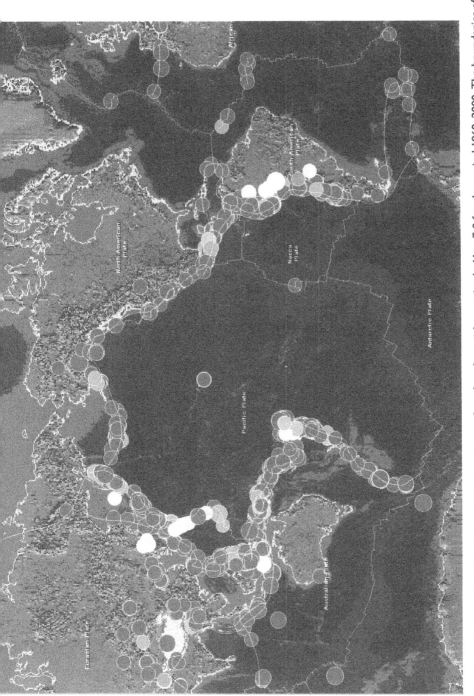

Figure 4.1 Geographical distribution of the epicentres of earthquakes with magnitude $M > 7.0$ for the period 1960–2000. The boundaries of the lithospheric plates are also shown.

quantity for classifying the dynamic characteristics of the ground (see Section 4.3.3). Since v_s is lower than the velocity v_p of P-waves, the latter are always the first to arrive at a station recording the seismic motion, followed by S-waves, which are associated with large amplitudes of motion.

When body waves reach the earth surface they are reflected back into the crust, and a vibration of the surface is initiated, which propagates through *surface waves*. Depending on the way these waves propagate along the earth surface, they are classified as *Rayleigh* waves or *Love* waves. Surface waves, along with S-waves, account for the strongest part of the seismic motion (i.e. these are the ones that may cause losses). P-waves are generally small amplitude and of interest to the seismologists only; they use the difference in arrival times between P and S-waves for determining the epicentre of an earthquake.

4.2.2 Measures of earthquakes

Designing against earthquakes presupposes that the phenomenon can be adequately quantified. There are two main ways for measuring the size (or strength) of earthquakes: One based on instrumental data, and one based on observation of the effects of earthquake motions on humans and structures; both are indispensable for hazard assessment and seismic design.

There are two types of instruments that can be used for recording earthquake motions:

- The *seismographs*, which record the *displacement* of the ground with time. These instruments are designed to magnify weak motions, so they can record motions caused by very distant earthquakes. Their recordings are of interest mainly for the seismologists, since they are used for locating earthquakes and characterizing their sources.
- The *accelerographs*, which record the acceleration of the ground with time. Until recently these instruments were recording (on film) whenever they were triggered by a minimum level of acceleration (e.g. 0.01 g), but more advanced instruments are currently available, which record in a digital form on reusable medium, hence they can operate continuously and save only records of interest; this has the extra advantage that the initial part of the motion is not lost. *Accelerograms* are the main type of earthquake record used for deriving design seismic actions.

Magnitude

The *magnitude* of an earthquake is a measure of the earthquake size or the source strength; usually, though not necessarily, the magnitude measures the amount of *energy* released by an earthquake. The Richter magnitude or *local* magnitude M_L is defined as the (base 10) logarithm of the maximum amplitude A (in μm) of the earthquake, corrected to a distance of 100 km; the correction is done by subtracting from $\log A$ the quantity $\log A_0$, where A_0 is (arbitrarily) defined as the earthquake that would produce an amplitude of 0.001 mm on a standard seismograph at a distance of 100 km from the source. M_L is empirically related to the *energy* E released at the source (i.e. at the fault) by the formula

$$\log E = 11.8 + 1.5M_L \tag{4.2}$$

where E is in ergs (1 erg $= 10^{-7}$ joules). It is worth pointing out that a unit increase in magnitude corresponds to an increase in energy by 32; hence, a magnitude 7 event releases 1,000 times more energy than a magnitude 5 event. $M_L = 5$ is practically the magnitude threshold for earthquakes that may cause damage to structures.

The instrument specific definition of M_L and the fact that it is limited to earthquakes recorded at distances of less than 1,000 km, have led to the definition of other magnitude measures, the most common of which is the *surface–wave magnitude M_s* defined by

$$M_s = \log\left(\frac{A}{T}\right)_{max} + 1.66 \log \Delta + 3.3 \tag{4.3}$$

where A is the amplitude, T the period of the ground motion, and Δ the *epicentral distance* (i.e. the distance from the site, in this case the recording station, to the epicentre). It is seen that this definition is independent of the instrument used (no need for A_0). M_s is determined with respect to the amplitude of Rayleigh waves with a period of about 20 sec. A similar definition exists for the *body–wave magnitude m_b*, determined by the maximum amplitude of P-wave motion. Another scale is based on the *seismic moment M_0* which is a description of the extent of deformation at the earthquake source; the *moment magnitude M_w* is defined as a simple function of $\log M_0$ (see e.g. Reiter, 1991).

Whenever magnitude is used for estimating seismic hazard (see Section 4.2.5), one should be particularly careful in identifying what type of magnitude is used in each earthquake catalogue, as all the previous definitions do *not* yield the same value, especially in the range of large magnitudes. A notable feature is the 'saturation' of all magnitude scales, with the exception of M_w (i.e. beyond a certain limit the scales stop increasing with increasing earthquake size). There is no upper or lower limit to magnitude, however, the largest size of an earthquake is limited by the strength of the rocks of the Earth's crust (Bolt, 1993). The largest earthquakes recorded in the 20th century had magnitude $M_L \cong 8.9$; the 1960 Chile earthquake had an $M_L = 8.3$, but a moment magnitude $M_w = 9.5$. The problem of saturation of wave amplitude-based scales is behind the current trend to use predominantly M_w as a measure of earthquakes; nevertheless M_s and even M_L are still widely used worldwide.

Intensity

Whereas the use of measurable quantities for characterizing earthquakes is obviously desirable, the fact remains that the instrumental record is less than 100 years old. Since the recurrence period of strong earthquakes (including design earthquakes) is significantly longer than 100 years, it is imperative to make some use of the *historical* record of earthquakes in seismic hazard analysis. For some regions of the world (the best example being China) historical records go back to more than a thousand years, but their completeness and quality vary greatly. The critical information that can be found in such records regards the effects of past earthquakes on humans and on structures.

The *(macroseismic) intensity* of an earthquake refers to the way an earthquake is felt at a specific site (i.e. its effects on humans, structures and the ground). Therefore, the intensity is a measure of the severity of ground shaking on the basis of observed effects in a certain area (rather than a measure of the energy release or the seismic moment). The major advantage of intensity is that it can be estimated from the historical records, therefore it is essentially the only viable tool in *historical seismicity*, and it can be estimated in all affected areas, including those where no instrumental records exist; hence it is also useful today as a complement to instrumental measurements. The major disadvantages of intensity are that it varies significantly within the area affected by an earthquake (note that an earthquake has one magnitude but several intensities), and its estimation involves substantial subjective judgement.

A major problem in estimating intensity is that similar structures respond differently to the same earthquake, due to several reasons whose discussion falls beyond the scope of this book. Hence the need for appropriately classifying the effects of damage (with at least some rough allowance for its statistics) and also for appropriately defining the extent of the areas for which a uniform intensity should be assumed. Typically these areas should correspond to a village or a relatively small town, or parts of a large city, but strict rules are difficult to set (European Seismological Commission, 1998).

Starting from the late 1800s, several *intensity scales* have been suggested. The ones most commonly used today are the *Modified Mercalli* intensity (I_{MM}), employed in the Americas, and the *Medvedev–Sponheur–Karnik (MSK)* intensity (I_{MSK}), widely used in Europe. Both scales have 12 degrees, and are generally equivalent (there is a small discrepancy at the lower end of the scales only). It used to be common to denote the degrees with Roman numerals (I–XII), primarily to discourage arithmetical manipulation, but the need for computer processing of intensity data has made it common nowadays to use normal (Arabic) numerals. Since 1992 the European Seismological Commission (ESC) has been developing an updated version of the MSK scale, called the 'European Macroseismic Scale' (ESC Working Group on Macroseismic Scales, 1998), which might be used extensively in the future.

All the aforementioned intensity scales share several common features, the most important one being that they are descriptive, in the sense that each degree on the scale is characterized by a set of 'diagnostics' referring to specific effects of an earthquake on humans, buildings, objects and the nature in general. As an example, a diagnostic referring to humans is 'many people find it difficult to stand, even outdoors' (intensity VIII), while a typical diagnostic referring to buildings is 'considerable damage in masonry structures built to withstand earthquakes' (intensity IX).

Once intensities have been assigned to several areas (defined as explained earlier in this section), then *isoseismal* maps showing the distribution of intensity in a larger area can be drawn. Figure 4.2 shows such a map drawn for the Los Angeles area following the 1994 Northridge earthquake (EERI, 1995). An interesting feature, quite common in such maps, is that the epicentre of the earthquake is not within the area where the maximum intensity was recorded.

4.2.3 Strong motions and path effects

However useful intensity maps may be, the definition of seismic loading for the purposes of structural analysis and design requires more refined information which can be provided by appropriate processing of strong ground motions.

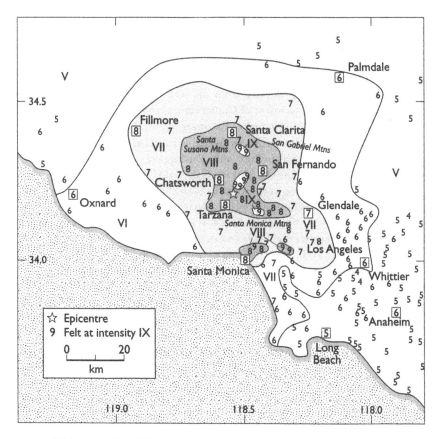

Figure 4.2 Distribution of I_{MM} in the epicentral region of the Northridge 1994 earthquake (EERI, 1995).

Strong motion records

Accelerograms of strong motions (i.e. time histories of acceleration) are recorded by accelerographs; these instruments record simultaneously the three components of the motion, two perpendicular horizontal (longitudinal, for instance N–S, and transverse, E–W), and one vertical. Before being used for 'engineering' purposes (e.g. for deriving response spectra, see Section 4.3.2), the records are corrected to remove frequency dependent instrument response and ambient noise. An example of corrected accelerogram from the 1971 San Fernando (S. California) earthquake is shown in Figure 4.3. Although this is not really a typical record, the observed difference in frequency content between the vertical and the horizontal components is indeed quite typical.

Accelerograms are arguably the most valuable information for deriving design seismic loads and it is fortunate that nowadays a very large number (several tens of thousands) of accelerograms are available; on the other hand, though, there are seismic areas for which the number of records is very low or even zero. Databanks of accelerograms have been compiled in many regions, particularly in the United States, Japan and Europe. One of the largest collections containing over 15,000 digitized and processed accelerograph records from all over the world (but mainly from the US), dating from 1933–1994, is available from the National Geophysical Data Centre in Boulder, Colorado. A number of American records can be downloaded directly from the web sites of the Strong Motion Data Centre of the US Department of Conservation, and from NISEE (National Information Service on Earthquake Engineering, University of California, Berkeley). In Europe, accelerograms are available from organizations such as the Institute of Engineering Seismology and Earthquake Engineering (ITSAK), Thessaloniki, Greece, and Servizio Sismico Nazionale (SSN), Rome, Italy.

The main purpose of using accelerograms is to characterize the strong ground motion, with a view to defining appropriate design loads. In this respect, the *Peak Ground Acceleration* (PGA, or simply A) (i.e. the highest value of the acceleration time history), is a parameter that has been extensively used in seismic hazard assessment (see Section 4.2.5). It is worth pointing out, though, that this is mainly due to its convenience, because otherwise the PGA is often a rather poor indicator of the destructiveness of the ground motion. The Peak ground Velocity (PGV, or simply V) and/or the Peak Ground Displacement (PGD or D) are better indicators of damage potential and have been used in some studies. Velocity and displacement time histories of the ground motion can be calculated by integration of the acceleration time history, but they are quite sensitive to the filtering procedure used in correcting the accelerograms. Hence, by far the most useful information that can be extracted from accelerograms is the response spectra, discussed in Section 4.3.2. One factor, though, that is not reflected in the spectra (which are plots of peak response) is the *duration* of the motion. This can be quite critical in certain cases, such as structures susceptible to strength degradation under reversed cyclic loading (i.e. change of the sign of the applied force or moment).

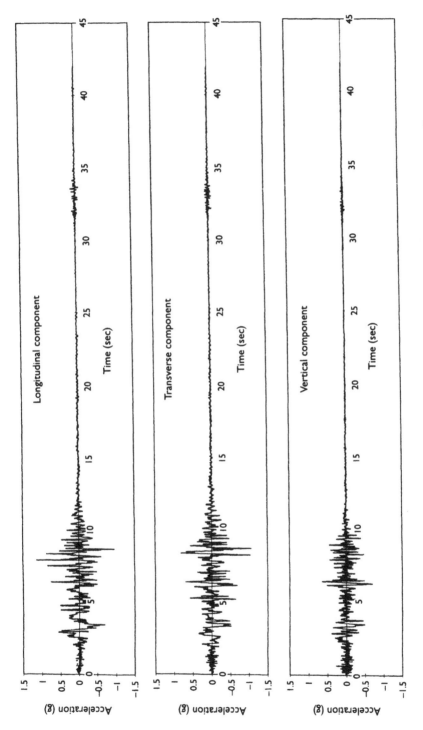

Figure 4.3 Three components of the accelerogram recorded during the 1971 San Fernando earthquake at the Pacoima Dam site.

Attenuation relationships

As the seismic waves propagate away from the source, their amplitude decreases; this results in the so-called *attenuation* of the ground motion. Attenuation is the reason why even the strongest motions cease to be damaging after a certain distance from the source. The previous statements should not be interpreted as meaning that the damage potential of a motion at, say, 100 or 200 km from the source is always lower than at a distance of say 10 or 20 km. Site effects (Section 4.2.4) can lead to quite the opposite effect, a notable example being that of the 1985 earthquake off the coast of Mexico whose most catastrophic effects (including about 10,000 fatalities) were recorded in Mexico City, 400 km away from the epicentre.

Attenuation relationships (i.e. models describing the values of strong motion parameters as a function of distance from the source) have been developed for magnitude, intensity (compare Figure 4.2), the strong motion peaks (PGA, PGV, PGD), and, more recently, spectral ordinates. The most commonly used one, particularly for defining seismic loads, is the relationship involving PGA. The typical form of such a relationship is

$$\log(A) = b_1 + b_2 M + b_2 \log R + b_4 R + \sigma P \tag{4.4}$$

where $R = (\Delta^2 + H_0^2)^{1/2}$, Δ being the epicentral distance, and H_0 can either coincide with the focal depth H, or just be a parameter to be defined by regression, together with the coefficients b_i. The parameter P is introduced to account for the significant uncertainty associated with all attenuation relationships; $P = 0$ if the mean (or 50-percentile) of PGA is sought, while $P = 1$ for calculating the mean plus one standard deviation (σ), which is the 84-percentile if a normal distribution of the residuals of $\log(A)$ is assumed. For design purposes either the 84-percentile or the 90-percentile of A is used.

As will be seen in the next section, attenuation relationships are essential in estimating seismic hazard and design seismic loads. Today there are several such relationships for several regions of the world, most of them referring to the US (especially the West Coast), Japan and Southern Europe. A comprehensive review of the attenuation relationships used in Europe can be found in Ambraseys and Bommer (1995) who suggested the following form of eqn (4.4) for horizontal PGA in Europe

$$\log(A) = -1.09 + 0.238M - \log R - 0.0005R + 0.28P \tag{4.5}$$

with $H_0 = 6$ km (if the actual H is used for H_0, the coefficients are markedly different). A comparison of eqn (4.5) with a more recent one suggested by Ambraseys and the relationship proposed by Joyner and Boore (1988) for western North America is shown in Figure 4.4. It is worth pointing out that differences among the predictions of the three equations are less than the scatter associated with them. It is also seen in Figure 4.4 that ground motion attenuation at relatively large distances is more pronounced in North America than in Europe. Equations

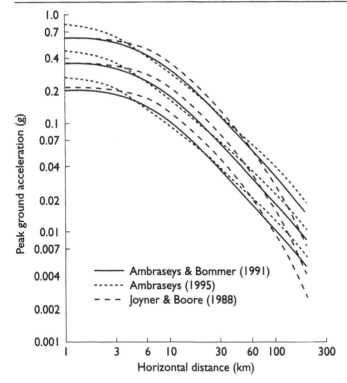

Figure 4.4 Comparison of attenuation relationships for PGA, for Europe and western North America for shallow earthquakes of magnitude 5, 6 and 7 (Ambraseys and Bommer, 1995).

similar to (4.5) have been developed for the *vertical* PGA (Ambraseys and Simpson, 1996), which is generally of the order of $\frac{2}{3}$ the corresponding horizontal acceleration (Newmark and Hall, 1982). However, in the near field (i.e. at distances from the source less than about 15 km) the ratio of the vertical to horizontal PGA may exceed unity, but falls off with distance (Ambraseys and Simpson, 1996).

Directivity effects

The source of the seismic waves (the fault rupture) is a moving source (i.e. the source travels along the fault at a certain velocity). The direction of the fault rupture has a strong influence on the resulting ground motion. If the fault rupture propagates towards a particular site the motion at that site will be stronger than at an equidistant site located opposite to the propagation of rupture. This phenomenon is called *directivity* and its effect is to produce the highest amplitude of motion together with the shortest duration in the direction of the rupture, and the smallest amplitudes but longest duration in the opposite direction.

Directivity effects have been observed in several earthquakes, a recent example being that of the Northridge earthquake, where the only extensive region with accelerations above 0.5 g was to the north of the epicentre (see Figure 4.2), consistent with the rupture propagation (EERI, 1995). Of particular concern with regard to the seismic behaviour of structures is the case of large amplitude and long period pulses in the acceleration time history due to directivity effects; these pulses are usually accompanied by large velocities and can be quite catastrophic.

4.2.4 Site and topography effects

The ground motion can be significantly affected by the properties and configuration of the layers underlying the earth's surface. The properties that most affect the amplitude of ground motion are the resistance to particle motion, called *impedance*, and the *soil damping* (or absorption). For most practical purposes the impedance can be defined as the product ρv_s where ρ is the density and v_s the previously defined (see eqn 4.1) shear wave velocity. The flow of energy (or energy flux) during the wave propagation is equal to $\rho v_s \dot{u}^2$; hence, when a seismic wave propagates through a region of decreasing impedance, the resistance of soil particles to motion decreases, and to preserve the total energy, the particle velocity and hence the amplitude of motion increases. It follows that assuming all other conditions remain the same, the seismic waves would have higher amplitude on soil (low ρ, low v_s) than on rock (high ρ, high v_s). On the other hand, damping is typically much higher on soft soils than on hard rock, therefore it tends to mitigate the adverse effect of low impedance in the former. As a result of the aforementioned effects, peak accelerations are generally not very different on sites classified as 'rock' and as 'soil' (or 'alluvium'); usually peak accelerations at the surface of soil deposits are slightly higher than on rock outcrops when these accelerations are small (less than 0.15 g), and smaller at higher acceleration levels. Peak velocities, though, as well as displacements, are always higher on softer soil sites.

The configuration of the layers underlying a site, for example, whether they are essentially horizontal or not, and whether there are variations of their properties along the (horizontal) length, may also significantly affect the amplitude of ground motion. A detailed discussion of the complicated phenomena involved can be found elsewhere (Finn, 1991; Reiter, 1991; Kramer, 1996). Here it will only be pointed out that the most adverse effect of layer configuration is *resonance*, particularly two-dimensional one that can appear in alluvial valleys. Resonance occurs whenever the predominant period of the ground motion practically coincides with the characteristic site period, which for a soil deposit of depth (to the bedrock) H is given by

$$T_s = \frac{4H}{v_s} \tag{4.6}$$

A lot of controversy prevailed until recently regarding the effect of *non-linear* soil response on the ground motion, the geotechnical engineers arguing that it is significant and the seismologists maintaining that existing evidence does not support this view. The main implication of nonlinearity is that when a soil layer becomes strongly inelastic the shear stress cannot increase significantly, hence the amplitude of motion ceases to increase. This is obviously a desirable effect regarding the response of structures, but it causes problems regarding the reliability of data (on v_s and similar quantities) measured from microtremor or other small amplitude testing. Quantitative evidence from recent earthquakes such as the 1985 Michoacan (Mexico) and the 1989 Loma Prieta (California), has clearly shown that much higher accelerations can be recorded on sites underlain by soft soil layers (such as the Mexico City clay and the San Francisco Bay mud), than on stiffer soil sites. Figure 4.5 reported by Finn, 1991 shows the reduction of the shear modulus G of clays characterized by different Plasticity Indices (PI) (note that the highest PI corresponds to the Mexico City clay). It is clear that for stiffer clays, with PI not exceeding about 40 or 50, G reduces significantly at relatively low shear strains, hence resulting in reduced amplification of the motion; similar behaviour is shown by other soil types, like sands. However, this is not the case with high PI clays which remain essentially elastic (G/G_{max} close to 1) for strains up to 0.1 per cent or even more. It is clear, therefore, that at least for this class of soils, the non-linear characteristics have a significant influence on the ground motion and should be accounted for in design.

Even more important than increasing peak accelerations, site effects are strongly influencing the shape of the response spectrum (see Section 4.3.3).

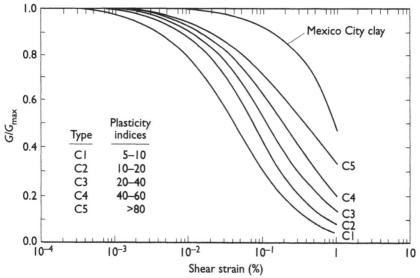

Figure 4.5 Reduction of normalized shear modulus for clays with different plasticity indices.

Topography effects

Topography of the site can also have a noticeable effect on amplification of ground motion. The strong motion shown in Figure 4.3 was recorded on a rocky ridge connected to the Pacoima Dam, and is characterized by a peak acceleration of 1.17 g, one of the highest ever recorded. Many people argued that this was mainly the result of a topographic amplification, although other interpretations were also suggested (Reiter, 1991).

The major parameter of the problem appears to be the steepness of the ridges; it can be shown that the displacement amplification at the crest of an essentially triangular hill is equal to $2/\nu$, where $\nu\pi$ is the angle formed by the ridges; therefore the amplification increases as the ridge becomes steeper. Observed amplifications at the crest (with respect to the base) range from 2 to 20, whereas theoretical predictions are generally much less (3 to 4), possibly due to the influence of three-dimensional effects and ridge to ridge interaction. Topography effects are discussed, among others, by Finn (1991) and Kramer (1996). Due to the complexity of the subject, it is generally considered as not mature enough to be included in code provisions. *The Recommendations of the French Association for Earthquake Engineering* (AFPS, 1990) appear to be the only document of regulatory character that has adopted rather detailed rules for the calculation of the topographic amplification factor.

Spatial variability of ground motion

While the smallest dimension of common structures such as buildings is usually small enough that the ground motion can be assumed to be the same along the entire plan of the structure, in elongated structures, such as long bridges and pipelines, a rather significant variability of the ground motion may occur, particularly whenever the large plan dimensions are combined with irregularities in the soil profile. The local spatial variation or *incoherence* of ground motion is mainly due to

- travelling wave effects, wherein non-vertical seismic waves reach different points of the structure at different times (time delay effect);
- scattering (reflection, refraction) of seismic waves caused by inhomogeneities along the travel path;
- local soil filtering and amplification of the motion.

The *coherency* of two ground motions is a measure of correlation of amplitudes and phase angles at different frequencies. Ground motions recorded by dense arrays of accelerographs have shown that coherency decreases with increasing distance and increasing frequency of motion (Clough and Penzien, 1993; Kramer, 1996).

4.2.5 Assessment of seismic hazard

Analysis of seismic hazard (resulting from strong motions) is the basis for defining seismic loading for design purposes, more particularly for deriving the *design response* spectrum, discussed in more detail in Section 4.3.2.

If seismic hazard is to be estimated in a deterministic way, an appropriate *earthquake* scenario has to be defined. This involves identifying the source (fault) which will give the most critical motion for the site under consideration, estimate the maximum magnitude that can be produced by this source, and then estimate the maximum PGA at the site using an appropriate attenuation relationship (similar to eqns 4.4, 4.5). This PGA can then be used for scaling or 'anchoring' a fixed spectral shape, with due allowance for site effects, in order to produce the design spectrum (see Sections 4.3.2, 4.3.4). Such a procedure (whereas not uncommon) suffers from various drawbacks. One problem is the difficulty in identifying the critical source (different sources can produce motions that may be critical for a particular type of structure), another one is the difficulty in predicting the 'maximum credible earthquake' associated with a source. Even if this earthquake is reliably estimated, it is generally uneconomical to design structures against it. These and other problems are the reason why today all major seismic hazard studies are carried out using a *probabilistic* approach.

The various components of a *probabilistic hazard analysis* are shown in Figure 4.6 (EERI Committee, 1989). The first step is the identification of all sources, which can be point sources or line sources (faults), or area sources. Then, for each type of source the recurrence of earthquakes has to be defined, mainly on the basis of historical data. Despite (or because of) its simplicity, the most commonly used recurrence relationship is the one proposed by Gutenberg and Richter back in 1944

$$\log(N) = a - bM \tag{4.7}$$

where N is the (cumulative) number of earthquakes greater than or equal to a given magnitude M, that are expected to occur during a specified period of time, typically taken equal to 1 year. The coefficients a and b have to be determined from regression analysis of available data. Usually an upper bound on magnitude is placed, based on the characteristics of the source and/or the maximum historical earthquake.

Design seismic loads for a structure are based on the ground motions having a desired probability of exceedance during the lifetime of the structure (about 50 years for usual buildings, higher for other types of structures); this probability is commonly taken equal to 10 per cent for buildings of usual importance. The probability p of an earthquake exceeding a certain magnitude M during the lifetime can be calculated if an appropriate statistical model is assumed, as shown in Figure 4.6(top left). For simplicity a *Poisson process* is assumed, wherein the various 'events' (i.e. that the magnitude M is exceeded within a certain time) are independent. This is equivalent to assuming that earthquake activity has no memory, which is not true, but the resulting error is not large. Using the definition of the Poisson distribution, this probability is

$$p = 1 - e^{-LN} \tag{4.8}$$

where L is the lifetime of the structure. Hazard assessment can then proceed by selecting a number of values of a strong motion parameter (e.g. A_i), calculate the

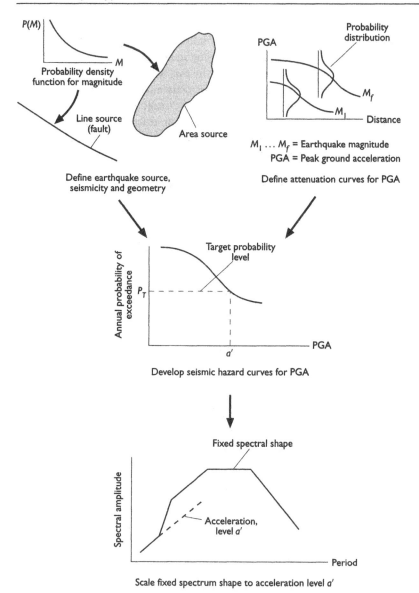

Figure 4.6 Development of the design spectrum on the basis of seismic hazard analysis for PGA (EERI, 1989).

corresponding magnitude M_i from an attenuation relationship (see Figure 4.6, top right) and then obtain the annual frequency N_i of earthquake with magnitude $\geq M_i$ by substituting M_i in the magnitude frequency eqn (4.7). The calculated value of N_i is then substituted in equation (4.8) to find the corresponding probability of exceedance. By repeating the procedure for an appropriate number of A_i a complete hazard curve as shown in Figure 4.6(middle) can be derived. The actual procedure is somewhat more complicated as the scatter in the attenuation relationship is also included in the analysis.

Results from such procedures are used to construct the hazard maps used as a basis for seismic codes. An example of such a map is shown in Figure 4.7; it provides the contours of the effective peak acceleration coefficient A_a for the United States (FEMA, 1995). This map was derived from similar maps showing the PGA's with a 10 per cent probability of exceedance in 50 years, after converting PGA to effective peak acceleration using procedures based in part on scientific knowledge and in part on judgement and compromise. For the purpose of defining design seismic actions, hazard maps such as that of Figure 4.7 are further simplified to include a limited number of *seismic zones* within which the value of A_a is considered as constant.

Response spectra for a target annual probability of exceedance P_T (e.g. 0.2 per cent) can be constructed by calculating the corresponding $A = a'$ from the curve of Figure 4.6(middle) and then anchor a fixed spectral shape to a', as shown in Figure 4.6(bottom), and further discussed in Section 4.3.2. Alternatively, a more complex procedure may be followed, whereby the attenuation relationships are developed for spectral ordinates (e.g. the spectral acceleration S_{pa}), rather than for PGA. These period dependent attenuation relationships are then used to construct the design spectrum period by period; this is called a hazard consistent or *uniform hazard spectrum* (EERI, 1989; Reiter, 1991).

4.3 DESIGN SEISMIC ACTIONS AND DETERMINATION OF ACTION EFFECTS

4.3.1 Design situations

The design seismic action or the *design earthquake* is a ground motion or a set of ground motions defined in a way appropriate for the design of engineering structures. Depending on the type and importance of the structure to be designed, the seismic action can be defined in different ways, i.e. as:

- a set of (equivalent) lateral forces;
- a response spectrum;
- a power spectrum;
- a set of acceleration time histories.

The foregoing can be defined either on the basis of a seismic code (most common case), or by carrying out a site specific seismic hazard analysis with due consideration of ground effects (see Sections 4.2.3–4.2.5). The scope of each procedure can be appreciated by considering the following four situations that might be faced by an engineer in practical design:

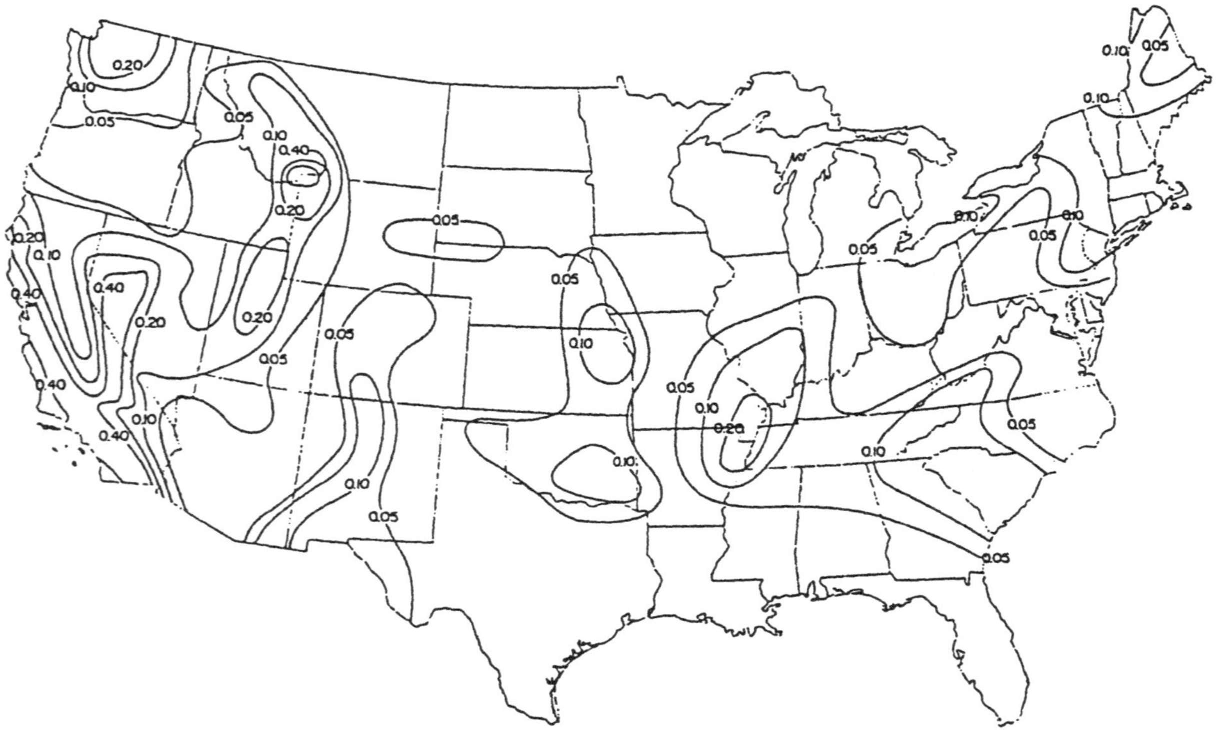

Figure 4.7 Contour map for effective peak acceleration coefficient A_a for the continental United States, from the 1994 NEHRP Provisions (FEMA 1995).

- For many building structures, and also for some 'small-scale' civil engineering structures (such as small bridges, viaducts, etc., and typical geotechnical structures such as retaining walls), the *equivalent lateral force* procedure can be used. The procedure is well documented in most current seismic codes, and will be described in Section 4.3.5 with specific reference to two major codes, the 1995 Eurocode 8 (EC8) and the 1997 UBC (American code).

- For buildings with configuration problems (irregular plan and/or elevation), for many types of medium bridges, and for many of the structures falling beyond the scope of this chapter, an elastic *dynamic analysis* has to be carried out, typically in the form of modal response spectrum analysis. The definition of the elastic spectrum (Section 4.3.2), its modifications due to site effects (Section 4.3.3), and its reduction to an inelastic design spectrum (Section 4.3.4), are some of the most important issues relating to seismic loading. Specific mention will be made in the aforementioned sections to the EC8 and the UBC spectra. In exceptional situations where a probabilistic approach is warranted, *power spectra* (Section 4.3.8) may be used instead of 'normal' response spectra.

- In cases such as the design of very important structures, or structures clearly falling outside the limits of the existing codes (e.g. structures with very high fundamental natural periods), a full *time history* analysis, typically in the inelastic range may be required. Note that there is no advantage in using this procedure for an elastic analysis of the structure which can be conveniently carried out (at essentially the same accuracy) using the modal superposition approach, the exception being structures where due to highly irregular geometry it is difficult to combine the modal contributions, or whenever the structural model includes critical frequency dependent parameters (Clough and Penzien, 1993). An appropriate selection and scaling of natural and/or artificial records has then to be made; a key point to be addressed is the correspondence between these records and the (code) design spectrum. The EC8 and the UBC recommendations will be referred to in Section 4.3.7 and it will be made clear that this type of procedure is more common in the case of assessment of existing structures which might not comply with current code requirements.

- Again for some exceptional cases, such as important structures whose construction cost is particularly high and/or the consequences of their failure particularly severe (a typical example being nuclear power plants), as well as in the case of construction in areas where a design spectrum or a code is not available, a site specific *seismic hazard assessment* study has to be made, typically using probabilistic techniques. Although normally the civil engineer will not carry out such a study, it is important that s/he realizes the main assumptions involved, and, more significantly, is capable of appropriately evaluating the results of such a study and making use of them for design purposes. A brief coverage of this procedure has already been given in the previous section (4.2.5).

4.3.2 Elastic spectra

Response spectra

A *response spectrum* (i.e. a plot of the peak response (to an input motion) as a function of the natural period) can be derived by analysing a series of Single Degree-Of-Freedom (SDOF) systems, as explained in Chapter 2 (Section 2.5) and, in more detail, in the literature (Newmark and Hall, 1982; Gupta, 1990; Clough and Penzien, 1993). The quantities typically plotted are the spectral pseudo-acceleration S_{pa}, pseudo-velocity S_{pv}, and displacement S_d, which are interrelated through the familiar expressions

$$S_{pa} = \omega S_{pv} = \omega^2 S_d \qquad (4.9)$$

Due to (4.9) the three spectral quantities can be plotted together on a log–log paper (see Figure 4.10). It should be recalled that S_{pa} and S_{pv} are *not* the actual response acceleration and velocity, respectively (see also Section 2.5). Nevertheless, S_{pa} is practically the same as the actual maximum acceleration for reasonable (i.e. not too high) values of damping, while S_{pv} is nearly the same as the actual velocity except in the very short and the long period range (Newmark and Hall, 1982). For design purposes, S_{pa} is more useful than the actual response acceleration, since the former can be used to calculate directly the maximum forces on the structure, as discussed in Section 4.3.6.

An example of response spectra, referring to the longitudinal (horizontal) component of the input motion of Figure 4.3, is given in Figure 4.8; for each spectrum five curves are plotted, corresponding to damping ratios from 0 to 20 per cent. It is first noted that for lower values of damping the variation of the spectral values with the natural period can be quite abrupt, whereas for high damping values the spectra become much smoother. An important piece of information provided by a spectrum is the range of periods for which the response of a structure is peaking. The S_{pa} curves in Figure 4.8(a) are typical in the sense that the peaks occur in the short period range, mainly from 0.2 to 0.5 sec; this is a common feature of motions recorded on rock sites. A second period range around 1 sec also shows some increase in the amplification, but significantly lower than that in the short period range. However, if the pseudo-velocity is used as the basis for identifying critical periods, it is seen in Figure 4.8(b) that the most critical range is that between 0.9 and 1.8 sec; the range of periods between 0.3 and 0.5 sec is also characterized by local peaks, but is less critical than the previous one. This illustrates an important problem in seismic design (i.e. the selection of the parameter which best characterizes the damageability of a particular ground motion). Many designers rely more on S_{pv} which is a direct measure of the seismic energy input, since for negligible damping the energy stored in an oscillator with mass m is equal to $\frac{1}{2} m S_{pv}^2$. On the other hand, recently developed *displacement based design* and *assessment* procedures are based on the displacement spectrum. Despite the aforementioned trends, all current codes base their design forces on S_{pa} spectra

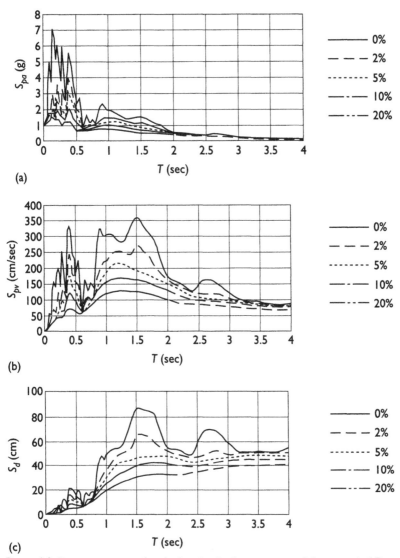

Figure 4.8 Response spectra for the longitudinal component of the record of Figure 4.3 (Pacoima Dam S16E): (a) pseudo-acceleration; (b) pseudo-velocity; (c) displacement.

(directly, or indirectly). Of course, due to eqn (4.9), S_{pv} and S_d curves can always be derived when S_{pa} is available.

The significant differences in the shape of response spectra derived from different ground motions are illustrated in Figure 4.9, which shows the 5 per cent damped S_{pa} spectra for three accelerograms recorded in three different parts of the world

(North America, Central America, and Southern Europe). The Pacoima Dam (California) S16E and the Kalamata (Greece) N10W records are from earthquakes with similar magnitude (6.6 and 6.2) and very close to the recording station (epicentral distances of 3 and 15 km). It is seen that, although the magnitude of the accelerations is significantly larger for the Pacoima record (see discussion of the topographic amplification effect in Section 4.2.4), the shape of the two spectra is quite similar, with peaks occurring in the short period range. On the other hand the Mexico City 1985 SCT transverse component, recorded during a magnitude 8.1 earthquake at a distance of 400 km, resulted in a significantly different spectral shape, wherein the critical period range is between 1.7 and 2.8 sec; the effect of soil conditions (very important in this case) is discussed in the next section. It is seen that the Mexico City record with a PGA of only 0.17 g, will be more critical for high rise buildings with $T > 1.7$ sec than the Pacoima record with a PGA of 1.17 g.

Fourier spectra

Although most engineering applications involve the aforementioned response spectra, a better understanding of the ground motion characteristics can be obtained from the *Fourier spectrum*, defined as

$$\ddot{V}_g(i\bar{\omega}) \equiv \int_{-\infty}^{\infty} \ddot{u}_g(t)\, e^{i\bar{\omega}t}\, dt \qquad (4.10)$$

where $\ddot{u}_g(t)$ is the ground acceleration time history and $\bar{\omega}$ is the circular frequency of a harmonic forcing function. It is then possible to express $\ddot{u}_g(t)$ through the superposition of a full spectrum of harmonics (Clough and Penzien, 1993). Common applications involve the *Fourier amplitude spectrum*, defined by

$$|\ddot{V}_g(i\bar{\omega})| = \left[\left(\int_0^{t_1} \ddot{u}_g(t) \cos \bar{\omega}t\, dt \right)^2 + \left(\int_0^{t_1} \ddot{u}_g(t) \sin \bar{\omega}t\, dt \right)^2 \right]^{1/2} \qquad (4.11)$$

where t_1 is the duration of the ground motion. Note that eqn (4.11) does not uniquely define a ground motion (as eqn 4.10 does) since the phase angles between pairs of harmonics have been lost in this definition.

Fourier spectra are commonly used to interpret phenomena associated with the transmission of seismic energy from the source to distant locations. A useful application of these spectra in the construction of simulated ground motions is briefly presented in Section 4.3.7.

Design spectra

For design purposes, it is clear that spectra smoother than those of Figures 4.9 and 4.10 are required, since a future motion is very unlikely to be identical to a previously recorded one, and also the exact periods of a structure are difficult to assess in practical situations (e.g. when stiff cladding or partition elements are

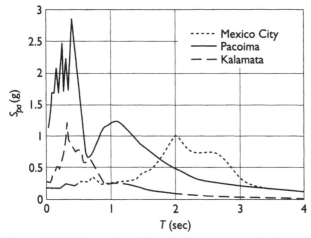

Figure 4.9 5% damped pseudo-acceleration spectra for three different ground motions.

present in steel or concrete frames). A smooth design spectrum encompasses a family of ground motions with the same overall intensity but possibly differing in the frequency content (particularly when two or more earthquake sources are considered) and in some details of the time sequences of motion that could critically affect the structural response. Smooth spectra for seismic design are generally derived from a statistical evaluation of actual spectra, and several alternative procedures are possible, as outlined in the following.

If the starting point is a pair (or a set of pairs) of M and R values (see Section 4.2.2), a number of records from earthquakes having characteristics falling within the desired range can be selected (whenever feasible, earthquakes from similar source mechanisms and site conditions should be used); the records are then scaled to a desired intensity (e.g. to the same PGA or PGV) and their spectra are calculated. A smoothed representation of the curve providing the desired percentile (e.g. 84 or 90) of the spectral values can be used for design.

By far the most common technique used today is the anchoring of a fixed spectral shape to a ground motion parameter such as the PGA, calculated using a probabilistic hazard analysis (see Figure 4.6). A well known spectral shape has been proposed by Newmark and Hall (1982), who noticed that when spectra are plotted on a log–log scale (see Figure 4.10) they are essentially scalar amplifications of A, V and D in their respective ('short'–'medium'–'long') period ranges. The amplification factors suggested by Newmark and Hall (1982) are summarized in Table 4.1 for some typical damping ratios; two values are given for each factor, one corresponding to the median (a log-normal distribution was assumed), and one to the 84 percentile (mean plus one standard deviation).

Using Table 4.1, the tripartite elastic response spectrum can be derived. If values of PGV and PGD are not available, they can be estimated from

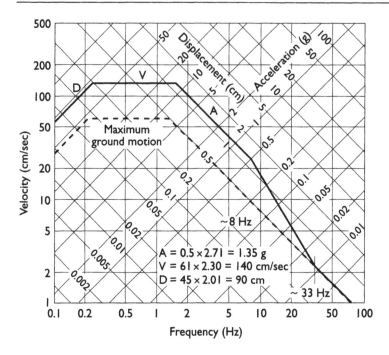

Figure 4.10 Elastic design spectrum corresponding to a PGA of 0.5 g, 5% damping, and one sigma cumulative probability (Newmark and Hall, 1982).

Table 4.1 Relative values of spectrum amplification factors (Newmark and Hall, 1982).

Percent of critical damping	Amplification factor for		
	Acceleration (A)	Velocity (V)	Displacement (D)
2	2.74 (3.66)*	2.03 (2.92)	1.63 (2.42)
5	2.12 (2.71)	1.65 (2.30)	1.39 (2.01)
10	1.64 (1.99)	1.37 (1.84)	1.20 (1.69)
20	1.17 (1.26)	1.08 (1.37)	1.01 (1.38)

* Median value (1 σ value).

$$V = c_1 \frac{A}{g}; \qquad D = c_2 \frac{(V)^2}{A} \qquad\qquad (4.12)$$

where the constants c_1 and c_2 should be calculated on the basis of statistical analysis of appropriately selected accelerograms. The calculated values of A, V, and D should be multiplied by the corresponding amplification factors from Table 4.1.

It is common to use amplification factors at the 84 percentile level with mean or median values of A, V, and D (Clough and Penzien, 1993).

The above method, which has influenced substantially the development of US and other codes, has the weakness that it ignores the fact that the spectral displacement tends to the PGD for very flexible structures (period tending to infinity). Until recently this had no practical consequences; however, in the case of displacement based design procedures which assume that structures respond well into the inelastic range (hence their effective periods can be quite long), an additional transition curve between the amplified displacement line and the constant PGD line might be necessary.

As an alternative to anchoring a fixed shape to a PGA and other ground motion parameters, spectra corresponding to a uniform probability of exceedance of their ordinates (uniform hazard spectra) can be constructed, as briefly discussed in Section 4.2.5. The effort required for their development is significantly higher than that associated with the previously described method.

Code spectra

Seismic codes typically specify pseudo-acceleration spectra only, consisting of a fixed shape to be anchored to a (design) PGA. Starting from the design PGA, it has long been argued that this should not correspond to the actually recorded peak acceleration, which might be associated with very short duration and high frequency pulses of the record, but should rather be representative of the effect of the acceleration on the structure. Hence, the concept of *Effective PGA* (usually denoted as EPA or A_{ef}) has been suggested. EPA can be calculated from the 5 per cent damping S_{pa} value in the region 0.1 to 0.5 sec, by dividing the average ordinate by an amplification factor of 2.5 (see Commentary to FEMA, 1995). The EPA is not the same as the PGA, and in fact when acceleration peaks are associated with very high frequencies, the EPA can be significantly lower than the PGA. The 1994 NEHRP Provisions (FEMA, 1995) also introduce the concepts of effective PGV (denoted as EPV) and the corresponding velocity related acceleration, which might control the design of longer period structures.

An indication of the uncertainty in the *shape* of the response spectrum is the difference between the median and 84 percentile values of the amplification factors given in Table 4.1; note that the coefficient of variation is higher for S_{pv} than for S_{pa}. It is worth pointing out that these high coefficients of variation were calculated for ground motions from one particular area (Western US) (i.e. for essentially the same geological and tectonic conditions).

A typical example of a code specified spectrum is shown in Figure 4.11, where the 5 per cent-damped elastic pseudo-acceleration spectrum of Eurocode 8 (CEN, 1994a) is plotted. The spectrum consists of four branches:

(i) An ascending linear branch ($A_1 B_1$ in Figure 4.11) described by the equation

$$S_e(T) = a_g S \left[1 + \frac{T}{T_B} (\eta \beta_0 - 1) \right] \tag{4.13}$$

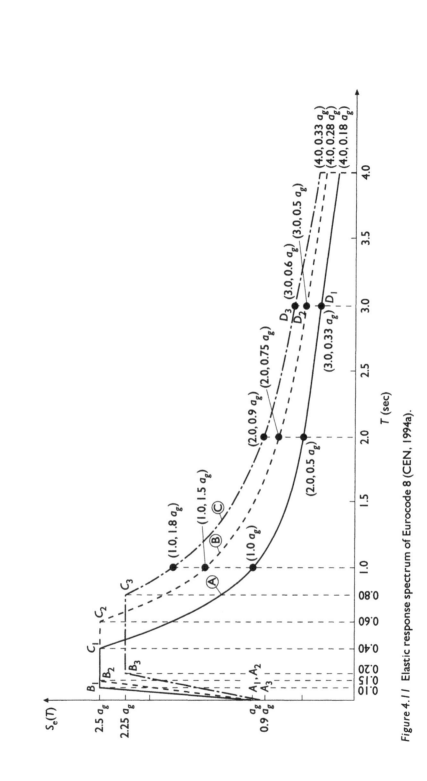

Figure 4.11 Elastic response spectrum of Eurocode 8 (CEN, 1994a).

where $\alpha_g \equiv A$ is the design PGA corresponding to a 10 per cent probability of being exceeded in 50 years (or a return period of 475 years); S is a soil parameter (see Section 4.3.3); β_0 is the spectral amplification factor taken equal to 2.5 (compare this with the values 2.1–2.7 in Table 4.1); and η is a damping correction factor given by

$$\eta = \sqrt{\frac{7}{2+\xi}} \geq 0.7 \tag{4.14}$$

and intended to account for viscous damping coefficients different from 5 per cent. The reference value of 5 per cent is generally appropriate for reinforced concrete (R/C) structures, but a lower value (3–4 per cent) is more appropriate for steel structures, and a somewhat higher value (about 6 per cent) is more appropriate for masonry structures. However, the approach adopted by the Eurocode is not to specify different damping ratios for different materials, but rather to include the effect of the difference in damping in the value of the force reduction factor (behaviour factor q) used for deriving the design seismic actions (see Section 4.3.4).

(ii) A flat branch ($B_1 C_1$ in Figure 4.11) defined by the constant value

$$S_e(T) = a_g S \eta \beta_0 \tag{4.15}$$

(iii) An exponentially descending branch ($C_1 D_1$ in Figure 4.11) defined by

$$S_e(T) = a_g S \eta \beta_0 \left(\frac{T_C}{T}\right)^{k_1} \tag{4.16}$$

The suggested value of k_1 is 1.0 (such values can be changed by the committees developing the 'national application documents', which will adopt the Eurocode as a national standard).

(iv) A second exponentially descending branch (beyond point D_1 in Figure 4.11) given by

$$S_e(T) = a_g S \eta \beta_0 \left(\frac{T_C}{T}\right)^{k_1} \left(\frac{T_D}{T}\right)^{k_2} \tag{4.17}$$

where $k_2 = 2.0$. The values of the periods T_B, T_C, and T_D (corresponding to points B_i, C_i, D_i in Figure 4.11) depend on the site conditions and are given in the next section.

The specification of two different descending branches is a feature unique to Eurocode 8 (EC8) and establishes a one to one correspondence with the Newmark–Hall spectrum (the region $T \geq T_D$ corresponds to the amplified displacement region, see left part of Figure 4.10). Furthermore, this is also an attempt to define a uniform hazard spectrum corresponding to a 50 per cent probability of exceedance. Note that the foregoing is the alternative approach to the one described in the previous section, where it was suggested to use the mean A in connection with the 84 or 90 percentiles of spectral amplifications.

As mentioned previously, the design PGA (α_g) corresponds to a 10 per cent probability of being exceeded in 50 years (or a return period of 475 years). This is the suggested probability for usual structures; for important structures, such as critical facilities, which should remain operational following the earthquake, lower probabilities of exceedance are appropriate. This is treated in a simple way in EC8 by specifying an *importance* factor γ_I which multiplies the seismic action (see eqn 4.29 in Section 4.3.5). For buildings γ_I ranges from 0.8 to 1.4, where the highest value corresponds to buildings of vital importance for civil protection (hospitals, power plants, fire stations), and the lowest value to buildings of minor importance (e.g. agricultural).

For the *vertical* response spectrum, EC8 recommends the use of the previously described spectrum for the horizontal motion, with the following modifications:

- for periods $T \leq 0.15$ sec the ordinates of the spectrum are multiplied by a factor of 0.7;
- for periods $T \geq 0.50$ sec the ordinates of the spectrum are multiplied by a factor of 0.5;
- for $0.15 < T < 0.50$ sec linear interpolation is used.

As mentioned in Section 4.2.3, the $\frac{2}{3} \cong 0.7$ factor is a reasonable value for the vertical to horizontal PGA ratio, but at distances from the source less than about 15 km it may exceed unity. The corresponding spectral acceleration ratios may also exceed one in the near field, but are typically less than one for intermediate and long periods (Ambraseys and Simpson, 1996).

The response spectrum specified in the American *Uniform Building Code*, UBC (International Conference of Building Officials, 1997) is similar to the first three branches of the Eurocode 8 spectrum. The ascending part starts from a value C_a, representing the design EPA value, while the flat part corresponds to a value of 2.5C_a, exactly as in the Eurocode. The exponential branch is defined by C_v/T, where C_v is an EPV dependent coefficient, identical to C_a for rock sites but higher for soil sites (C_a and C_v are given in Table 4.4 of the next section). The corner periods (see points B and C in Figure 4.11) in the UBC spectrum are $T_C = C_v/2.5C_a$ and $T_B = 0.2T_C$.

A unique feature, first introduced in the 1997 edition of UBC, is the specification of *near source factors* N_a and N_v, given in Table 4.2, which account for the fact that

Table 4.2 Near source factors in 1997 UBC, N_a/N_v.

Seismic source definition	Closest distance to known seismic source			
	≤ 2 km	5 km	10 km	≥ 15 km
$M_w \geq 7$ and SR ≥ 5	1.5/2.0	1.2/1.6	1.0/1.2	1.0/1.0
All other cases	1.3/1.6	1.0/1.2	1.0/1.0	1.0/1.0
$M_w \leq 6.5$ and SR ≤ 2	1.0/1.0	1.0/1.0	1.0/1.0	1.0/1.0

M_w: moment magnitude; SR: slip rate (mm/year).

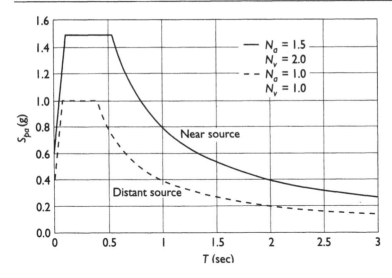

Figure 4.12 UBC elastic spectra corresponding to different distances from the source.

ground motions are significantly stronger near the earthquake source; this has long been recognized but not explicitly accounted for in previous codes. It is believed that these effects are significant for large earthquakes only, hence the N-factors of Table 4.2 are only applicable to the highest seismic zone in the US. Two typical UBC spectra for Zone 4 (highest) are shown in Figure 4.12. Both correspond to faults capable of producing large magnitude events ($M \geq 7$) and have high rate of seismic activity (slip rate ≥ 5 mm/y). However, one spectrum corresponds to a source which is very nearby (within 2 km), hence it is capable of producing significant near source effects, whereas the other corresponds to a source at least 15 km away from the site, for which no near source effects are expected.

The vertical component is defined in UBC by scaling the horizontal one by the $\frac{2}{3}$ factor, but where the near source factor $N_a > 1.0$, site specific response spectra should be used.

4.3.3 Site specific spectra

As already discussed in Section 4.2.4, the properties and configuration of the layers underlying the Earth's surface can significantly affect the seismic motion. As local site conditions influence the frequency content of surface motions, their effect is particularly important with respect to the response spectrum characteristics, i.e. for the same motion at the bedrock significantly different spectra can be calculated for the motions at the surface, depending on the characteristics of the soil layers. The general trend is that as the predominant period of the site increases (i.e. as the soil becomes softer) the peak, as well as the transition from the (approximately) flat to the exponential branch of the spectrum (compare Figure 4.11, 4.12) occur at longer periods; these periods are close to, but not necessarily the same as, the predominant period of the site. Referring to Figure 4.9, it is seen that the response spectrum for the accelerogram from Mexico City, recorded at a station (SCT building) underlain by about 40 m of soft clay (having an average shear wave velocity of only 75 m/sec) has its peaks in the range around 2 sec, whereas the other two motions recorded on much firmer soils are characterized by peaks at much shorter periods (around 0.5 sec). Until relatively recently, it was thought that for sites consisting of soft to medium clays the amplification of the acceleration in the short period range tends to be somewhat lower than the corresponding values for rock and stiff soils (Commentary to NEHRP Provisions, FEMA, 1995).

Recognizing the aforementioned trends, EC8 defines the site specific elastic response spectrum by modifying the basic shape of Figure 4.11 in two ways:

- by increasing the corner periods T_B and T_C in the case of softer soils;
- by decreasing the value of S_{pa} in the short period range for softer soils (soil factor S).

Of particular practical importance is the way soils are classified (into three classes in EC8), for design purposes; classification must be precise enough to avoid ambiguities, but also simple enough to avoid the need for costly detailed geotechnical investigations in the case of usual structures. The best indicator is probably the shear wave velocity of a soil layer, which captures the effect of both stiffness (through the shear modulus G) and density, as shown by eqn (4.1). In addition to this, the depth of each layer for which a constant v_s can be assumed is also of importance, while site amplification is further influenced by soil damping and the geometry (configuration) of the subsurface.

In situ measurements of the v_s profile by *in-hole geophysical methods* such as downhole or cross-hole tests (see description in Kramer, 1996) are strongly recommended for important structures and/or high seismicity. In other cases, empirical correlations of v_s with other geotechnical properties, typically the *cone penetration resistance*, may be used. The difference between the small strain values of v_s (as measured by *in situ* tests) and the strain values anticipated during the design earthquake must be taken into account.

The basic values of the site dependent parameters, along with the rest of data required for the construction of the EC8 spectrum (Figure 4.11) are summarized in Table 4.3. In the final version of EC8 it has been agreed to modify the values of

Table 4.3 Values of the parameters describing the EC8 elastic response spectrum.

Subsoil class	S	β_0	k_1	k_2	T_B(sec)	T_C(sec)	T_D(sec)
A	1.0	2.5	1.0	2.0	0.10	0.20	3.0
B	1.0	2.5	1.0	2.0	0.15	0.60	3.0
C	0.9	2.5	1.0	2.0	0.20	0.80	3.0

Table 4.3 in line (though not in full compliance) with the provisions of the new US codes, briefly discussed in the following.

Data from recent earthquakes, in particular the 1985 Mexico earthquake and the 1989 Loma Prieta (North California) earthquake, have indicated that accelerations on soft soils are larger (sometimes much larger) than on nearby rock sites; this is related to the high level of strain at which soft clay nonlinearity occurs, as discussed in Section 4.2.4. Moreover, soil-to-rock amplification factors for S_{pa} at long periods can be significantly higher than those adopted by EC8 and previous American codes (Borcherdt, 1994; FEMA, 1995). As expected (due to soil non-linearity effects), the spectral amplifications are higher for motions with low PGA and lower for higher PGA.

The 1997 UBC adopts the recommendations initially included in the 1994 NEHRP Provisions (FEMA, 1995), that are based on the foregoing consider-ations. The seismic coefficients C_a and C_v used for the definition of the response spectrum depend both on soil conditions and on the level of the design PGA. The site classification scheme adopted by NEHRP and UBC is quite simple, as only the shear wave velocity in the uppermost 30 m (the typical maximum depth of boring in geotechnical investigations) of the soil are used. As an alternative to v_s, geotechnical parameters such as the standard penetration resistance (for co-hesionless soils) or the untrained shear strength (for cohesive soils) can be used, but this will usually lead to more conservative results (FEMA 1995, 1997a). The site dependent seismic coefficients of the 1997 UBC are given in Table 4.4, where the definition of each soil profile type is also included; note that the Z-factor in the Table is the seismic zone coefficient, which for practical purposes can be seen as a PGA value (expressed in terms of g). The paramount effect of soil conditions on the C-values (particularly on C_v, which defines the response spectrum at longer periods), is clear from Table 4.4. Note that the maximum soil to rock ampli-fication factors for S_{pa} (calculated as the ratio of C-values corresponding to soils S_E and S_A) range from 4.1 to 1.3 (corresponding to $Z = 0.075$ and 0.40, respec-tively) for the short-period coefficient C_a, and from 4.3 to 3.0 for the long period coefficient C_v. Note also the upper bound of $0.36N_a$ imposed on C_a in the highest seismic zone, for the case of soft soils; this should be interpreted as the maximum acceleration that such soils are deemed to be able to transmit (due to non-linear effects).

Comparisons between UBC and EC8 response spectra for various site conditions show that for both 'intermediate' (S_C and S_D) and 'soft' (S_E and S_F) soils the UBC spectra result in higher S_{pa}-values than EC8 for PGA's up to 0.2 g, whereas this is not generally the case for 0.3 g. On the other hand, EC8 appears to be more conservative for rock sites.

Finally, with respect to *vertical* motion response spectra, it appears that they are less influenced by site conditions than horizontal spectra. Nevertheless, for short periods both horizontal and vertical spectra for soft sites are characterized by smaller amplification than for stiff sites; the opposite trend appears at intermediate and long periods (Ambraseys and Simpson, 1996).

Table 4.4a Seismic coefficients C_a of the 1997 UBC.

Soil type	v_s (m/sec)	Z = 0.075	Z = 0.15	Z = 0.2	Z = 0.3	Z = 0.4
S_A (hard rock)	>1,500	0.06	0.12	0.16	0.24	$0.32N_a$
S_B (rock)	760–1,500	0.08	0.15	0.20	0.30	$0.40N_a$
S_C (very dense soil)	360–760	0.09	0.18	0.24	0.33	$0.40N_a$
S_D (stiff soil)	180–360	0.12	0.22	0.28	0.36	$0.40N_a$
S_E (soft soil)	<180	0.19	0.30	0.34	0.36	$0.36N_a$
S_F (special[1])		See footnote 1 below Table 4.4b				

Table 4.4b Seismic coefficients C_v of the 1997 UBC.

Soil type	v_s (m/sec)	Z = 0.075	Z = 0.15	Z = 0.2	Z = 0.3	Z = 0.4
S_A (hard rock)	>1,500	0.06	0.12	0.16	0.24	$0.32N_v$
S_B (rock)	760–1,500	0.08	0.15	0.20	0.30	$0.40N_v$
S_C (very dense soil)	360–760	0.13	0.25	0.32	0.45	$0.56N_v$
S_D (stiff soil)	180–360	0.18	0.32	0.40	0.54	$0.64N_v$
S_E (soft soil)	<180	0.26	0.50	0.64	0.84	$0.96N_v$
S_F (special[1])		See footnote 1				

[1] Soil with v_s < 180 and large thickness (S_E has limited thickness); requires site specific geotechnical investigation.

4.3.4 Inelastic spectra and design spectra

For the vast majority of engineering structures it is not economically feasible to design them to withstand the seismic actions corresponding to a return period of about 500 years (the design earthquake in many modern codes, see Section 4.3.2) without developing inelastic deformations. This has long been recognized (Newmark and Hall, 1982), but the complications arising from the need to account in a simple and practical way for the inelastic response of a structure to the design earthquake without carrying out a proper non-linear analysis, are still a matter of controversy, as well as the subject of current research. The powerful modal analysis procedures, although strictly applicable to elastically responding structures only, are nevertheless used for analysing structures expected to develop significant amounts of inelastic deformation when subjected to the design earthquake. It is clear that such a procedure is not really rigorous, and there are situations (particularly in bridge design) that a full inelastic dynamic analysis is required by codes (see Section 4.3.7); however, due to its relative simplicity, this 'equivalent' modal analysis still forms the basis of most current code procedures. The basis of this type of analysis is the *inelastic spectrum* derived for nonlinear SDOF systems,

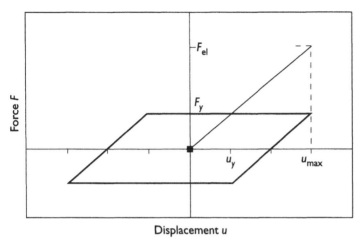

Displacement u

Figure 4.13 Elastoplastic response corresponding to a particular ductility factor ($\mu = 2$).

which is discussed in the following with a view to clarifying its role in seismic design.

Inelastic spectra

The general procedure for analysing SDOF systems with elastoplastic behaviour is presented in Chapter 2 (Section 2.2.3). Application of this procedure results in the calculation of the maximum (inelastic) displacement of the system u_{max} to a particular earthquake motion. This displacement can then be used to calculate the (displacement) ductility factor of the SDOF structure

$$\mu = \frac{u_{max}}{u_y} \tag{4.18}$$

where u_y is the yield displacement of the structure (i.e. the displacement corresponding to the yield force $F_y = k u_y$ (k is the elastic stiffness of the SDOF system)). The ductility factor of eqn (4.18) is a useful indicator of the amount of inelasticity expected to develop in a structure subjected to a given motion. For instance, a ductility factor of 3 (see Figure 4.13) means that the inelastic (plastic) displacement will be equal to twice the yield displacement. Moreover, for an elastoplastic system, the *energy* dissipated during a full symmetric cycle (peak amplitudes of u_{max} and $-u_{max}$) is equal to $4 k_0 u_y^2 (\mu - 1)$, as can be inferred from the geometry of the elastoplastic loop in Figure 4.13. Given a set of yield resistances (F_y), inelastic response spectra can be calculated indicating the ductility demand corresponding to each value of F_y; these are called *constant strength spectra*.

On the basis of the foregoing considerations, the need arises to design a structure to respond to a given earthquake excitation within a desired level of inelastic behaviour (i.e. not exceeding a target ductility). It is therefore particularly useful

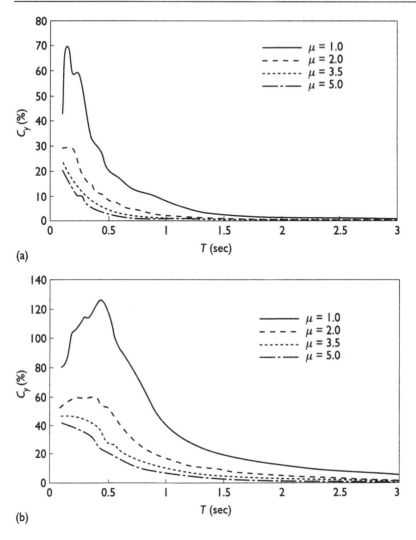

Figure 4.14 Mean elastic and inelastic strength spectra for various ductility levels: (a) records on rock sites; (b) records on alluvium sites. Records are from earthquakes in Greece, and are scaled to the maximum spectrum intensity in each soil type.

to construct response spectra corresponding to specific values of the ductility factor μ. This can be done either by interpolating between (closely spaced) constant strength spectral curves, or by iteratively adjusting the level of F_y (for each period) in order to match as closely as desired the target ductility value. Examples of such spectra calculated for appropriately selected sets of ground motions typical of earthquakes in Southern Europe are shown in Figure 4.14 (Kappos,

1999), for two site conditions ('rock' and 'alluvium'), and four ductility factors: 1 (elastic behaviour), 2 (low ductility), 3.5 (medium ductility) and 5 (high ductility). Note that the shape of inelastic spectra is generally different from that of the corresponding elastic spectra; they are much smoother than the latter, and smoothness increases with the ductility level. For $\mu \geq 3.5$ the strength requirement decreases monotonically with the period, regardless of soil conditions. Inelastic behaviour appears to be more effective in reducing the maximum elastic acceleration in the case of motions recorded on rock, but in all cases elastic force reduction is very significant in the medium and long period range. Also of practical significance is the observation that for $\mu \geq 3.5$ inelastic strength demands are just slightly influenced by the ductility level, for both rock and alluvium; the implication of this is that for relatively small changes in the strength of medium and high ductility structures, the increase in the required ductility is significant.

Design spectra

Seismic codes still rely upon the concept of *inelastic spectrum* for specifying design actions (forces) to be used for elastic modal analysis of structures which are expected to respond inelastically to the design earthquake. This is a rather crude approximation and errors tend to increase as the level of inelasticity (or target ductility μ) and the fundamental natural period (or the number of storeys) increase (Anagnostopoulos *et al.*, 1978; Krawinkler and Nassar, 1992).

Since for design purposes several ground motions with different characteristics have to be taken into account, an average inelastic response spectrum has to be used, and this would generally involve considerable work. Hence, several attempts have been made to construct (inelastic) design spectra directly from the corresponding elastic spectra, by appropriate modification of the latter. The typical way to do this is to divide the ordinates of the elastic response spectrum by a factor which depends on the type of inelastic behaviour (e.g. elastoplastic, stiffness degrading, etc.) and the damping (typically 5 per cent is used for the design spectra, as mentioned in Section 4.3.2), in addition to the period; i.e. for a given hysteretic behaviour and damping ratio

$$R(T) = \frac{S_{el}(T)}{S_{in}(T)} \tag{4.19}$$

where the subscripts 'el' and 'in' refer to the ordinates of the elastic and inelastic response spectrum, respectively. Note that in eqn (4.19) T is the fundamental period of the structure before yielding, often referred to as the elastic period. This period is *not* the effective or the predominant period of the *inelastically* responding structure (particularly when the plastic deformations are significant), hence it should not be forgotten that plotting S_{in} as a function of the initial T is merely a convention.

Newmark and Hall (1982) have noted the following characteristics of inelastic response spectra:

- For periods longer than about 0.5 sec, the displacements of the inelastic systems are very close to those of the elastic systems; referring to Figure 4.13, it can be shown that in this case the yield force in the inelastic system is $F_y = F_{el}/\mu$, where F_{el} is the force in the elastic system corresponding to the displacement u_{max}. Given that the maximum force in an elastoplastic system is its mass times the pseudo-acceleration ($F_y = mS_{pa}$) the corresponding R-factor defined from eqn (4.19) is in this case constant (i.e. $R = \mu$).
- For periods between about 0.12 and 0.5 sec the energy stored in the inelastic system (the area under the monotonic $F - u$ curve from 0 to u_{max} in Figure 4.13) is roughly the same as the area stored by an elastic system with the same initial stiffness (but smaller maximum displacement); by equating the areas under the two curves it can be shown that in this case

$$F_y = \frac{F_{el}}{\sqrt{2\mu - 1}} \tag{4.20}$$

or $R = \sqrt{2\mu - 1}$

- For periods less than 0.03 sec the force (or acceleration) is the same for elastic and inelastic systems (i.e. $F_y = F_{el}$). This leaves a transition range from 0.03 to 0.12 sec, wherein a linear decrease from F_{el} to the value given by eqn (4.20) is assumed for F_y; this is equivalent to R varying from 1 to $\sqrt{2\mu - 1}$.

Following the Newmark–Hall proposal for inelastic spectra construction, a number of studies, some of them based on more extensive databases of records, have appeared. A review of most proposals regarding the R-factor can be found in Miranda and Bertero (1994), wherefrom Figure 4.15 has been reproduced. It is seen that although all proposals for the R-factor follow a similar trend, differences up to about 40 per cent can result between them.

Another critical issue regarding the use of design spectra is the feasibility of capturing the inelastic response of a Multiple Degree-of-Freedom (MDOF) system using spectra that have been derived from SDOF system analysis. More specifically, the question arises whether an MDOF system designed for a base shear derived from an inelastic response spectrum corresponding to a target ductility μ, will develop an (equivalent) ductility of this order when subjected to earthquakes compatible with the aforementioned spectrum. Both earlier (e.g. Anagnostopoulos et al. 1978) and more recent (e.g. Krawinkler and Nassar, 1992) studies have indicated that the danger exists that the ductility factors for the MDOF system may significantly exceed the target ductility (i.e. the one for which the inelastic spectrum for the SDOF system has been constructed). The critical aspect of the problem is the type of inelastic mechanism that forms in the MDOF system, which depends largely on the philosophy adopted for design (see Section 4.4). If a soft storey mechanism develops, the ductility demands for the MDOF system are much higher than those for the corresponding SDOF system; on the

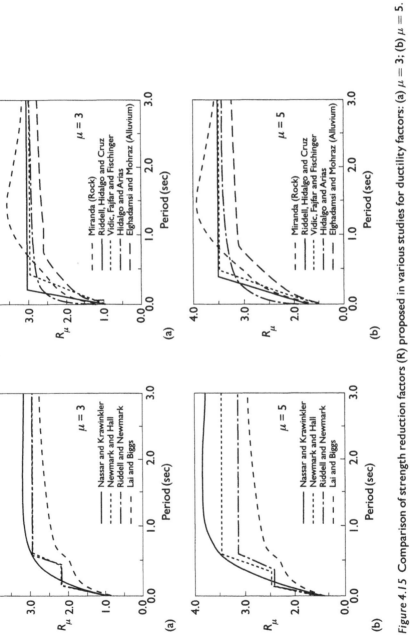

Figure 4.15 Comparison of strength reduction factors (R) proposed in various studies for ductility factors: (a) $\mu = 3$; (b) $\mu = 5$.

other hand, the differences are much smaller when a mechanism involving primarily beam hinging forms. Moreover, the increase in the ductility demands in the MDOF system is larger for increased target ductility factors and for longer fundamental periods (i.e. for taller buildings). The need therefore arises for modifying the (inelastic) design spectrum in the long period range to remedy the previous situation. Newmark and Hall (1982) have suggested lowering the exponent (k_1 in eqns 4.16, 4.17) of the period dependent term giving spectral accelerations in the long period range ($T > 1$ sec) from 1 to $\frac{2}{3}$; this has been adopted by several seismic codes but Krawinkler and Nassar (1992) have found that it is only valid for well-designed structures (i.e. those forming beam mechanisms).

Code spectra

The design spectrum in Eurocode 8 is defined by eqns (4.13–4.17), with the following modifications:

- the term $\eta\beta_0$ is substituted by β_0/q, where the so-called *behaviour factor q* is analogous to the R-factor of eqn (4.19).
- the exponents $k_1 = 1.0$ and $k_2 = 2.0$ are replaced by $k_{d1} = \frac{2}{3}$ and $k_{d2} = \frac{5}{3}$, respectively;
- a cut-off value of $0.2\alpha_g$ for the design acceleration is introduced.

The introduction of the reduced k_d exponents in combination with the cut-off of $0.2\alpha_g$, results in a substantial increase in the design forces for long period structures, such as tall buildings or long span bridges. This is generally in line with the remarks made previously for such structures, although no particular justification appears to exist for specifying a constant minimum seismic force (the cut-off value).

Design spectra in the American codes are similarly derived from the corresponding elastic spectra (i.e. factors similar to q are specified for reducing the elastic spectrum ordinates and/or the elastic base shear). They are called *response modification factors* (R) in the NEHRP [National Earthquake Hazard Reduction Program] Provisions (FEMA 1995, 1997a), whereas they are referred to simply as the R *coefficients* in UBC. It is deemed that the term *response reduction* factor (or force reduction factor) offers a clearer indication of the nature of this factor, which plays a paramount role in seismic design, and is discussed in more detail in the next section. Unlike EC8, the American UBC specifies a lower bound to the design base shear equal to 90 per cent of the value used in the equivalent (static) lateral force procedure (Section 4.3.5); this appears to be mainly due to historical reasons, as lateral force design has long prevailed, whereas modal analysis was traditionally restricted to 'special' structures. Similarly to EC8, the UBC specifies a minimum base shear (see Section 4.3.5), lower than the EC8 one.

Force reduction factors

The force reduction factor can be defined as the ratio of the elastic strength demand (i.e. the strength that would be required in the structure if it were to respond elasti-

Table 4.5 Seismic force reduction factors for high ductility R/C structures.

	Symbol	Frame	Structural wall	Frame wall
Eurocode 8 (ULS)	q	5	4–5	4.5–5
UBC[1] (ULS)	R	8.5	4.5–5.5	8.5
NZS 4203[2] (ULS)	μ	$\leq 6^3$	$\leq 5^4$	≤ 5–6^4
Japan[4] (Level 2 earthquake)	$1/D_s$	2.2–3.3	1.8–2.5	2.0–2.9

[1] R factor must be reduced by a reliability/redundancy factor of between 1 and 1.5.
[2] The structural performance factor S_p also applies, in addition to μ, hence the values in the table are typically increased by 50%.
[3] Depending on the mechanism of inelastic deformation.
[4] Depending on the aspect ratio and coupling.
[5] The factor D_s is calculated for each storey separately, rather than the building as a whole.

cally to the design earthquake), to the inelastic strength demand (i.e. the strength required in the structure for it to respond beyond the elastic range but within the selected ductility (and/or displacement) limits). If the elastic strength demand is denoted as F_{el} and the inelastic (design) strength demand as F_d, it follows that the reduction factor

$$R = F_{el}/F_d \tag{4.21}$$

Differences in the numerical values of the force reduction factors specified in various codes for the same type of structure can be quite substantial. The values specified for high ductility R/C frames in four leading codes are summarized in Table 4.5 (Booth *et al.*, 1998); it is seen that the reduction factor is equal to 8.5 in UBC, 5 in EC8, and ≤ 3.3 in the Japanese Code (whose conceptual basis is generally different from that of the other three codes). It should be noted, however, that if appropriate adjustments are made to these values to account for the different partial safety factors used in each code (for loads, as well as for member resistances), differences become smaller.

The value of the reduction factor depends on the *ductility* of the structure (which relates to the detailing of the structural members), but also on the *strength reserves* that normally exist in a structure (depending mainly on its *redundancy* and on the *overstrength* of individual members), as well as on the (effective) *damping* of the structure; all these factors directly affect the *energy dissipation* capacity of a structure. Bertero (1989) suggested a definition of the force reduction factor along the foregoing lines, i.e.

$$R = R_{\mu}R_sR_{\xi} \tag{4.22}$$

where R_{μ} is the ductility dependent component, R_s the overstrength dependent component, and R_{ξ} the damping dependent component of the reduction factor; the latter is of interest mainly in the case of structures with supplemental damping devices (see Section 4.4.4). A detailed discussion of possible procedures for quantifying R_μ and R_s can be found elsewhere (Fischinger and Fajfar, 1994; Kappos, 1999). These and other studies have indicated significant values for the overstrength component R_s (at least 1.5) for both R/C and steel structures. This is particularly important from the design point of view, since ductile detailing requirements can be relaxed for structures possessing substantial overstrength.

The concept expressed by eqn (4.22) is not explicitly recognized in Eurocode 8. Nevertheless, if the ratio of the EC8 elastic spectra of eqns (4.13–4.17) to the inelastic (design) spectra resulting from the aforementioned modifications is calculated, the resulting R-factor (eqn 4.19) is period dependent (i.e. $R < q$ for both short and long period structures, and $R = q$ only for the intermediate period (from T_B to T_C) structures).

Contrary to the EC8 approach, the American codes specify essentially period independent values of the R factors, something that has been criticized in the past (Miranda and Bertero, 1994). Although a proposal has been made by SEAOC (1996) to include a two component $(R_\mu R_s)$ reduction factor in the UBC, this has not been done in the 1997 edition, which, however, does include a redundancy factor $(\rho \le 1.5)$, intended as a lower bound, below which a penalty (an increase of up to 50 per cent) is applied with regard to seismic force levels in structures lacking redundancy. Some other national codes have adopted expressions for R that explicitly differentiate between the ductility and the overstrength component of R (Fischinger and Fajfar, 1994).

Design displacements

In addition to the determination of the 'inelastic' forces expected in a structure, it is also necessary to have an estimate of the inelastic displacements under the design and/or the serviceability earthquake; these are typically required for checking that the code-specified drift limits are not exceeded. Based on the previous discussion, it is reasonable to assume that elastic and inelastic displacements are about the same (except for short period structures), and calculate the latter by simply amplifying the (elastic) displacements, calculated for the factored seismic loading (corresponding to F_{el}/R), by the reduction factor (R) used for forces. This is indeed the recommended procedure in Eurocode 8: Under the serviceability earthquake (inelastic) drifts are calculated as $q\Delta_{el}/\nu$, where Δ_{el} is the drift calculated on the basis of the design seismic forces and ν is a factor intended to account for the lower intensity of the serviceability earthquake (for buildings $\nu = 2.0$ to 2.5).

The corresponding procedure in the UBC (ICBO, 1997) is to estimate drifts under the design earthquake as $0.7R\Delta_{el}$ (i.e. the amplification factor for inelastic displacements is 30 per cent lower than the reduction factor (R) for forces). Although the UBC background document (SEAOC, 1996) claims that this is a better 'average' value of the inelastic drift, this is a point of rather considerable controversy.

4.3.5 Equivalent lateral force procedures

Until very recently seismic design of most structures was based on a static analysis using a set of lateral (horizontal) forces assumed to represent the actual (dynamic) earthquake loading. In the absence of commercial software appropriate for dynamic analysis of three-dimensional structures, as well as of the expertise for using whatever software of this type was available, most codes of practice clearly promoted the simpler static procedure. However, the last 10 to 15 years were marked by a massive introduction of more advanced software packages, running on increasingly more powerful hardware; this was probably the main reason for a change of attitude, both from the practising engineer's and the code drafter's point of view. As a consequence, in modern codes, such as the EC8, dynamic analysis (Section 4.3.6) is adopted as the reference method, and its application is compulsory in many cases of practical interest.

The typical procedure in the equivalent static analysis method is the determination of an appropriate value of the base shear in terms of the structure mass and the design earthquake intensity, properly reduced for inelastic effects, along the lines discussed in the previous section. The base shear is then used for estimating a set of lateral forces distributed along the structure following (more or less) the fundamental mode of vibration. Since the base shear itself is also calculated on the basis of the fundamental period, it is clear that the application of the equivalent lateral force method should be restricted to structures whose dynamic response is governed by the fundamental mode.

The Eurocode 8 procedure

The method is referred to as 'simplified modal response spectrum analysis', rather than as 'equivalent static analysis', and is restricted to structures that are not significantly affected by higher modes and/or stiffness irregularities.

The *base shear* (sum of horizontal loads) is calculated from

$$V_b = S_d(T_1)W \tag{4.23}$$

where $S_d(T_1)$ is the ordinate of the design spectrum (see Section 4.3.4) corresponding to the fundamental period T_1 of the structure, and W is the gravity load contributing to inertial forces; this is taken as the permanent loading (G) and a portion $\psi_E Q$ of the variable (live) loading Q. The fundamental period T_1 can be estimated either from a proper eigenvalue analysis (see Section 2.3.2), or from Rayleigh's method, or from empirical formulae included in the code.

The *lateral forces* corresponding to the base shear of eqn (4.23) are calculated assuming (conservatively) that the effective mass of the fundamental mode is the entire mass of the structure; hence

$$F_i = V_b \frac{s_i W_i}{\sum s_j W_j} \tag{4.24}$$

where F_i is the horizontal force acting on storey i, s_i, s_j are the displacements of the masses m_i, m_j in the fundamental mode shape, and W_i, W_j are the weights corresponding to the previous masses. It is permitted by the code to avoid the calculation of the fundamental mode shape and assume instead that it is increasing linearly with the height of the building, hence s_k in eqn (4.24) are substituted by z_k, the heights of the masses m_k (typically the heights of the storeys) above the foundation level. The forces F_i are then used for a standard static analysis of the building, which can be based on two planar models.

In order to cover uncertainties in the distribution of mass and stiffness (of 'non-structural' elements), as well as the spatial variability of ground motion, an *accidental eccentricity* of the loads F_i with respect to the mass centre C_M of the storey has to be introduced in the analysis; this is equal to

$$e_{1i} = \pm 0.05 L_i \qquad (4.25)$$

where L_i is the floor dimension perpendicular to the direction of force F_i. The eccentricity e_1 is additional to any existing eccentricity e_0 between the stiffness centre C_S and the mass centre C_M at any storey. Instead of applying the forces at an eccentricity from C_M, it is usually more convenient to consider a torsional moment $M_t = F_i(e_0 + e_1)$, or simply $F_i e_1$ if a three-dimensional model is used, acting at the mass centre.

While the aforementioned eccentricities e_0 and e_1 are present in both static and dynamic analysis, an additional complication arises when the former is used. It is known that static analysis underestimates dynamic torsion effects (Chopra, 1995), hence EC8 requires consideration of an additional eccentricity e_2 to account for the dynamic effect of simultaneous translational and torsional vibrations. Appropriate (rather complicated) expressions for e_2 as a function of the geometry and the stiffness of a storey are given in EC8 1–2 (CEN, 1994b).

The *load combination* involving the seismic loading is

$$\sum G_{kj}{}' +{}' \sum \psi_{2i} Q_{ki}{}' +{}' \gamma_I E_d \qquad (4.26)$$

where ' + ' means 'to be combined with', $\sum$ implies the combined effect of several actions of the same type (permanent or 'dead' G, variable or imposed Q), G_{kj} is the characteristic (upper 5 per cent fractile) value of the permanent action j, $\psi_{2i} Q_{ki}$ is the 'quasi-permanent' value of the variable action, γ_I the importance factor (Section 4.3.2), and E_d the design value of the seismic action.

The UBC 1997 procedure

The method is applicable to all buildings in the low seismicity zone (Zone 1) and usual structures in seismic Zone 2, regular structures with a height up to 73 m, and irregular structures having no more than five storeys.

The design base shear is

$$V_b = \frac{C_v I}{R T_1} W \leq \frac{2.5 C_a I}{R} W \tag{4.27}$$

where W is the 'seismic' dead load, including the total dead load and applicable portions of other loads (partition load of at least $0.5\ kN/m^2$, permanent equipment, etc.), and the factors C_a, C_v, I and R, that define the design spectrum were discussed in the previous section. Two lower bounds to V_b are set in UBC

- For all seismic zones

$$V_b \geq 0.11 C_a I W \tag{4.28a}$$

- For seismic Zone 4 (highest)

$$V_b \geq \frac{0.8 Z N_v I}{R} W \tag{4.28b}$$

Note that this is to account for near source effects, as discussed in Section 4.3.2. The fundamental period T_1 is calculated using the same procedures as in EC8. The distribution of lateral forces along the height of the structure is given by

$$F_i = \frac{(V_b - F_t) z_i W_i}{\sum z_j W_j} \tag{4.29}$$

It is seen that this is the same as the simplified version of eqn (4.24), wherein the heights z_i replace the mode shape amplitudes s_i, with the exception that part of the total base shear is applied as a concentrated force at the top, $F_t = 0.07 T_1 V_b$, which need not exceed 25 per cent of the total base shear and may be taken as zero for $T_1 \leq 0.7$ sec. The top force F_t is a simple way of accounting for the effect of higher modes on the force pattern and is important for tall buildings only.

The $0.05L$ accidental eccentricity discussed previously for EC8 is also specified in UBC, but no provisions are included regarding the 'dynamic' eccentricity (e_2 in EC8).

The following combinations involving the seismic loading E are specified in UBC (the notation for loads has been changed here to facilitate comparison with EC8)

$$1.2G + E + f_1 Q_1 + f_2 Q_2 \tag{4.30a}$$

$$0.9G \pm E \tag{4.30b}$$

where Q_1 is the live load (its factor f_1 is typically equal to 0.5) and Q_2 the snow load (f_2 is 0.2 or 0.7 depending on the roof configuration). The load factors in eqns (4.30) should be increased by 10 per cent for the design of R/C and masonry structures.

The UBC earthquake loading E is calculated as

$$E = \rho E_b + E_v \tag{4.31}$$

where E_h is the load due to the horizontal component (corresponding to the base shear of eqn 4.27), ρ is the redundancy factor described previously, and E_v is the load effect resulting from the vertical component of the ground motion, accounted for by adding an extra permanent load (additional to G), equal to $0.5C_aIG$. This is an interesting difference between the two codes, since EC8 requires consideration of the vertical component in special cases only (i.e. horizontal cantilever members, long span (>20 m) members, prestressed concrete members, and beams supporting columns); in all these cases the vertical component can be considered locally (for the members under consideration and their associated supporting members).

4.3.6 Modal analysis procedures

For the purposes of seismic design the method is almost invariably applied in combination with the design response spectrum, and is typically referred to as '*modal response spectrum analysis*'. Its field of applicability covers essentially all cases for which the equivalent static analysis is not appropriate (i.e. cases where modes other than the fundamental one affect significantly the response of the structure). There are a few cases where modal analysis is not deemed appropriate and a full dynamic (time history) analysis is required, a notable example being the design of base isolated bridges to EC8 Part 2 (CEN, 1994c). Detailed presentations of the modal response spectrum analysis can be found elsewhere (Gupta, 1990; Clough and Penzien, 1993; Chopra, 1995).

Review of the procedure

In modal analysis involving lumped mass systems, the (elastic) force vector $\mathbf{f}_n$ for the nth mode, calculated on the basis of the response spectrum, is

$$\mathbf{f}_n = \mathbf{m}\boldsymbol{\phi}_n \frac{L_n}{M_n} S_{pan} \tag{4.32}$$

where $\mathbf{m}$ is the mass matrix, $\boldsymbol{\phi}_n$ is the nth mode shape vector, L_n is the earthquake excitation factor (depending on the mass distribution and the corresponding mode shape), M_n is the generalized mass (see also Section 2.3.2), and S_{pan} is the spectral pseudo-acceleration corresponding to the period T_n of the nth mode. Note that the forces f_{in} are acting on the (lumped) masses m_i; in the common case of buildings with floor diaphragms, m_i is the mass of the ith storey and f_{in} the nth mode force acting on this mass.

The corresponding maximum *base shear* for the nth mode is given by

$$V_{0n} = \frac{L_n^2}{M_n} S_{pan} \tag{4.33}$$

The *displacements* for the *n*th mode can be calculated from

$$\mathbf{u}_n = \boldsymbol{\phi}_n \frac{L_n}{M_n \omega_n^2} S_{pan} \tag{4.34}$$

where $\omega_n = 2\pi/T_n$ is the circular frequency of the *n*th mode. Recall that S_{pa}/ω^2 equals S_d, the spectral displacement (Section 4.3.2).

Since the response of a structure results from the contribution of all modes, and since modal maxima generally do not occur simultaneously, it is customary to combine the action effects S_i from the individual modes in a statistical way. The most commonly adopted procedure is the Square Root of the Sum of Squares (SRSS) combination that is:

$$S_{i,\max} \cong \sqrt{S_{i1}^2 + S_{i2}^2 + S_{i3}^2 + \cdots} \tag{4.35}$$

where $S_{i,\max}$ is the probable maximum value of the action effect (force or displacement), and the subscripts $1, 2, 3, \ldots$ refer to the first, second, third $\ldots$ mode; a sufficient number of modes should be considered in estimating $S_{i,\max}$ (see code criteria in the next subsections). Note also that action effects due to earthquake should always be taken with alternate sign (i.e. both as positive and negative). Equation (4.35) gives reasonable values in many practical cases, but is generally unconservative when two or more modes are closely spaced (i.e. their periods are close to each other); this is often the case in three-dimensional structures susceptible to torsional effects. In these cases more refined combination rules, such as the Complete Quadratic Combination (CQC) (Wilson *et al.*, 1981) are appropriate.

A significant shortcoming of modal response spectrum analysis is that it is not possible to define exactly the *simultaneous* values of forces, for instance the axial loading corresponding to the maximum moment in a column section, and vice versa. Therefore, in addition to the approximation of modal combination (eqn 4.35), it is customary to assume that the probable maxima of the various action effects (M, N, V) for a given earthquake action (e.g. a response spectrum in a particular direction) occur simultaneously; this is usually, but not necessarily, conservative, with regard to design of members. The problem of combining moments (generally pairs of moments) and axial loads in the design of columns for (biaxial) bending and axial force (M_x, M_y, N) is discussed in more detail elsewhere (Gupta, 1990; Penelis and Kappos, 1997).

The Eurocode 8 procedure

The basis of the method is the design response spectrum discussed in Section 4.3.4; this has to be applied along two, properly identified, perpendicular axes of the structure.

The criterion for the required number of modes to be included in the analysis is two-fold:

- the sum of the effective modal masses (L_n^2/M_n, see eqn 4.33) of the considered modes should amount to at least 90 per cent of the total mass of the structure;
- all modes with effective mass greater than 5 per cent of the total mass should be considered.

The modal action effects should be combined using the SRSS approach (eqn 4.35), unless the periods of two of the considered modes differ by less than 10 per cent, in which case the CQC approach should be used.

The accidental eccentricity e_1 (eqn 4.25) could be considered in buildings either by displacing the location of the mass of each storey diaphragm by e_1 or (more conveniently) by introducing an equivalent torsional moment, exactly as in the case of equivalent static analysis.

The simultaneous action of the two horizontal components should be taken into account; this is also required in equivalent static analysis. Since peak values do not occur at the same time in both directions (x and y), the simultaneous action can be modelled either:

- by an SRSS combination (compare eqn 4.35) of the 'x' and 'y' action effects; or
- by considering the combinations

$$S_{Ex}` + ' 0.30 S_{Ey} \quad \text{and} \quad S_{Ey}` + ' 0.30 S_{Ex} \tag{4.36}$$

where S_{Ex} are the action effects due to the application of the seismic action along the selected x-axis of the structure, and S_{Ey} the corresponding effects for the seismic action applied along the y-axis.

Both procedures are statistical ones, and both introduce small errors on the safe, as well as the unsafe side (Penelis and Kappos, 1997).

In the case of elongated structures, such as bridges exceeding about 600 m, the *spatial variability* of the ground motion should be given due consideration (see also Section 4.2.4). Methods for accounting for spatial variability are described in the (informative) Annex D to EC8 Part 2 (CEN, 1994c).

The UBC 1997 procedure

There are two differences in the modal analysis procedure specified in UBC, compared to the previously described EC8 procedure:

- The elastic, rather than the design, response spectrum is used for estimating action effects; the resulting *displacements* are directly used for design. Recall that in EC8 displacements are calculated by scaling the values resulting from the design spectrum (which includes $1/q$) by the q-factor.
- The elastic forces calculated as above are then scaled down to account for inelastic and related effects. This is done by adjusting them to 90 per cent of

the base shear used in equivalent static analysis (eqns 4.27, 4.28) in the case of regular structures, or 100 per cent this base shear in the case of irregular structures.

Modal combinations, torsional effects, and orthogonal (x and y) effects are treated in the same fashion as in EC8.

4.3.7 Time history representations

Time history analysis is used for design purposes only as an exception (see Section 4.3.1), and almost exclusively whenever non-linear effects are to be considered explicitly, rather than through the R-factor approach. When acceleration time histories are used for design, it is imperative that they actually correspond to the design earthquake for the site under consideration, which means that the envelope of the response spectra of the accelerograms used should reasonably match the elastic design spectrum for the site (no reduction through R-factors).

Several options are available for selecting an appropriate set of design accelerograms:

- use of records from actual earthquakes, which generally have to be scaled to the design earthquake intensity;
- use of artificial accelerograms generated so as to match the (target) elastic response spectrum; this is sometimes referred to as the 'engineering method';
- use of simulated accelerograms generated by modelling the source and travel path mechanisms of the design earthquake ('seismological method').

Each option has its own merits and limitations, as discussed in the following.

Selection of recorded accelerograms

This can be the ideal solution whenever an extensive database of acceleration time histories is available, containing records from earthquakes with a large range of characteristics. Then, a selection can be made of records matching the source parameters (focal mechanism and depth, distance from source), travel path, magnitude, peak ground motion parameters (A, V, D), and duration, for the site under consideration. Note that, with the possible exception of major projects (such as the design of critical facilities) in areas where abundant data exist, such as the US and Japan, the foregoing is a rather over-ambitious procedure, since not only an adequate database of records is required, but also a complete characterization of the seismic hazard at the site.

A more pragmatic approach would involve the following main parameters to be considered when selecting natural records:

- site conditions;
- magnitude;

- distance to source (or epicentral distance);
- closeness to the site under consideration.

It is beyond the scope of this book to discuss in detail these parameters and their relation to the characteristics of the ground motion. In a practical context, the need becomes clear of using a minimum number of criteria (ideally only one) for selecting ground motion records for the purpose of time history analysis. An interesting proposal in this respect is the use of the A/V ratio (Zhu *et al.*, 1988), which is a simple parameter, easy to calculate from the commonly available values of $A(\equiv\mathrm{PGA})$ and $V(\equiv\mathrm{PGV})$, and correlates well with the M–R relationship, as well as with site conditions.

Another possible criterion is to select ground motion records whose spectra (S_{pa} or preferably S_{pv}) are peaking in the vicinity of the fundamental period of the structure under consideration, irrespective of their other characteristics, which is generally a conservative approach.

Scaling of recorded accelerograms

Whenever a careful selection of natural accelerograms has been made, for instance on the basis of (M, R) pairs within a narrow range, one might argue that these could be directly used for design purposes. In fact, if these records are used for analysis, significant variability in the calculated response is found; Shome *et al.* (1998) reported dispersions of about 50 per cent to 60 per cent in the inelastic peak interstorey drifts of medium rise steel frames subjected to sets of motions, each corresponding to a narrow magnitude range (e.g. 6.5–7.0) and distance range (e.g. 50–70 km). This significant variability is attributed to the very different characteristics of ground motions at a given location resulting from an earthquake of a given M, and is a clear indication of the effect of neglecting the other important parameters characterizing the ground motion. This points to the need for scaling (or *normalizing*) the selected earthquake accelerograms before using them for time history analysis. In addition to the foregoing considerations, scaling is also necessary whenever different limit states (serviceability, ultimate, etc.) have to be considered, since it is generally impractical to select different sets of records for each limit state.

The most commonly applied scaling procedure is based on the PGA (i.e. all records used for design are scaled to the same PGA). Unfortunately this convenient procedure is one of the most unsatisfactory ones, with the exception of structures with very low periods (not exceeding about 0.2 sec). As discussed in Section 4.3.2, the spectral ordinates are proportional to the PGA over the short period range only, whereas for longer periods (covering most of the usual civil engineering structures) they are proportional to the PGV, and for very long periods (more than about 3.0 sec) they are proportional to the PGD. The peak ground parameters have indeed been used as scaling factors, and so have the integrals of their squared values and their root-mean-square values (Nau and Hall, 1984). All these values are ground motion dependent only (i.e. they are not correlated in any way to the characteristics of the structure to be designed).

A sensible choice of scaling parameters accounting for the characteristics of both the record and the structure, are the spectral values, either those of the response spectrum or of the Fourier spectrum. Since spectral ordinates vary with T, a critical question is which range of the spectra should be considered for deriving a scaling parameter. An early (1952), but still quite popular, proposal is Housner's *spectrum intensity*, SI_v (see Housner 1970), which is the area under the S_{pv} spectrum

$$SI_v = \int_{T_a}^{T_b} S_{pv}(T,\xi)\,dT \tag{4.37}$$

with $T_a = 0.1$ sec, $T_b = 2.5$ sec, and $\xi = 20$ per cent. The reason for selecting these period limits is that they were deemed to represent the range of typical periods of buildings at the time; it is understood that SI_v is intended to be an overall measure of the 'damageability' of a ground motion with respect to a population of structures. Whenever a particular structure is to be designed or assessed, a condensation of the limits suggested by Housner is appropriate. Kappos (1991) suggested a modified Housner intensity based on $T_a = 0.8T_1$ and $T_b = 1.2T_1$, where T_1 is the fundamental period of the structure, calculated using the average of the SI values from the 5 per cent and 10 per cent velocity spectra. Martinez-Rueda (1998) suggested values of $T_a = T_y$ and $T_b = T_h$, where T_y and T_h are the fundamental periods calculated at yield and in the post yield (hardening) range; these periods are rather difficult to calculate for actual structures. Using a different approach, Shome *et al.* (1998) have suggested scaling of accelerograms selected for a given (M, R) pair to the median S_a value corresponding to the fundamental period of the structure T_1. All these definitions make the set of records structure dependent which is reasonable, but not particularly convenient if time history analysis is to be performed in several design projects.

Use of simulated ground motions

In the 'engineering method', artificial accelerograms are generated so as to match the (target) elastic response spectrum, hence they are typically called *spectrum compatible* motions. Depending on the availability of appropriate recorded motions, the starting point of the method could be either:

- A numerically derived time history generated by superimposing sinusoidal components with pseudo-random phase angles, which are then multiplied by a deterministic intensity function (envelope of the time history) selected on the basis of the characteristics of the design earthquake (see Clough and Penzien, 1993; CEN, 1994c; Hu *et al.* 1996); or
- An actual acceleration record having the desired seismological features.

The selected record is then processed iteratively by multiplying the Fourier amplitudes (see eqn 4.11) by the corresponding average of the ratios of target S_{pv} values to the S_{pv} values calculated for the initially selected record, with a view to better matching the target spectrum.

In the 'seismological method', simulated accelerograms are generated by modelling the source and travel path mechanisms. The method generally involves two steps (Hu *et al.*, 1996):

- define the ground motion at the site due to an 'element' of earthquake source or fault rupture; planar sources are divided into a number of elements
- sum up the contributions of motions due to all elements, in the time domain.

A detailed discussion of this method, which is less common than the previous one, falls outside the scope of this book. References to the pertinent literature can be found, for instance, in Clough and Penzien (1993) and Hu *et al.* (1996).

Code treatment

All the aforementioned types of accelerograms are generally allowed as input for time history analysis in EC8, which, however, appears to promote spectrum compatible records, generated using the elastic response spectrum as the target. The duration of the records must be consistent with the characteristics (*M*, *R*, etc.) of the earthquake underlying the establishment of the design α_g. A minimum of five records is required for time history analysis, which should be enough to provide a stable statistical measure of the response; additional rules are given in EC8 regarding the allowable difference between the mean spectrum of these records and the code spectrum.

Whereas spectrum compatible records are an attractive choice of dynamic input, in the sense that scaling is not required and code requirements are imposed in a rather straightforward way, care is required in their construction to avoid over conservatism as well as inconsistencies. Referring to the previously described EC8 procedure, it is emphasized that it is the *mean* of the response spectra of artificial motions that should match the design spectrum, rather than each individual spectrum. In practice what is commonly done is that the elastic design spectrum is used as the target for all records (i.e. each spectrum matches closely the design one); this is inconsistent with the very nature of the design spectrum which does not represent a particular ground motion but rather envelopes the spectra of several motions generated from different sources and at different distances from the site. As shown by Naeim and Lew (1995), design spectrum compatible motions may represent velocities, displacements, and energy content which are very unrealistic; as a result their use in inelastic time history analysis may lead to unreliable estimates of design displacement demands.

Eurocode 8 also allows the use of recorded (natural) accelerograms, as well as of accelerograms generated by simulation of the source and travel path effects (*seismological method*). A minimum of three records is required, to be scaled to the design

PGA. As discussed in the previous section, this type of scaling, albeit convenient, is one of the most unreliable ones. A more sophisticated procedure based on matching of spectra is included in EC8 Part-2, Bridges (CEN, 1994c).

The 1997 UBC recommends the use of actual recorded accelerograms as input for time history analysis; these should be selected from at least three different events, with due consideration of magnitude, source distance and mechanisms, that should be consistent with the design earthquake. In three-dimensional analysis, pairs of records are required (i.e. a minimum of 3 pairs). Simulated time histories are allowed whenever three appropriate recorded motions are not available (to make up the total number of records required for design). For each pair of records the SRSS of the 5 per cent-damped site specific spectra is first constructed. The accelerograms are then scaled in such a way that the average value of the SRSS spectra does not fall below 1.4 times the 5 per cent-damped design spectrum, for the period range $0.2T_1$ to $1.5T_1$. Note that in two-dimensional building models T_3 (third mode period) is close to $0.2T_1$, while $1.5T_1$ is a reasonable estimate of the post-yield period of the structure. If only three time history analyses are performed the maximum response parameters are used for design, while if seven (or more) analyses are carried out, the average response parameters can be used. If the analysis is elastic, the response parameters can be scaled to the design base shear level, as in modal analysis (see Section 4.3.6). If a non-linear time history analysis is performed, the resulting response parameters (forces and displacements) can be directly used.

4.3.8 Power spectrum analysis

Although treatment of the ground motion as a random process is a very reasonable approach given the uncertainties involved in seismic wave propagation, the difficulties in calculating the response of MDOF structures to a non-deterministic input (particularly when some account for inelasticity must be made) make the application of stochastic dynamics to practical seismic design almost prohibitive. In fact, among the leading codes, EC8 is the only one including some provision for this type of analysis.

The basis of the procedure is the power spectrum (i.e. the *power spectral density* of the acceleration time history that is considered as a random process). As explained in Section 10.2, if a stationary process $x(t)$ has zero mean value and is gaussian, its power spectral density $S_x(\omega)$ completely characterizes the process, since other properties can be calculated from it, for instance the autocorrelation function is related to $S_x(\omega)$ through the Fourier integral. It is often assumed for convenience that the ground motion does possess the previous characteristics, which significantly simplifies the analysis. Models for stationary processes can be found in the literature (e.g. Hu *et al.*, 1996); one of the most commonly adopted for the ground acceleration is the modified Kanai–Tajimi model proposed by Clough and Penzien, whose power spectral density is given by

$$S(\omega) = S_0 \frac{1 + 4\xi_g^2(\omega/\omega_g)^2}{[1 - (\omega/\omega_g)^2]^2 + 4\xi_g^2(\omega/\omega_g)^2} \cdot \frac{(\omega/\omega_1)^4}{[1 - (\omega/\omega_1)^2]^2 + 4\xi_1^2(\omega/\omega_1)^2}$$

(4.38)

where S_0 is the intensity of the ground motion, ω its frequency, ω_g and ξ_g are the frequency and damping ratio of the soil, and ω_1, ξ_1 are parameters selected to produce the desired filtering of very low frequencies (high frequencies are filtered out by the first multiplier of S_0, known as the Kanai–Tajimi filter). It is seen that eqn (4.38) describes a *filtered white noise* type of random process.

General procedures, based on modal superposition, for calculating the response of MDOF structures subjected to ground motion described by a power spectrum such as that of eqn (4.38), can be found in Clough and Penzien (1993), while a presentation of the EC8 procedure for stochastic analysis of structures, including some suggested simplifications, is given by Di Paola and La Mendola (1992).

EC8 requires the use of power spectra compatible with the elastic response spectrum described in Section 4.3.2, within ±10 per cent over the range of periods from 0.2 sec to 3.5 sec, but provides no such spectrum. Some procedures for relating a power spectrum to a response spectrum are given, for instance, in Hu *et al.* (1996). A simple proposal for an EC8 spectrum compatible power spectral density can be found in Di Paola and La Mendola (1992).

4.4 CONCEPTUAL DESIGN FOR EARTHQUAKES

4.4.1 Basic principles

The objective of seismic design is to ensure that a structure behaves satisfactorily when subjected to earthquake loading. As is the case with most loading types, the anticipated behaviour or *performance levels* for the structure are different for different levels of the loading. Ideally, and taking into account the large uncertainty associated with earthquake loading, several levels of performance should be considered in design, each one corresponding to a different probability of exceedance of the seismic loading. Similarly to gravity load design, the structure should remain serviceable under 'frequent' earthquakes (SLS) and 'safe' under the ULS earthquake. Recent events, such as the 1994 Northridge earthquake and the 1995 Great Hanshin (Kobe) earthquake, have shown that whereas structures built in industrialized countries aware of the seismic risk are in general adequately safe, the *cost of damage* inflicted in these structures by earthquakes, as well as the indirect cost resulting from business disruption, need for relocation, etc. can be difficult to tolerate. This points to the need to address the problem of designing a structure for a set of performance objectives (limit states), recently referred to as *Performance Based Design* (PBD) (Fajfar and Krawinkler, 1997).

The intent of current seismic codes is usually to produce building designs capable of achieving two or three performance objectives: to resist minor earthquakes without significant damage, moderate earthquakes with repairable damage, and major earthquakes without collapse. However, as a rule, design checks are only explicitly performed for one performance objective, typically for the ULS (corresponding to either life safety or no collapse requirements). The Eurocodes recognize explicitly two limit states (ULS and SLS), whereas, interestingly, the American code (ICBO, 1997) explicitly states that 'the purpose of the provisions is primarily to safeguard against major structural failures and loss of life, not to limit damage or maintain function'.

One problem with all existing seismic codes is that their criteria for evaluating adequacy of performance are not always directly tied to specific measures of performance. Moreover, the actions used for checking these criteria are typically based on one design earthquake (the ULS action); of course, as discussed in Section 4.3.4, in serviceability related checks the lower intensity of the SLS earthquake is implicitly accounted for (ν factor in EC8, resulting in the SLS displacements being 1/2 to 1/2.5 the ULS displacements).

The intent of PBD is to ensure that structures perform at appropriate levels for all earthquakes, and is deemed to provide engineers with the ability to design structures capable of providing controlled and predictable performance for multiple performance objectives (Fajfar and Krawinkler, 1997). The difficulty, of course, lies in the quantification of this attractive concept.

One recent attempt to quantify performance levels and corresponding hazard levels has been done in the new *NEHRP Guidelines for Seismic Rehabilitation (strengthening) of Buildings* (FEMA 1997b). The hazard levels are expressed by probabilities of exceedance in 50 years, or the corresponding mean return periods T_r; that is:

- 50 per cent/50 year ($T_r = 72$ year);
- 20 per cent/50 year ($T_r = 225$ year);
- 10 per cent/50 year ($T_r = 475$ year);
- 2 per cent/50 year ($T_r = 2{,}475$ year).

Structural performance levels are quantified both in a qualitative sense (description of the type of damage associated with each one) and in a quantitative sense. At building level a convenient global measure of damage is the *interstorey drift*. Appropriate drift values are suggested by FEMA (1997b) for various types of structural systems; significantly higher values are applicable to flexible and ductile structures, such as frames, compared to stiffer and more brittle structures, such as walls (particularly masonry walls). Requirements (performance levels) are also included in FEMA (1997b) for non-structural elements (partitions, cladding, mechanical and electrical installations, plumbing, contents and finishings). It is pointed out that depending on the type of structure to be designed (or assessed), different combinations of hazard and performance levels would be appropriate. For a normal structure (e.g. an apartment building) the 'immediate occupancy' level would normally be associated with an earthquake with a 50 per cent/50 year probability, 'life safety' with a 10 per cent/50 year probability earthquake (the usual 'design' earthquake in current codes), and 'collapse prevention' with an earthquake having a 2 per cent/50 year probability of exceedance.

Explicit checks of performance can also be made at local level (i.e. for each member in the structure), in which case appropriate limits for local deformation quantities, such as rotations of plastic hinge zones are required; tabulated values of such quantities for various types of members can be found in FEMA (1997b), but it has to be stressed that these issues are still the subject of current research. Local ductility of members (e.g. plastic rotation capacity of R/C or steel beams and columns) is dependent on appropriate design and detailing, but also on quality control, particularly at the construction phase. Moreover, exceeding the available ductility capacity at one or even a few critical regions does not necessarily mean (incipient) collapse of the structure, particularly when the latter is characterized by high redundancy and ability to redistribute loading.

Recent approaches, such as those briefly outlined previously, are essentially deterministic procedures, since uncertainty is explicitly accounted for only in the case of the seismic input (spectral accelerations are adjusted to the target probability of exceedance selected for the performance level that is being checked). However, the real issue is the reliability (or the probability of failure) of the structure when subjected to a particular earthquake. This is only marginally addressed in EC8, where it is stated that *target reliabilities* for the 'no collapse requirement' and the 'damage limitation requirement' should be established by national authorities for different types of structures, on the basis of the consequences of failure. Unfortunately, these target reliabilities are not given, even as 'boxed' (i.e. indicative) values in EC8 or other codes. In a recent study (Wen *et al.* 1996) addressing this issue, suggested target 50-year probabilities range from 30 per cent to 50 per cent for the SLS and from 4 per cent to 6 per cent for the ULS; both values refer to a specific performance criterion, i.e. exceeding a drift limit.

4.4.2 Configuration issues

The selection of the configuration of the structure (i.e. of the arrangement of the structural system as well as of the non-structural elements and their connection to the former) is arguably the most critical step in the seismic design procedure. It is clear from the distribution of damage in earthquake struck regions that structures with a reasonably regular and symmetric configuration perform consistently better than structures with irregular configuration (Arnold and Reitherman, 1982). An irregular configuration is characterized by one or more of the following problems:

- The plan of the building includes large re-entrant angles (L-shaped, C-shaped, H-shaped plans; see Figure 4.16a).
- The distribution in plan of stiff members, such as walls, is not symmetric with respect to the mass centre (*M* in Figure 4.16b).

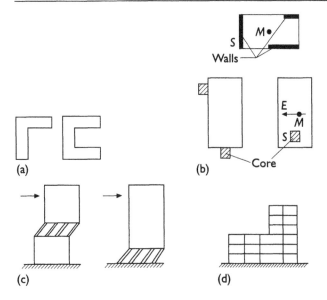

Figure 4.16 Irregular configurations in plan (a), (b) and in elevation (c), (d).

- The distribution in plan and/or in elevation of the mass of the structure is not reasonably uniform.
- The aspect ratio (height to length) of a building is high (more than about 4).
- There are abrupt changes in lateral load resistance along the height of the building (see Figure 4.16c).
- There are abrupt changes in lateral stiffness along the height of the building, due to termination of stiff elements (such as walls or heavy partitions) and/or due to the presence of setbacks; see Figure 4.16c, d.

There are several reasons for avoiding problematic configurations; they have to do with:

- our inability to accurately predict the (inelastic) response of irregular structures subjected to strong earthquakes;
- the tendency of damage to concentrate in the weakest parts of a structure; this is true, regardless of whether dynamic or other refined analysis has been used in the design;
- the increased cost required for providing to an irregular structure the same seismic resistance as in a similar regular structure.

Some of the problems mentioned previously, particularly the ones related to irregularities in plan, can often be tackled effectively by splitting a building into smaller parts separated by *seismic gaps*, so that each individual part becomes a regular structure. Seismic gaps should account for a substantial part of the anticipated relative movement of the adjacent parts during the earthquake, hence they are generally wider than standard construction (expansion) joints.

Whereas the advantages of regular configurations have relatively long been recognized, quantification of regularity requirements is a critical issue that can not be deemed as been resolved so far. An important consideration is that most of the irregular buildings damaged by previous earthquakes were designed on the basis of rough methods and the level of detailing required by the then applicable codes was quite low, while poor construction practices often made it even lower. Experimental studies involving irregular structures such as frames with setbacks designed and detailed to modern codes (Wood, 1992), tested on the shaking table, have clearly indicated that irregular R/C structures can indeed perform adequately, even when subjected to earthquakes significantly stronger than the one they were designed for. It has to be pointed out, though, that the foregoing tests involved application of unidirectional earthquake input, hence they did not address the problem of *torsion*.

Code criteria for regularity tend to be conservative but the consequences of a building being classified as irregular are typically not grave. In the UBC the presence of irregularities affects the analysis procedure (compulsory use of multimodal analysis), but it does not affect the value of the response modification factor R; in contrast, the EC8 q-factor is reduced by 20 per cent for irregular structures.

4.4.3 Failure mechanisms and capacity design

If the structure is allowed to behave inelastically during the design earthquake (Section 4.3.4), it is obvious that it will respond even further into the inelastic range whenever a stronger event (having a lower probability of exceedance) occurs. The requirement under such a rarer event would normally be that the structure does not collapse and/or does not sustain damage that would jeopardize human life. Given that it is very unlikely that the response of a structure close to failure can be analysed (particularly in the framework of a practical design), it has long been accepted that the main goal of seismic design should be to ensure that the collapse (or failure) mechanism of the structure is a favourite one, so that the structure could displace well into the inelastic range without falling down in part or entirely.

A typical illustration of the above concepts is made in Figure 4.17 that shows two generic plastic mechanisms for a multistorey frame. In the first one (Figure 4.17a) the design strength of all beams has been exceeded at a certain level of the lateral loading (roughly corresponding to the design earthquake) and 'plastic hinges' have formed at the beam critical sections. This is also the case at the base of the columns, but not in any other column section. In contrast, the plastic mechanism shown in Figure 4.17b is characterized by column hinging both at the top and bottom of the ground storey columns; this is commonly referred to as a 'soft storey' or 'weak storey' mechanism. It is clear from the kinematics of the two mechanisms that for the same top displacement (δ_u) the ratio of the required plastic

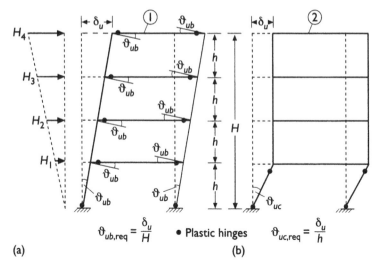

$$\vartheta_{ub,req} = \frac{\delta_u}{H} \quad \bullet \text{ Plastic hinges} \quad \vartheta_{uc,req} = \frac{\delta_u}{h}$$

(a) (b)

Figure 4.17 Favourable and unfavourable collapse mechanisms in buildings: (a) beam mechanism (favourable); (b) column mechanism (unfavourable).

hinge rotations is equal to the ratio of the total height to the storey height (i.e. in the four storey structure of Figure 4.17 the plastic rotation of the columns in the mechanism of Figure 4.17b is four times that of the members in the beam mechanism (Figure 4.17a)). It is clear that the ductility requirements, expressed here by the plastic hinge rotations, are higher in the column sidesway mechanism, and the difference increases with the height of the frame. As discussed in detail in the literature (for instance by Penelis and Kappos, 1997, for R/C structures, and by Bruneau *et al.*, 1998, for steel structures), the *available* ductility of members subjected to compressive axial loading (columns) is lower than that of beams, while the second order (P-Δ) effects that may lead to physical collapse of the structure are also more critical in the column sidesway mechanism. These are the main reasons why the plastic mechanism involving mainly hinges in the beams is considered a favourable one, whereas the mechanism involving hinges at both the top and bottom of columns is an unfavourable one. A practical way to avoid the formation of the latter mechanism is to ensure that the beams at a beam–column joint are stronger than the columns.

Provisions to materialize the previously described concept are included in most modern codes and form part of the so-called *capacity design* of a structure subjected to seismic loading. Capacity design is essentially a procedure for imposing on a structure the desired member strength hierarchy and eventually achieving a failure mechanism involving inelastic response in members that can conveniently (and reliably) be detailed to develop inelastic deformations. Most seismic codes recognize this principle, albeit to a varying degree of clarity, and the degree to which capacity design is incorporated in each code also varies significantly.

Capacity design generally dominates the response of structures that heavily rely on the development of inelastic deformations to ensure a satisfactory seismic performance, while structures that are designed for relatively high seismic forces, hence are not required to develop significant inelastic deformations, are much less controlled by capacity design considerations. This interrelationship between the required ductility (and, inversely, the level of design force) and the degree to which capacity design affects a structure are also recognized by most codes.

4.4.4 Passive and active control

Although the concepts of inelastic spectra and behaviour factors, coupled with capacity design principles, clearly dominate current seismic codes, it has to be emphasized that they do not represent the only conceptual framework available for seismic design. Furthermore, an engineer should fully realize that designing a structure on the basis of these concepts means that under earthquakes of an intensity equal to or exceeding that of the design event, damage to the structure could be both substantial and extending into a large part of the structure. Perhaps more importantly, formation of a favourable mechanism does not guarantee that interstorey drifts and/or floor accelerations will be low enough to prevent extensive damage to the non-structural elements and the content of the building. These and other concerns have led to the development of alternative conceptual frameworks for seismic design, currently referred to as 'passive' and 'active' control of the seismic response of the structure. By far the most practical approach is passive control that incorporates the fundamental ideas of *seismic isolation* and provision of *supplemental damping*. These will be discussed in the remainder of this section, followed by a brief reference to the idea of active control.

Seismic isolation and passive control

Isolating a structure from the shaking ground is a rather old concept, but it is only since the 1970s that practical isolation systems have been developed and used for earthquake protection of buildings and bridges. The concept was initially referred to as *base isolation* but at present the term *seismic isolation* prevails, in view of the fact that the isolating devices do not have to be always located at the base of the structure.

There are two interrelated ideas behind developing a seismic isolation system: the first one is to make the structure much more flexible than it is, by altering the way it rests on the ground, hence shift it to the long period range of the response spectrum that is typically characterized by reduced accelerations and consequently reduced inertial forces; the second is to introduce some kind of 'fuse' between the structure and the ground, whereby the amount of base shear to be transferred from the shaking ground to the structure is controlled by the strength of the fuse. By making the structure more flexible, one might achieve lower seismic forces, but displacements tend to increase. It is therefore essential to also control the amount of horizontal displacement of the isolated structure and an efficient way to do this is by increasing its damping (refer to Figure 4.8 for the effect of damping on seismic response spectra). This type of structural response control is referred to as *passive control*.

Currently used isolation systems are based on the concept of flexible supports which can either remain essentially elastic (linear isolation) or enter the inelastic range (non-linear isolation) upon exceeding a certain level of horizontal shear (Skinner *et al.*, 1993). The basic elements included in a seismic isolation system are:

- Horizontally flexible supporting devices (*isolators*) located either between the structure and its foundation or at a higher level in the structure; in buildings the flexible supports are commonly located at the superstructure–foundation interface, whereas in bridges they are located at the top of the piers and abutments.
- A *supplemental damping* device (or energy dissipator) for reducing the relative horizontal displacement between the superstructure and substructure (i.e. the portion of the structure below the isolators).
- Some means for controlling displacements at service levels of lateral loading (i.e. wind loading and SLS or smaller earthquake loading).

Today there are many types of isolators including, among others, rubber (elastomeric) bearings, roller bearings, sliding plates, rocking structures, cable supports, sleeved piles, helical springs, and air cushions. Detailed descriptions of the various isolating devices can be found in the massive literature available, which includes two recently published books dealing exclusively with this topic (Skinner *et al.*, 1993; Naeim and Kelly, 1999) and chapters on seismic isolation included in books of broader scope (Booth, 1994; Hu *et al.*, 1996; Priestley *et al.*, 1996).

Supplemental damping devices can be of different types, including

- *Hysteretic* dampers, wherein energy dissipation is taking place by yielding of metals such as lead and mild steel, which have hysteresis loops very close to elastoplastic. A popular isolator that incorporates a damping device is the *lead–rubber* bearing, shown in Figure 4.18, which is an elastomeric bearing (layers of rubber reinforced with thin steel plates to increase the vertical stiffness) with a lead core which provides both damping (after yield) and resistance to service lateral loads.
- *Viscous* dampers, such as the oil dampers commonly used in the motor industry, but also newer devices such as shear panels containing high viscocity fluids that have recently been developed in Japan. These mechanical devices are separate from the isolators.
- *Frictional* dampers based on the concept of friction between different materials, for instance stainless steel and PTFE (Teflon). Such systems have a number of advantages, but (unlike the previous ones) they need to be supplemented by a restoring force mechanism (i.e. a means for returning the isolated

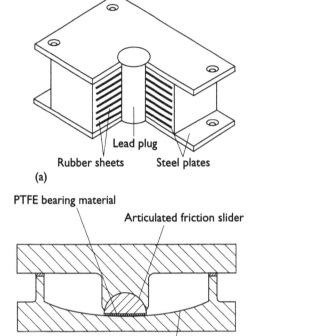

Lead plug

Rubber sheets   Steel plates

(a)

PTFE bearing material

Articulated friction slider

Spherical concave surface of hard
dense chrome over steel

(b)

Figure 4.18 Two commonly used isolating devices: (a) the lead–rubber bearing; (b) the friction pendulum system.

structure to its initial position after a strong earthquake). An efficient system in this category is the *friction pendulum* (Figure 4.18b), wherein the sliding surface of the bearing is concave, hence the restoring force is provided by the horizontal component of the weight of the structure itself.

Control of displacements under service horizontal loading can be obtained in several ways. Specially manufactured elastomers have a high rigidity at low strains, typically three to four times that at higher strains, and so do the aforementioned lead–rubber bearings. Alternatively, fuse-type sacrificial elements such as steel pins can be used, designed to fail at a desirable level of lateral loading; these elements should be replaced after each earthquake motion exceeding that level.

In a seismically isolated structure the largest part of the lateral displacement takes place at the location of the isolators. So long as this displacement can take place, the drifts in the superstructure can remain very low, hence damage to both structural and non-structural elements is minimal. Failure of such a system can occur due to rollover (instability by falling over) of the bearings at large displacements, exceedance of their shear strain capacity, or buckling of the bearings (at low strains). Special restrainers (such as steel angles) can be provided close to the bearings to prevent them from toppling over.

A critical point in passive control systems is that whereas isolator damping is always reducing the displacements of the structure that are controlled by the fundamental mode, it tends to increase floor accelerations caused by higher modes. This might be very important in structures where protection of secondary systems (equipment and non-structural elements) is the main reason for using seismic isolation. Seismic attack on secondary systems is frequency selective and it is possible to design isolation systems that reduce the response of such systems more than that of the primary structural system. A related issue is that in non-linear isolation systems (which are used in the majority of applications), control of the amount of base shear through the strength and the stiffness of the isolators does not guarantee control of the storey shear distribution along the height of the building. Whenever higher mode response is not adequately controlled, 'bulged' distributions of storey shear can result and in extreme cases the shear in the upper half of the structure may exceed the base shear (Skinner *et al.*, 1993). The foregoing are clear indications of the need for a reliable dynamic analysis when dealing with isolated structures.

The first design guidelines for seismic isolation were issued in California in 1986, and have been subject to several revisions; they were incorporated first (1991) as an appendix and later (since 1994) as a formal part of the UBC. A critical review of code provisions for seismic isolation can be found in Naeim and Kelly (1999). The current versions of UBC (ICBO, 1997) and NEHRP (FEMA, 1997a) contain provisions that are essentially identical, with the exception of the definition of design earthquake (see Section 4.3.2). These provisions include both the equivalent lateral force and the dynamic analysis procedures for seismically isolated buildings, but the restrictions for the former are such that in most practical cases the dynamic approach has to be applied. Two sets of verifications are required: The first one is for the design earthquake (10 per cent/50 year probability), under which the structure is required to remain essentially elastic. The second one is a stronger event (10 per cent/250 year probability) for which the isolation system should be designed and tested, while all building separations and utilities that cross the isolation interface should be designed to accommodate the forces and displacements associated with this seismic input. Whereas simplified methods based on the equivalent SDOF are available (see, among others, Skinner *et al.*, 1993) and can efficiently be used for preliminary design, most seismically isolated structures are currently designed using time history analysis. In the current Eurocode package, provisions for seismic isolation are only included in the bridge part EC8-2 (CEN, 1994c). However, currently (2000) such provisions are being developed for buildings and will be incorporated in the final (EN) version of EC8.

The main reason why isolation is not widely used today (particularly in buildings) is the concern regarding initial cost of the project (i.e. that in most cases a seismically isolated building costs 1 per cent to 5 per cent more than the corresponding conventional one). Such a comparison is strictly not valid as it completely ignores cost/benefit issues relating to future savings due to much lower level of damage in the isolated structure, which can be substantial (Mayes *et al.* 1990). On an initial cost basis, isolation can offer more economical solutions if the design force level is high (e.g. important structures in high seismicity areas) or if, as a result of using isolation to control damage, the structure is detailed for less ductility than in the case of conventional buildings; the latter is an option that is not explicitly recognized by current seismic codes. Finally, the isolation solution can become attractive when it leads to lower cost of insurance (i.e. lower premiums or no mandatory insurance against earthquakes in high seismicity areas).

Active control

Whereas in passive control specially provided devices absorb most of the energy input into the structure, the devices used in *active* control introduce an energy (or force) source into the structure. Active control systems have been developed during the last two decades for reducing the response of buildings (particularly tall ones) to wind and earthquake loading.

In a structure subjected to seismic loading and incorporating an active control system, the ground motion and/or the structure's response have to be monitored with appropriate sensors during the earthquake. Records from the sensors are then fed into a *controller* (computer) that activates devices for modifying the structure's response continuously during its excitation. These devices are either hydraulic actuators acting against masses in a direction that opposes that of the earthquake forces or they change the dynamic properties of the structure in order to reduce its response.

This is an attractive concept, but when applied to massive civil engineering structures such as tall buildings (instead of mechanical engineering structures) several practical problems arise, for instance the provision of adequate reaction systems to resist the large control forces produced by the actuators. Another serious problem is that since active control systems depend on power supply, it has to be ensured that this supply will not be interrupted during a strong earthquake (as it often happens), otherwise the whole system will remain idle exactly at the time that it will be required to function.

Currently used active control systems include *active mass drivers, active tendons* (wherein tension in the prestressed tendons is varied during the earthquake excitation in a way to reduce the structure's response), *active adjustable stiffness systems* (joints between the braces and the structure are either engaged or disengaged by closing or opening a control valve), and *pulse generators* (systems of pneumatic actuators and nozzles). Combinations of the above systems have also been suggested (Soong *et al.* 1991), offering some advantages.

Despite the attractiveness of the concept and the high quality interdisciplinary research carried out over the last two decades, the practical application of active control systems still remains very limited, mainly in full-scale demonstration projects.

In addition to purely passive and purely active control systems, *hybrid* systems have also been suggested (Soong *et al.*, 1991), that complement each other (for instance, a passive damper can reduce the force that has to be reduced by the active controller), hence producing an effective protecting system.

4.5 REFERENCES

AFPS [French Association for Earthquake Engineering] (1990) *Recommendations pour la redaction de règles rélatives aux ouvgages et installations à réaliser dans les régions sujettes aux séismes*, Presses de l'ENPC, Paris.

Ambraseys, N. N. and Bommer, J. J. (1995) 'Attenuation relationships for use in Europe: an overview', paper given at *5th SECED Conference 'European Seismic Design Practice'* Chester, October 1995, Amsterdam, Balkema, pp. 67–74.

Ambraseys, N. N. and Simpson, K. A. (1996) 'Prediction of vertical response spectra in Europe', *Earthquake Engineering Structural Dynamics*, **25**(4): 401–12.

Anagnostopoulos, S. A., Haviland, R. W. and Biggs, J. M. (1978) 'Use of inelastic spectra in aseismic design', *Journal Struct. Division, ASCE*, **104**(ST1): 95–109.

Arnold, C. and Reitherman, R. (1982) *Building Configuration and Seismic Design*, John Wiley, New York.

Bertero, V. V. (1989) 'State-of-the-art report: Ductility based structural design', paper given at *9th World Conference on Earthquake Engineering*, Tokyo–Kyoto, Japan, August 1988, Maruzen, Tokyo, VIII, pp. 673–86.

Booth, E. (ed.) (1994) *Concrete Structures in Earthquake Regions: Design and Analysis*, Longman, London.

Booth, E., Kappos, A. J. and Park, R. (1998) 'A critical review of international practice on seismic design of reinforced concrete buildings', *The Structural Engineer*, **76**(11): 213–20.

Bolt, B. A. (1993) *Earthquakes*, W. H. Freeman, New York.

Borcherdt, R. D. (1994) 'Estimates of site-dependent response spectra for design (methodology and justification)', *Earthquake Spectra*, **10**(4): 617–53.

Bruneau, M., Uang, C-M. and Whittaker, A. S. (1998) *Ductile Design of Steel Structures*, McGraw-Hill, New York.

CEN (1994a) Techn. Comm. 250/SC8 Eurocode 8, *Design Provisions for Earthquake Resistance of Structures: Part 1, General Rules – Seismic Actions and General Requirements for Structures* (ENV 1998–1–1), CEN, Brussels.

CEN (1994b) Techn. Comm. 250/SC8 Eurocode 8, *Design Provisions for Earthquake Resistance of Structures: Part 1, General Rules – General Rules for Buildings* (ENV 1998–1–2). CEN, Brussels.

CEN (1994c) Techn. Comm. 250/SC8 Eurocode 8, *Design Provisions for Earthquake Resistance of Structures: Part 2, Bridges* (ENV 1998–2). CEN, Brussels.

Chopra, A. K. (1995) *The Dynamics of Structures: Theory and Applications to Earthquake Engineering*, Prentice-Hall, Englewood Cliffs, NJ.

Clough, R. W. and Penzien, J. (1993) *Dynamics of Structures*, 2nd edn, McGraw-Hill, New York.

Di Paola, M. and La Mendola, L. (1992) 'Dynamics of structures under seismic input motion (Eurocode 8)', *European Earthquake Engineering*, 6(2), 36–44.

EERI [Earthquake Engineering Research Institute, US] Committee on Seismic Risk (1989) 'The basics of seismic risk analysis', *Earthquake Spectra*, 5(4): 675–702.

EERI (1995) 'Northridge Earthquake of January 17, 1994 Reconnaissance Report – Vols. 1 and 2, *Earthquake Spectra*, Suppl. to Vol. 11, April.

ESC (European Seismological Commission) (Working Group Macroseismic Scales) (1998) *European Macroseismic Scale 1998 (EMS-98)*, Vol. 15, Cahiers du Centre Européen de Géodynamique et de Seismologie, Luxembourg.

Fajfar, P. and Krawinkler, H. (eds) (1997) 'Seismic Design Methodologies for the Next Generation of Codes', (paper given at International Workshop, Bled, Slovenia, June 1997, Balkema, Rotterdam.

FEMA [Federal Emergency Management Agency] (1995, 1997a) *NEHRP Recommended Provisions for Seismic regulations for New Buildings (1994, 1997 Edns): Part 1 – Provisions; Part 2 – Commentary*, BSSC, Washington, DC.

FEMA (1997b) *NEHRP Guidelines for the Seismic Rehabilitation of Buildings*, FEMA-273, Washington, DC.

Finn, W. D. L. (1991) 'Geotechnical engineering aspects of microzonation', paper given at 4th International Conference Seismic Zonation, Stanford, CA, pp. 199–259.

Fischinger, M. and Fajfar, P. (1994) 'Seismic force reduction factors', paper given at 17th Regional European Seminar on Earthquake Engineering, Haifa, Israel, September 1993 (Rutenberg, A., ed.), Balkema, Rotterdam, pp. 279–96.

Gupta, A. K. (1990) *Response Spectrum Method in Seismic Analysis and Design of Structures*. Blackwell Scientific Publications, Boston.

Housner, G. W. (1970) 'Strong ground motion', in R. L Wiegel (ed.) *Earthquake Engineering*, Prentice-Hall, New Jersey, pp. 75–91.

Hu, Y.-X., Liu, S-C. and Dong, W. (1996) *Earthquake Engineering*. E & FN Spon, London.

International Conference of Building Officials (ICBO) (1997) *Uniform Building Code – 1997 Edition, Vol. 2: Structural Engineering Design Provisions*, ICBO, Whittier, CA.

Joyner, W. B. and Boore, D. M. (1988) 'Measurement, characterization, and prediction of strong ground motion', given at Earthquake Engineering and Structure Dynamics, Geotechnical Division ASCE, Park City, UT, pp. 43–102.

Kappos, A. J. (1991) 'Analytical prediction of the collapse earthquake for R/C buildings: suggested methodology', *Earthq. Engineering Struct. Dynamics*, 20(2): 167–76.

Kappos, A. J. (1999) 'Evaluation of behaviour factors on the basis of ductility and overstrength studies', *Engineering Structures*, 21(9): 823–35.

Kramer, S. L. (1996) *Geotechnical Earthquake Engineering*, Prentice-Hall, Englewood Cliffs, NJ.

Krawinkler, H. and Nassar, A. A. (1992) 'Seismic design based on ductility and cumulative damage demands and capacities', in: P. Fajfar & H. Krawinkler (eds.) *Nonlinear Seismic Analysis and Design of Reinforced Concrete Buildings*, Elsevier Applied Science, London & New York, pp. 23–39.

Martinez-Rueda, J. E. (1998) 'Scaling procedure for natural accelerograms based on a system of spectrum intensity scales', *Earthquake Spectra*, 14(1): 135–52.

Mayes, R. L., Jones, L. R., and Kelly, T. E. (1990) 'The economics of seismic isolation in buildings'. *Earthquake Spectra*, 6(2): 245–63.

Miranda, E. and Bertero, V. V. (1994) 'Evaluation of strength reduction factors for earthquake-resistant design', *Earthquake Spectra*, 10(2): 357–79.

Naeim, F. and Kelly J. M. (1999) *Design of Seismic Isolated Structures: From Theory to Practice*, John Wiley, New York.

Naeim, F. and Lew, M. (1995) 'On the use of design spectrum compatible time-histories', *Earthquake Spectra*, **11**(1): 111—27.

Nau, J. M. and Hall, W. F. (1984) 'Scaling methods for earthquake response spectra', *J. Struct. Engineering*, ASCE, **110**(7): 1533–48.

Newmark, N. M. and Hall, J. W. (1982) *Earthquake Spectra and Design*. EERI, Berkeley, CA.

Penelis, G. G. and Kappos, A. J. (1997) *Earthquake Resistant Concrete Structures*. E & FN SPON (Chapman & Hall), London.

Priestley, M. J. N., Seible, F. and Calvi G. M. (1996) *Seismic Design and Retrofit of Bridges*, John Wiley, New York.

Reiter, L. (1991) *Earthquake Hazard Analysis*, Columbia University Press, New York.

SEAOC Seismology Committee (1996) *Recommended Lateral Force Requirements and Commentary*, Structural Engineers Association of California, Sacramento, CA.

Shome, N., Cornell, C. A., Bazzuro, P. and Carballo, J. E. (1998) 'Earthquakes, records, and nonlinear responses', *Earthquake Spectra*, **14**(3), 469–500.

Skinner, R. I., Robinson, W. H. and McVerry, G. H. (1993) *An Introduction to Seismic Isolation*, John Wiley, Chichester, UK.

Soong, T. T., Masri, S. F. and Housner, G. W. (1991) 'An overview of active structural control under seismic loads', *Earthquake Spectra*, **7**(3): 483–505.

Tso, W. K., Zhu, A. C. and Heidebrecht, A. C. (1992) 'Engineering implication of ground motion A/V ratio', *Soil Dynamics Earthquake Engineering*, **11**(3), 133–44.

Wen, Y. K., Collin, K. R., Han, S. W. and Elwood, K. J. (1996) 'Dual-level designs of buildings under seismic loads', *Structural Safety*, **18**(2–3): 195–224.

Wilson, E. L., Der Kiureghian, A. and Bayo, E. P. (1981) 'A replacement for the SRSS method in seismic analysis' (Short Communication), *Earthquake Engineering Structure Dynamics*, **9**(2): 187–94.

Wood, S. L. (1992) 'Seismic response of R/C frames with irregular profiles', *Journal Structural Engineering, ASCE*, **118**(2): 545–66.

Zhu, T. J., Tso, W. K., and Heidebrecht, A.C. (1988) 'Effect of peak ground a/v ratio on structural damage', *Journal Struct. Engineering, ASCE*, **114**(5): 1019–37.

Chapter 5

Wave loading

Torgeir Moan

5.1 INTRODUCTION

Wave loading is important for ships, fixed and compliant offshore structures, floating bridges and airports. The focus here will be on marine civil engineering structures such as fixed and compliant platforms and buoyant bridges, as shown in Figure 5.1.

Wave load effects are required for design checks of ultimate, accidental and fatigue limit states (ISO, 1994). Ultimate and accidental limit states are often governing and are based on extreme load effects. For structures in extratropical climates, fatigue may also be an important design criterion and require an estimate of the number of stress ranges at different magnitudes for the service life. It is noted that the main contribution to fatigue damage is caused by load effects which are of the order of 15–25 per cent of the extreme load effects in the service life and hence by waves with periods in the range 2 to 8 sec.

Wave load effects for design are commonly determined by quasi-static analysis methods when the structures or structural modes have natural periods in the lower range of wave periods, while an intrinsically dynamic analysis approach is required for structures with natural period above the wave period. Typical ranges of natural periods for some marine structures are displayed in Figure 5.2.

However, besides wave loads with period equal to the wave period, in the range of 2–3 to 20 sec, the presence of certain non-linear features of the loading may cause steady state loads with a period which is a fraction $\frac{1}{2}, \frac{1}{3}, \ldots$ or a multiple 2, 3, ... of the wave period. Such steady state loads, as well as wave impact loads and other transient loads, may cause inertia and damping effects which need to be accounted for by using a proper dynamic analysis methodology, also for platforms with natural periods below the wave period. For such structures dynamic effects on fatigue loads will be most important.

For situations where dynamic effects need to be considered, the long-term stochastic character of the wave loading and its effects is important to recognize. At the same time, design analyses should be made as simple as possible, not least to avoid errors.

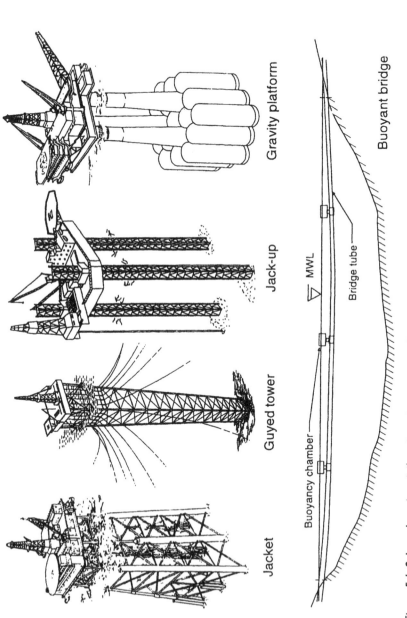

Jacket   Guyed tower   Jack-up   Gravity platform

Buoyancy chamber

▽ MWL

Bridge tube

Buoyant bridge

Figure 5.1 Selected marine civil engineering structures (Moon et al., 1990).

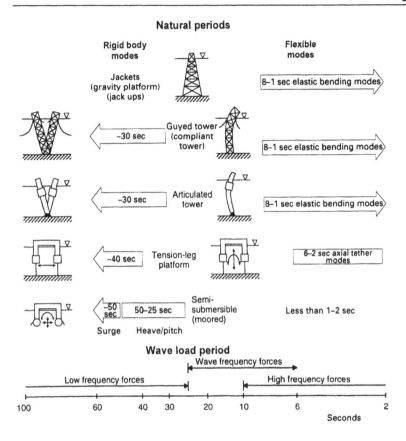

Figure 5.2 Natural periods of marine structures and wave excitation periods.

With this background in mind, the present chapter addresses characteristic features of wave loading and associated dynamic load effects, primarily for marine civil engineering structures.

5.2 WAVE AND CURRENT CONDITIONS

5.2.1 General

Surface water waves may be generated by wind, tidal bore, earthquake or landslides. Internal water waves may be generated at a boundary between water layers of different densities, and are not likely to occur together with wind generated surface waves.

The focus here is on oscillatory wind generated surface waves. Waves developed in an area may endure after the wind ceases and propagate to another area – as

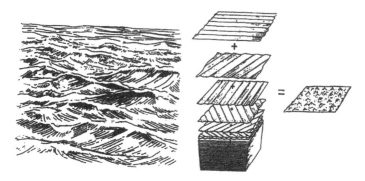

Figure 5.3 Surface elevation as a superposition of regular waves with different height, frequency and direction.

swell with decaying intensity and slowly changing form. Long period swell travels a very long distance as long-crested waves.

Wind-generated waves consist of a large number of wavelets of different heights, periods and directions superimposed on one another (Figure 5.3). Although regular waves are not found in real seas they can closely model some swell conditions. They also provide the basic components in irregular waves and are commonly used to establish wave conditions for design. Regular waves are therefore first characterized in terms of dynamic pressure, particle velocity and accelerations in Section 5.2.2. Then, irregular waves are dealt with in Section 5.2.3, while their long-term variability is briefly treated in Section 5.2.4.

The current velocity, in general, is composed of two components, namely, wind driven (v_{cwi}) and tide driven (v_{ct}) components. In addition, coastal and ocean currents may occur. Also, eddy currents, currents generated over steep slopes, currents caused by storm surge and internal waves, should be considered. Very little information about their surface velocity and velocity distribution is available. The wind current is commonly put equal to 1.0–2.0 per cent of the 'sustained' wind velocity 10 m above the sea surface. The surface value of the tidal current v_{cto} in the North Sea may be in the range of 0.2 to 0.5 m/s. The variation of current speed over time is slow in comparison with the natural periods of a structure. Hence, the current velocity is taken to be constant.

In other areas (e.g. Brazilian waters and the Gulf of Mexico), higher current velocities may be experienced. If the contribution from current velocity on the (drag) load is significant, local measurements at the actual offshore site should be performed.

Relevant variation of the current velocity over water depth is shown in Figure 5.4.

Wave and current interact. When the current is constant in time and space, the wave appears to travel on the current. In a stationary axis system this results in a Doppler shift in the wave period–wavelength relationship.

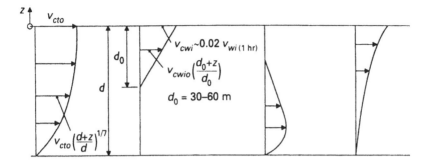

Figure 5.4 Current velocity profile.

5.2.2 Regular waves

General

Based on the assumption of an inviscid, irrational and incompressible fluid, the wave problem may be formulated in terms of a velocity potential Φ such that the velocity vector is given as: $v = \{\partial\Phi/\partial x, \partial\Phi/\partial y, \partial\Phi/\partial z\}^T$. The velocity potential should fulfil the Laplace equation (see e.g. Clauss *et al.*, 1991)

$$\nabla^2\Phi = 0 \qquad (5.1)$$

and the following boundary conditions:

1. *Kinematic boundary conditions:* No flow through the sea bottom:

$$\frac{\partial\Phi}{\partial n} = 0 \qquad \text{on sea bottom} \qquad (5.2a)$$

where $\partial/\partial n$ denotes the derivative normal to the sea bottom. In deep water, an alternative formulation of this condition is:

$$\nabla\Phi \rightarrow 0 \qquad \text{when } z \rightarrow \infty \qquad (5.2b)$$

If a body is present, a 'no flow through the body' criterion must also be satisfied:

$$\frac{\partial\Phi}{\partial n} = v \cdot n \qquad \text{on the body surface} \qquad (5.2c)$$

Here v and n denote the velocity and normal vector of the body present, respectively. Further, a fluid particle on the free surface is assumed to remain on the free surface. This is expressed in the kinematic free surface condition:

$$\frac{\partial\zeta}{\partial t} + \frac{\partial\Phi}{\partial x}\frac{\partial\zeta}{\partial x} + \frac{\partial\Phi}{\partial y}\frac{\partial\zeta}{\partial y} - \frac{\partial\Phi}{\partial z} = 0 \qquad \text{on } z = \zeta(x, y, z, t) \qquad (5.2d)$$

where ζ denotes the instantaneous free surface elevation.

2. *Dynamic boundary conditions:* On the free surface, the pressure is to be equal to the

atmospheric pressure. This is expressed by use of Bernoulli's equation:

$$g\zeta + \frac{\partial \Phi}{\partial t} + \frac{1}{2}\left(\left(\frac{\partial \Phi}{\partial x}\right)^2 + \left(\frac{\partial \Phi}{\partial y}\right)^2 + \left(\frac{\partial \Phi}{\partial z}\right)^2\right) = 0 \quad \text{on } z = \zeta(x, y, z, t) \quad (5.2e)$$

It is noted that the free surface conditions are non-linear, and that they are to be fulfilled on a free surface which is not known until the problem is solved.

Linear theory

The Airy theory is based on a linearization (i.e. Φ is supposed to be proportional to the wave amplitude), the wave elevation amplitude ζ_a is small (i.e. derivatives of ζ are zero) and the velocity square terms in eqn (5.2e) are neglected. This also means that the free surface boundary conditions can be satisfied on $z = 0$ instead of $z = \zeta$.

The solution of the linearized problem, as obtained by separation of variables (e.g. Clauss et al., 1991), may be written as

$$\Phi_0 = \frac{g\zeta_a}{\omega} \cdot \frac{\cosh[k(z+d)]}{\cosh(kd)} \cos(\omega t - kx) \qquad \text{for } z \geq -d \qquad (5.3)$$

with the circular frequency:

$$\omega = \frac{2\pi}{T} = \sqrt{kg \tan(kh)} \qquad (5.4)$$

and the wave number:

$$k = 2\pi/\lambda \qquad (\lambda = \text{wavelength}) \qquad (5.5)$$

The wave elevation is given by:

$$\zeta = \zeta_a \sin(\omega t - kx) \qquad (5.6)$$

Equation (5.6) represents a wave propagating along the positive x-axis. The linearized dynamic pressure is

$$p_D = -\rho \frac{\partial \Phi_0}{\partial t} = \rho g \zeta_a \frac{\cosh[k(z+d)]}{\cosh(kd)} \sin(\omega t - kx) \qquad (5.7)$$

and the velocities and accelerations in the x and z directions are:

$$v_x = \frac{\partial \Phi_0}{\partial x} = \omega \zeta_a \frac{\cosh[k(z+d)]}{\sinh(kd)} \sin(\omega t - kx) \qquad (5.8)$$

$$v_z = \frac{\partial \Phi_0}{\partial z} = \omega \zeta_a \frac{\sinh[k(z+d)]}{\sinh(kd)} \cos(\omega t - kx) \qquad (5.9)$$

$$a_x = \frac{\partial v_x}{\partial t} = \omega^2 \zeta_a \frac{\cosh[k(z+d)]}{\sinh(kd)} \cos(\omega t - kx) \qquad (5.10)$$

$$a_z = \frac{\partial v_z}{\partial t} = -\omega^2 \zeta_a \frac{\sinh[k(z+d)]}{\sinh(kd)} \sin(\omega t - kx) \qquad (5.11)$$

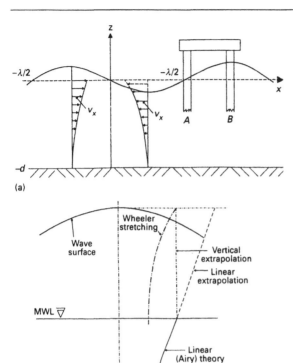

Figure 5.5 Wave elevation and kinematics; (a) Linear theory; (b) wave crest kinematics.

The horizontal velocity and acceleration are seen to have their absolute maximum values at crest/trough and wave nodes, respectively. The wave elevation, dynamic pressure and horizontal velocity are in phase, while the horizontal acceleration is 90° out of phase. Moreover, it is seen that the kinematics (e.g. horizontal velocity) at locations half a wavelength apart is in opposite phase. These phase relationships are of considerable significance for calculation of wave loads on structures (Figure 5.5). Further details about the Airy theory may be found, for example, in Clauss *et al.*, 1991.

Modifications of the kinematics of linear theory

The linear theory is valid only for small values of wave amplitudes. Particle velocities in the crest region will especially be subject to significant uncertainties, which will affect drag forces, which are proportional to velocity squared, and other loads which depend upon crest kinematics. Rather than extrapolating, for example, the particle velocity according to the exponential variation of eqn (5.8), various empirical modifications of the linear theory have been proposed to improve the accuracy. One alternative is to use a linear extrapolation or simply use a constant velocity equal to that at the Mean Water Level (MWL) in the crest. Linear extrapolation of the kinematics above MWL is obtained by, for example, replacing eqn (5.8) by

$$v_x(x, z, t) = v_x(x, 0, t) + z\frac{\partial v_x}{\partial z}(x, 0, t) \qquad \text{for } 0 \le z \le \zeta \tag{5.12}$$

A more sophisticated approach commonly used is the so-called Wheeler stretching (Wheeler, 1970). This modification introduces a new vertical co-ordinate which moves together with the free surface. The velocity potential Φ and the corresponding kinematics can then be obtained by introducing a co-ordinate z_ℓ instead of z in eqn (5.8); with $z_\ell = d(z - \zeta)/(d + \zeta)$. This means that the kinematic quantities have the same size and vertical distribution, only now with the free surface as starting point instead of the MWL, as shown in Figure 5.5b. Gudmestad (1993) recently reviewed various engineering approximations to wave kinematics and compared them with experimental results.

A deficiency of the original and modified Airy theory is that it provides symmetric waves while extreme waves are known to be asymmetric (i.e. with a larger crest than trough). Higher order wave theories have been proposed to better represent the shape and kinematics of the waves (see e.g. Clauss *et al.*, 1991).

Higher order wave theory

The linear wave theory represents a first order approximation of the free surface conditions, which means that errors will become large as the waves become higher (i.e. as ζ_a/λ increases), because of the neglected higher order terms. This deficiency can be improved by introducing higher order terms. Commonly this is done by means of perturbation theory. Wave elevation and velocity potential are then expanded into power series, with α being a small perturbation parameter, so that the significance of additional terms decreases with their order (see e.g. Sarpkaya and Isaacson, 1981)

$$\zeta = \alpha\zeta^{(1)} + \alpha^2\zeta^{(2)} + \alpha^3\zeta^{(3)} + \cdots + \alpha^n\zeta^{(n)} \tag{5.13}$$

$$\Phi = \alpha\Phi^{(1)} + \alpha^2\Phi^{(2)} + \alpha^3\Phi^{(3)} + \cdots + \alpha^n\Phi^{(n)} \tag{5.14}$$

Each individual potential $\Phi^{(i)}$ satisfies both the Laplace equation and the non-linear boundary conditions with successive refinement. At the free surface, the velocity potential is expanded as a Taylor series about the still water level to obtain successive approximations of higher order wave theories:

$$\Phi(x, z, t) = \Phi(x, 0, t) + \zeta\frac{\partial\Phi(x, 0, t)}{\partial z} + \frac{\zeta^2}{2!}\frac{\partial^2\Phi(x, 0, t)}{\partial z^2} \tag{5.15}$$

The perturbation parameter (α) turns out to be (ζ_a/λ). The second order expansion of the surface elevation is

$$\zeta = \zeta_a \sin(\omega t - kx) - \tfrac{1}{2}\zeta_a^2 k f^{(2)}(\chi)\cos[2(\omega t - kx)] \qquad (5.16)$$

where $f^{(2)}(\chi)$ is a function of χ (e.g. Sarpkaya and Isaacson, 1981). Equation (5.16) shows that the crest becomes more peaked while the troughs become more shallow.

The effect of higher order wave theory on the kinematics depends upon wave height $(H/(gT^2))$ and water depth $(d/gT^2))$ parameters. For high waves in deep water the Airy theory yields larger particle velocities than Stokes higher order theory.

Alternative wave theories based, for example, on stream function instead of velocity potential are discussed, for example, by Sarpkaya and Isaacson (1981).

It is noted that the Stokes theory still depends upon the limitation of the assumed small non-linearities within the perturbation theory.

When a current is present, the kinematics corresponds to a superposition of horizontal current and wave particle velocities.

5.2.3 Wave kinematics of irregular waves in short-term periods

Linear theory

During a suitably short-term period of time (from half an hour to some hours) the *sea surface elevation* is commonly assumed to be a zero mean, stationary and ergodic Gaussian process (e.g. Kinsman, 1965). An interpretation of this process is a linear combination of independent and arbitrarily distributed random disturbances. In strong wind generated waves non-linearities in the wave process tend to disturb the Gaussian character. The Gaussian process is completely specified in terms of autocorrelation function of the surface elevation or the three-dimensional wave spectral density. Due to the unique relationship between wave frequency and wave number for water waves, a two-dimensional spectral density suffices (see e.g. Kinsman, 1965; Sigbjørnsson, 1979).

In the *time domain* the wave elevation may be described by a sum of long crested waves specified by linear theory, with different amplitude (a_{ik}), frequency (ω_i), wave number (k_i), direction relative to the x-axis (θ_k) and phase angle (ε_{ik}) as follows:

$$\zeta(x,t) = \sum_{k=1}^{K}\sum_{i=1}^{N} a_{ik}\sin[\omega_i t - k_i(x\cos\theta_k + y\sin\theta_k) + \varepsilon_{ik}] \qquad (5.17)$$

If $\theta_k = 0$ (eqn (5.17)) expresses an irregular wave propagating along the x-axis. For another period of the same sea state, the coefficients (ε_{ik}) will be different while the distribution of a_{ik} over ω_i and θ_k will be 'the same'. The distribution of a_{ik} over ω_i and θ_k is a 'deterministic' measure of that sea state, while the phase angle ε_{ik} appears to be uniformly distributed over $(-\pi, \pi)$.

The amplitude a_{ik} may be expressed by the two-dimensional energy spectrum:

$$a_{ik} = \sqrt{2S_\zeta(\omega_i, \theta_k)\,\Delta\omega\,\Delta\theta} \tag{5.18}$$

The two-dimensional (directionality frequency) spectral density $S_\zeta(\omega, \theta)$ is conveniently expressed by

$$S_\zeta(\omega, \theta) = S_\zeta(\omega)D(\theta, \omega) \tag{5.19}$$

where $S_\zeta(\omega)$ is the one-dimensional spectral density that can be estimated from observations of $\zeta(t)$ at a given location, by a Fourier transform of the autocorrelation function of the $\zeta(t)$ process (see Chapter 10). $D(\theta, \omega)$ is the so-called spreading function.

Various analytical formulations for the wave spectrum are applied (as discussed e.g. by Price and Bishop, 1974). In developing seas the JONSWAP spectrum (Hasselman *et al.*, 1973) is recommended and frequently used. For fully developed seas, the Pierson–Moskowitz spectrum (see e.g. Gran, 1992) is relevant. Wind sea and swell have different peak periods and a combined sea state may have a two-peaked spectrum (as proposed e.g. by Torsethaugen, 1996). It should be noted that much of the wave energy is concentrated in a narrow frequency band close to the peak(s) of the spectrum. Moreover there is a significant difference in the spectral amplitudes for high frequencies, implied by different models.

The JONSWAP spectrum is parameterized in the following form:

$$S_\zeta(\omega) = 0.3125 H_{m0}^2 T_p(\omega/\omega_p)^{-5} \exp(-1.25(\omega/\omega_p)^{-4})$$

$$\times (1 - 0.287 \ln \gamma)\,\gamma \exp(-0.5((\omega/\omega_p - 1)/\sigma)^2) \tag{5.20}$$

where H_{m0} and $T_p = 2\pi/\omega_p$ are the significant wave height and spectral peak period, respectively, $\sigma = 0.07$ for $\omega \le \omega_p$ and $\sigma = 0.09$ for $\omega > \omega_p$. The peakedness parameter γ depends upon $T_p/\sqrt{H_{m0}}$ and varies in the range from 1 to 7.

While the spreading function $D(\theta, \omega)$ generally is frequency dependent, it is usually approximated by

$$D(\theta, \omega) = D(\theta) = C\cos^n(\theta - \theta_0) \qquad \text{for } -\frac{\pi}{2} \le \theta - \theta_0 \le \frac{\pi}{2} \tag{5.21}$$

where θ_0 denotes the mean wave direction and C is a normalization factor to ensure that the integral of $D(\cdot)$ over θ is unity, and n normally varies between 2 and 8.

The kinematics (particle velocities, accelerations, pressures) for irregular waves are then obtained by superposition of the kinematics based on linear (Airy) theory for each regular wave. It is noted that there is no phase lag in the kinematics in the vertical direction.

In the *frequency domain* the kinematics are described by spectral densities. Hence, the following cross-spectral density can be derived from the wave number spectral density (Sigbjørnsson, 1979)

$$S_{\zeta_m \zeta_n}(\omega) = \int_{-\pi}^{\pi} S_\zeta(\omega, \theta) \exp[ik(\omega)(\Delta x \cos\theta + \Delta y \sin\theta)] \, d\theta \qquad (5.22)$$

where ζ_m and ζ_n are the wave amplitudes at points m and n with co-ordinates (x_m, y_m) and (x_n, y_n), respectively, $\Delta x = x_m - x_n$, $\Delta y = y_m - y_n$ and $i = \sqrt{-1}$.

The probabilistic description of the wave kinematics in terms of the particle velocities and accelerations is commonly achieved by applying the principle of superposition of independent and arbitrarily distributed disturbances and the Airy wave theory. Then the frequency cross-spectral density of, for example, the water particle velocity v_x may be expressed as follows, applying eqns (5.8) and (5.22)

$$S_{v_x(m)v_x(n)}(\omega) = S_\zeta(\omega)\omega^2 \frac{\cosh[k(z_m + d)]\cosh[k(z_n + d)]}{\sinh^2(kd)}$$

$$\times \int_{-\pi/2}^{\pi/2} D(\theta) \begin{bmatrix} \cos^2\theta & \sin\theta\cos\theta \\ \sin\theta\cos\theta & \sin^2\theta \end{bmatrix}$$

$$\times \exp[ik(\Delta x \cos\theta + \Delta y \sin\theta)] \, d\theta \qquad (5.23)$$

Analogous expressions hold for the frequency cross-spectral densities of acceleration, and acceleration and velocity.

Higher order irregular wave theory

To reduce the deficiencies of the linear theory, especially in predicting extreme values, a consistent second-order or higher order irregular wave theory, analogous to the higher order regular wave theories mentioned, may be established. However, in current engineering practice, improved kinematics is obtained by modification of the linear theory (e.g. Gudmestad, 1993). It should be noted that this formulation does not represent the asymmetry in wave elevation nor the non-linear interaction between individual waves in an irregular wave process.

The sea surface elevation is not a perfect Gaussian process (see e.g. Longuet-Higgins, 1963; Haver and Moan, 1983; Vinje and Haver, 1994). In the same way as a finite regular wave is not perfectly sinusoidal (i.e. the crest is larger than the trough), the random sea elevation is skewed and has more kurtosis than a Gaussian process. Vinje and Haver (1994) found that the skewness depends upon H_{m0} and T_p according to $\gamma_1 = 34.4 H_{m0}/(gT_p^2)$ and kurtosis $\gamma_2 = 3.0 + 3\gamma_1^2$. Non-Gaussian surface elevation may be generated by a second order (irregular) wave model for instance based on Stokes' expansion (see e.g. Longuet-Higgins, 1963), or by transformation of a Gaussian process by a Hermite expansion (see e.g. Winterstein, 1988).

5.2.4 Wave kinematics of irregular waves in long-term periods

The non-stationary sea state in a long term period (i.e. of some years duration), can be assumed to consist of a sequence of short term sea states (i.e. stationary zero mean ergodic processes), each completely described by the spectral density (see definitions in Chapter 10). For a given analytical model of the spectrum (e.g. JONSWAP or Pierson–Moskowitz), the spectral parameters H_{m0}, T_p, γ, θ_0, etc. completely specify the sea state. By expressing the magnitude of these parameters and possibly the current and wind velocity and direction in probabilistic measures, the long term process is described. For extratropical regions, like the North Sea, the joint probability density of the parameters is applied towards this aim (see e.g. Haver, 1980). A Weibull distribution is then commonly used to describe the marginal distribution of H_{m0}, while the conditional distribution of T_p given H_{m0} is often taken to be a log-normal distribution. For tropical areas subject to hurricanes, the long term wave climate can be described by storms arriving in a sequence (e.g. Jahns and Wheeler, 1972).

Data for the long-term model of the waves can be generated (i) by direct observation of wave condition; (ii) hindcasting based on wind data.

The probabilistic description of the *wave kinematics* in terms of the particle velocities and accelerations is commonly achieved by applying the principle of superposition of independent and arbitrarily distributed disturbances and the Airy or modified Airy wave theory.

5.3 HYDRODYNAMIC LOADING

5.3.1 General

In general, the effects of waves and currents on marine structures are obtained as vector superposition of all forces on the individual structural elements. If relevant, the subsequent response (e.g. the motion of the structure) also needs to be considered. To calculate hydrodynamic forces, it is necessary to integrate the pressure field over the wetted surface of the structure. The main force components are (Clauss *et al.*, 1991; Faltinsen, 1990):

- Froude–Krylov force – pressure effects due to undisturbed incident waves;
- hydrodynamic 'added' mass and potential damping force – pressure effects due to relative acceleration and velocity between water particles and structural components in an ideal fluid;
- viscous drag force – pressure effect due to relative velocity between water particles and structural components.

The Froude–Krylov (FK) force acting on a submerged body in a wave field may be obtained by integrating the pressure p on the surface S

$$F_{FK} = -\int_{\text{surface}(S)} p\mathbf{n}\,ds = -\int_{\text{volume}(\forall)} \nabla p\,d\forall = \int_{\forall} \rho \frac{d\mathbf{v}}{dt} d\forall \approx \int_{\forall} \rho \frac{\partial \mathbf{v}}{\partial t} d\forall \quad (5.24)$$

when the basic surface integral expression, first, is transformed to a volume integral by applying the Gauss theorem, then Euler's equation for an incompressible, inviscid and irrotational fluid is introduced and finally the convective term of acceleration is ignored. **n** denotes a normal to the surface, **v** is the particle velocity of the fluid and ρ denotes the density.

For a slender body, the water particle acceleration changes only slightly within the structure and may be substituted by the acceleration at the component axis, to yield

$$F_{\text{FK}} = \rho \forall \, \frac{\partial \mathbf{v}}{\partial t} \tag{5.25}$$

The hydrodynamic (added) mass force acting on a body is obtained by integration of the pressure field arising from the relative acceleration between the structural component and fluid over the wetted surface. This force can be determined by accelerating the body in a fluid at rest, and can generally be written as

$$F_A = C_A \rho \forall \, \mathbf{a} \tag{5.26}$$

In general, C_A depends upon the flow conditions and the location of the body. It is frequency dependent for bodies at or close to the surface, whereas it is independent of frequency for submerged, slender bodies. Data may be found, for example, in Clauss *et al.* (1991). For a submerged, slender cylinder C_A is equal to 1.0.

The viscous (drag) force per unit length normal to a member may be written as

$$\mathbf{F}_D = \tfrac{1}{2} C_D \rho A \, v_n |v_n| \tag{5.27}$$

where v_n is the velocity normal to the axis of the member with projected cross-section of A. Drag coefficients may be found, for example, in Clauss *et al.* (1991).

The load formulation applicable depends upon the flow condition, as measured, for example, by the Keulegan–Carpenter number (KC) and the Reynolds number (Re). KC is defined as $\text{KC} = vT/D$. (v is the maximum horizontal wave particle velocity, T is the wave period and D is the diameter of the structure). For KC smaller than 2, potential theory applies, while viscous effects should be included for KC larger than 2. Re is defined as $\text{Re} = v \cdot D/\nu$, where v and D are given above and ν is the kinematic viscosity of water, $\nu = 1.11 \times 10^{-6}\,\text{m}^2/\text{s}$.

For slender structures, F_{FK} and F_A are approximated by a single inertia term, and the viscous force makes up the drag term F_D. In this case, it was assumed that the water particle velocity and acceleration in the region of the structure do not differ significantly from the values at the cylinder axis. This assumption is only acceptable when the diameter, D, of the structure is small compared with the wave length, λ (i.e. for $D/\lambda < 0.2$). The loading on slender members is further discussed in Section 5.3.2.

With larger structural diameters, the incident wave is significantly disturbed by the structure. Assuming linear wave theory, the steady state wave field then results from the interference of the incident wave and the body, and may be derived from the superposition of the potentials of the undisturbed incident wave and an induced wave field of the same frequency, generated by and radiating from the body. Here viscous forces are of less significance, since the ratio of wave height to structural diameter remains sufficiently small. According to potential theory, the pressure distribution and the corresponding forces can be calculated from the velocity potential, as discussed in Section 5.3.3.

When the wave acts upon a structure, the latter will be set in motion, which will set up waves radiating away from it. Reaction forces are then set up in the fluid that are proportional to acceleration and velocity of the structure, respectively. These are inertia (added mass) and potential damping forces due to wave generation. In addition, viscous (drag) forces are set up. This issue is treated in Section 5.3.4. Finally, Section 5.3.5 deals with particular transient wave loading phenomena such as wave slamming and ringing.

5.3.2 Steady-state loading on slender structures

If the characteristic dimension (e.g. the diameter D of a circular structural component) is small relative to the wavelength λ (i.e. $D/\lambda < 0.2$), there is little alteration of the incident wave when it passes the structure. The wave does not 'see' such a slender structure: as diffraction and reflection phenomena are negligible, the structure is said to be 'hydrodynamically transparent'. With relatively small dimensions, local variations of particle velocity and acceleration in the region of the structural element are small enough to be ignored, and values are calculated at the position of the structural element as a whole.

For slender members which are fixed the force per unit length q_n, normal to the member, is most often calculated by the extended, empirical Morison formula (e.g. Clauss $et\,al.$, 1991):

$$q_n = q_I + q_D = \rho\,d\forall\,a_n + C_A\rho\,d\forall\,a_n + \tfrac{1}{2}C_D\rho\,dA\,v_n|v_n|$$
$$= C_M\rho\,d\forall\,a_n + \tfrac{1}{2}C_D\rho\,dA\,v_n|v_n| = q_I + q_D \tag{5.28}$$

where ρ is the density of the fluid, C_M and C_D are the inertia and drag force coefficients, respectively, and v_n and a_n are, respectively, the wave particle velocity and acceleration perpendicular to the member. dA and $d\forall$ are the exposed area and displaced water of unit length. For a circular member, with a diameter D, $dA = D$ and $d\forall = \pi D^2/4$.

The first term results from the FK and hydrodynamic mass force while the second terms $q_D = \tfrac{1}{2}C_D\rho\,dA\,v_n|v_n|$ in eqn (5.28) is due to the viscous drag term and downstream wake.

The v_n and a_n for a design wave are obtained directly from the kinematics for a regular wave. In the case of random waves, v_n and a_n are obtained by superimposing the kinematics for all regular waves that constitute the random wave history.

For a vertical cylinder in deep water, the total integrated forces q_D and q_I are equal for a wave height to diameter ratio of about 10.

A crucial issue in applying Morison's equation is the determination of C_D and C_M. Extensive data from laboratory experiments indicate a general range of 0.6 to 1.2 for C_D and 1.2 to 2.0 for C_M, depending upon flow conditions (as measured by the KC, Re numbers) and surface roughness (Sarpkaya and Isaacson, 1981). When applying hydrodynamic coefficients to calculate loading on platforms consisting of many members, additional uncertainties are encountered and should actually be reflected in the coefficients.

Under such circumstances the coefficients are chosen so as to adequately represent the loading in view of the wave kinematics formulation used.

The API (1993/1997) recommendation for calculating loads on jacket platforms may serve to illustrate this point. The key points in this procedure for calculating extreme load effects are:

- regular wave with appropriate height (e.g. corresponding to 100 year return period) and period;
- wave kinematics according to two-dimensional Stokes fifth order (or other Stokes type) methods and appropriate correction factors for shortcrested seas and current shielding or blockage (i.e. the effect of the structure on the kinematics);
- the input current velocity profile, which refers to MWL, is modified by stretching to provide current velocities over the total wetted surface;
- the effective diameter of the member is calculated by $D = D_c + 2t$, where D_c is the clean outer diameter and t is the thickness of the marine growth;
- drag and inertia coefficients for calculation of global loads are selected as:

smooth cylinders: $C_D = 0.65$, $C_M = 1.6$
rough cylinders: $C_D = 1.05$, $C_M = 1.2$

Members located 2 m above MWL may be considered smooth and those below are considered to be rough.

The hydrodynamic coefficients were calibrated to fit in-service measurements (Heideman and Weaver, 1992).

The relevant hydrodynamic loads for fatigue analyses correspond to more moderate waves (i.e. with smaller KC numbers than for extreme waves). The implication may be to apply the same C_D for smooth cylinders and reduce the C_D for rough cylinders and increase C_M to 2.0 (API, 1993/1997).

Morison's equation accounts for in-line drag and inertia forces, but not for the 'out of plane' (plane formed by the velocity vector and member axis) *lift force* due to periodic, asymmetric vortex shedding from the downstream side of a member. Due to their high frequency, random phasing and oscillatory (with zero mean) nature, lift forces are not correlated across the entire structure. Their effect on global loads can therefore be ignored while they may have to be considered for local loads. Morison's equation also ignores axial FK, added mass and drag forces, which will be of increasing importance with increasing diameter to member length.

For (dynamic) spectral or time domain analysis of surface piercing framed structures in random Gaussian waves and use of modified Airy (Wheeler) kinematics with no account of kinematics factor, the hydrodynamic coefficients should, in absence of more detailed documentation, be taken to be (NORSOK N-003, 1999) $C_D = 1.0$ and $C_M = 2.0$.

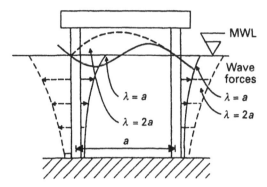

Figure 5.6 Effect of phase angle on forces in regular waves.

These values apply both for stochastic analysis of extreme and fatigue action effects. It is noted that the increased value of especially C_M is to account for the non-symmetry of wave surface elevation in severe wave conditions.

The presence of a *current* will change the wave height (and, hence, the spectral density for the sea elevation), the type of flow orbits (and, hence, in principle the wave force coefficients) as well add a contribution to the sea particle velocity (Sarpkaya and Isaacson, 1981). In many cases, the effect of current is implicit in observed wave data. In such cases, the effect of current on wave height should not be considered. The current velocity is added vectorially to wave particle velocities. With a 100 year surface current velocity of the order 0.5 to 2.0 m/s, and a maximum wave particle velocity (in a 30 m high wave) of the order 7 to 9 m/s, the current contributes significantly to the hydrodynamic loading, due to the quadratic form of F_D.

The cyclic character of waves implies that there is a phase angle between the wave forces on different members, as illustrated in Figure 5.6.

5.3.3 Steady-state loading on large volume structures

As mentioned in Section 5.3.1, the accuracy of Morison's equation will diminish when D/λ increases beyond 0.2.

Consider, for instance, a vertical cylinder with a diameter $D = 2R$, resting on the seabed and piercing the surface. The incident potential Φ_0, given by eqn (5.3), is known.

The radiation potential Φ_7 is solved from a boundary value problem in terms of the Laplace differential equation in the fluid domain and appropriate boundary conditions. The boundary conditions consist of the conditions at the ocean bottom, the free surface and the surface of the structure as well as a radiation condition far from the structure. It is demonstrated, for example, in Clauss *et al.* (1991) that Φ_7

can be expressed in a polar co-ordinate system by a product of a function in z and a function in r (radial co-ordinate).

Once the velocity potential ($\Phi = \Phi_0 + \Phi_7$) is known, the pressure on the surface of the structure can be calculated from the linearized Bernoulli's equation ($p = -\rho\, \partial\Phi/\partial t$) and the horizontal and vertical forces may be determined by integrating the pressure.

For a vertical cylinder with diameter $D = 2R$, a closed form solution often named the MacCamy and Fuchs (1954) approach, can be obtained.

The horizontal force q in the x-direction per unit axial length of the cylinder is computed as:

$$q = 4\rho \frac{g\zeta_a}{k} \frac{\cosh k[(z+d)]}{\cosh kd} A_1(kR)\cos(\omega t - \varepsilon_1) = \rho C_M\, d\forall\, a_x \tag{5.29}$$

where $A_1(kR) = [J_1'(kR)^2 + Y_1'(kR)^2]^{-1/2}$, $\varepsilon_1 = \tanh^{-1}[J_1'(kR)/Y_1'(kR)]$. $J_1'(kR)$ and $Y_1'(ka)$ are the derivatives of first order Bessel functions of the first and second kind, respectively. $d\forall$ is the volume (πR^2) of the cylinder per unit length. The horizontal force may be expressed in terms of an effective inertia coefficient C_M and a horizontal water particle acceleration component a_x at the centre of the section of the cylinder and at an elevation z corresponding to the inertia term of the Morison equation. Hence, a_x is given by eqn (5.10). It is noted that the horizontal wave force is phase-shifted with respect to the acceleration. It is seen that

$$C_M = \frac{4A_1(kR)}{\pi(kR)^2} \tag{5.30}$$

As shown in Figure 5.7, C_M is approximately equal to the slender body value of 2.0 for $kR \leq 0.1$.

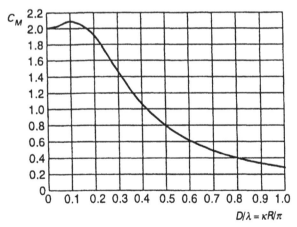

Figure 5.7 Effective inertia coefficient versus diffraction parameter for a large diameter vertical cylinder, piercing the water surface.

It is noted that the diffraction effects will significantly reduce C_M when waves with a period of, say, 5 sec act upon columns with a diameter that exceeds 8 m. This issue is important when the structure has natural periods around 5 sec.

Analytical solutions to several other cases of simple geometry for offshore structures can be developed (see e.g. Gran, 1992). These generally require that the member be far removed from a boundary, particularly the free surface. Some of these members include spheres, horizontal cylinders, bottom seated hemispheres and bottom seated half-cylinders. For more complex cases numerical methods have been proposed to obtain wave diffraction solutions. These methods include boundary element, finite fluid element, conformal mapping, and hybrid techniques. The solutions have received many experimental verifications and practical applications (see e.g. Clauss *et al.*, 1991, and Faltinsen, 1990).

Wave diffraction solutions do not include viscous actions. When body members are relatively slender and have sharp edges, viscous effects may be important and should be added to the diffraction forces determined.

Wave loads on structures composed of large volume parts and slender members may be computed by a combination of wave diffraction theory and Morison's equation. Parts of the structure may be modelled both by boundary elements to represent the potential hydrodynamic loads and beams to represent the viscous drag loads. The modifications of velocities and accelerations as well as surface elevation (wave enhancement) due to the large volume parts should, however, be accounted for when using Morison's equation. This situation may arise in connection with caissons of gravity structures, strong interaction between large columns, non-vertical sides near the water plane and other features. The results from boundary element methods should be carefully checked for surface-piercing bodies to ensure that irregular frequencies are avoided. Moreover, estimates of loads for novel structural shapes need to be checked by model tests. Model tests have also been carried out systematically to establish Morison-type formulation for inertia forces on gravity structures (see e.g. Moan *et al.* 1976).

5.3.4 Effect of motions

As mentioned above, when the structure moves, as a result of excitation forces, inertia (added mass) and potential damping forces are generated. If the structure moves, the *total* inertia force acting on a slender member of the structure, may then be established as the same FK force as that acting on a fixed structure, together with the added mass force associated with the relative acceleration between fluid and structure. The drag force may be established by replacing the particle velocity in eqn (5.28) with the relative velocity. Hence, the force normal to the axis of the member may be written as

$$q_n = \rho \forall\, a_n + C_A \rho\, d\forall\, (a_n - \ddot{x}_n) + \tfrac{1}{2} C_D \rho\, dA\, (v_n - \dot{x}_n)|v_n - \dot{x}_n|$$

$$= C_M \rho \forall\, a_n - C_A \rho\, d\forall\, \ddot{x}_n + \tfrac{1}{2} C_D \rho\, dA\, (v_n - \dot{x}_n)|v_n - \dot{x}_n| \qquad (5.31)$$

Equation (5.31) is particularly relevant in connection with analysis of motions of floating structures and structural dynamics of bottom supported structures. In the latter case the relative velocity term in eqn (5.31) should be used with caution. The amplitude of the structures motion needs to be equal to the member diameter to set up the fluid flow for which eqn (5.31) is valid. Otherwise, using eqn (5.31) may overestimate the damping and hence lead to non-conservative load effects.

Analogous considerations apply to large volume structures (large cross-section dimension relative to the wavelength). However, in that case the FK and added mass loads both need to be determined by analytical or numerical methods, as mentioned in Section 5.3.3. The added mass and damping contributions are then determined by introducing a potential Φ_j for each of the six rigid body modes as well as possible flexible modes.

For the structures considered herein wave kinematics is commonly determined with reference to the initial position of the structure. When motion amplitudes become large (i.e. of the order of the wave amplitude), the position of the structure may be updated, when excitation forces are determined.

5.3.5 Non-linear wave loading

Slender bodies

The drag force in Morison's equation, eqn (5.31), is non-linear in particle velocity. The particle velocity is proportional to wave height according to linear theory. Moreover, the fact that the drag force is non-linear will introduce higher order harmonics in the force associated with a regular, periodic wave. The drag and inertia force on, for example, a horizontal member caused by a regular wave with particle velocity $v_x = \sin(\omega t)$ and acceleration $a_x = \cos(\omega t)$ (Mo, 1983; Mo and Moan, 1984) are:

$$q_D \propto v_x|v_x| = \sin(\omega t)|\sin(\omega t)|$$
$$= 0.85 \sin(\omega t) + 0.17 \sin(3\omega t) - 0.02 \sin(5\omega t) \ldots \tag{5.32}$$

$$q_I \propto a_x = \cos(\omega t) \tag{5.33}$$

When a harmonic wave of finite height passes a structure, forces on a horizontal or a segment of a vertical member in the 'splash zone' may vary in time as indicated in Figure 5.8. Clearly, by expanding these forces in Fourier series, it is observed that there will be higher order harmonic components in the overall excitation of the structure. This effect will be more pronounced when drag forces are predominant because they attain their maxima at maximum and minimum wave elevation. Also, drag forces are more important in an extreme seaway than in a moderate one.

To illustrate this point more explicitly, consider a cylinder piercing the wave surface. When the velocity is assumed to be constant above the MWL and equal to the velocity at the MWL (vertical extrapolation in Section 5.2.2), the drag and

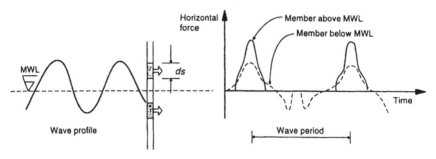

Figure 5.8 Schematic force–time history for a horizontal member or segments of a vertical member above and below the mean water level.

inertia forces can be shown to be (Mo, 1983):

$$q_D \propto v_x |v_x| H[\zeta(t)] = \sin(\omega t) |\sin(\omega t)| H[\sin(\omega t)]$$
$$= 0.25 + 0.424 \sin(\omega t) - 0.25 \cos(2\omega t)$$
$$- 0.085 \sin(3\omega t) \dots \tag{5.34}$$

$$q_I \propto a_x H[\zeta(t)] = \cos(\omega t) H[\sin(\omega t)]$$
$$= 0.5 \cos(\omega t) + 0.424 \sin(2\omega t)$$
$$+ 0.170 \sin(4\omega t) \dots \tag{5.35}$$

where $H[\cdot]$ is the Heaviside unit function defined by $H[x]$: (0 for $x < 0$; $\frac{1}{2}$ for $x = 0$; 1 for $x > 0$). This fact will also be reflected in the probabilistic description of the complete Morison equation.

Moreover, if a current velocity v_c is added vectorially to the wave particle velocity in eqn (5.28), the nature of the (drag) force will be affected. Consider, for simplicity, two wave components with velocity amplitudes of v_{x1} and v_{x2}, respectively, together with a current v_c. The dragforce during that part of the wave cycle for which the dragforce is positive may be obtained as:

$$q_D \propto (v_{x1} \cos \beta_1 + v_{x2} \cos \beta_1 + v_c)^2 = v_c^2 + \frac{v_{x1}^2 + v_{x2}^2}{2}$$
$$+ \tfrac{1}{2} v_{x1}^2 \cos(2\beta_1) + \tfrac{1}{2} v_{x2}^2 \cos(2\beta_2)$$
$$+ 2v_c(v_{x1} \cos \beta_1 + v_{x2} \cos \beta_2)$$
$$+ v_{x1} v_{x2}(\cos(\beta_1 + \beta_2)$$
$$+ \cos(\beta_1 - \beta_2)) \tag{5.36}$$

in which β_i is $\omega_i t + \varepsilon_i$; and ε_i is a (random) phase angle.

Clearly this (drift) expansion can be extended to comprise all frequencies ω_i in the random sea.

This expression shows that apparent wave force frequencies will have the original wave frequencies enhanced by the current, resulting in force components with frequencies equal to a difference, sum and double frequencies of the wave components.

Components containing the difference frequencies lead to long period forces which may be critical for rigid body modes of behaviour. The terms with $\omega_1 + \omega_2$ lead to high frequency forces, which may cause dynamic effects in bottom supported platforms.

The non-linearity in Morison's equation may be linearized to facilitate efficient response analysis. Linearization may, for instance, be achieved:

- deterministically by requiring that the same energy be dissipated per wave cycle for the linear and non-linear model;
- stochastically by assuming that the particle velocity follows a Gaussian distribution and finding the linearization that minimizes the expected mean square error.

Linearization by consideration of energy dissipation for a single wave-component corresponds to taking the first term in the Fourier expansion, eqn (5.32), and ignoring high-frequency terms (see e.g. chapter 2, Almar-Næss, 1985). Stochastic linearization of $K_D v(t)|v(t)|$ yields $K_D(\sqrt{8/\pi}\sigma_v)v(t)$ where σ_v is the standard deviation of $v(t)$ (see e.g. Leira, 1987). Stochastic linearization yields accurate estimates of loads when used to determine response variance, which is relevant for fatigue analysis, but needs to be used with caution in estimation of extreme values.

As mentioned in section 5.2.3, particle velocities and accelerations are Gaussian processes in the time domain. In the frequency domain the kinematics is described by spectral densities (e.g. eqn (5.23)). The forces (eqn (5.28)) on slender members may also be expressed in the frequency domain by the cross-spectral density of the load intensity at two locations m and n. This topic is thoroughly treated by Borgman (1972). It is seen that the spectral density has peaks at the wave frequency as well as at multiples of the wave frequency as displayed by the Fourier expansion, eqns (5.34), (5.35).

Clearly, a linearization of eqn (5.36), which yields:

$$q_{D_L} \propto c(v_{x1} \cos \beta_1 + v_{x2} \cos \beta_2 + v_c) \tag{5.37}$$

where c is a constant, ignores the higher order components.

Large volume structures

Higher order terms in the potential theory to account for finite wave elevation also cause time-variant sum and frequency forces on large volume structures in (irregular) waves. For instance, the second order term of the surface elevation (in e.g. eqn (5.16) for the deterministic wave) and the quadratic velocity terms in Bernoulli's equation (eqn (5.2e)) based on the first order potential will contribute second order force components. The term in Bernoulli's equation is somewhat analogous to the velocity squared term in the Morison equation.

The purpose of the higher order theory is to approximate more accurately the boundary conditions (i.e. the zero normal flow condition at the instantaneous position of the body and the pressure condition at the free boundary). Such higher order excitation forces are commonly derived by a perturbation method, with the following assumptions: variables, x such as wave height, velocity potential, dynamic pressures and motions of the structure are expanded in a series of a small perturbation parameters α

$$x = x^{(0)} + \alpha x^{(1)} + \alpha^2 x^{(2)} \tag{5.38}$$

where $x^{(0)}$ represents the stillwater condition and $x^{(1)}$ corresponds to the first order (linear) approximation. It is noted in particular that the different terms of quadratic velocity potential $\Phi^{(2)} = \Phi_0^{(2)} + \Phi_7^{(2)} + \Sigma \Phi_j^2$ are quadratic functions of the first order potentials $\Phi_0^{(1)} = \Phi_0$, $\Phi_7^{(1)}$ and $\Phi_j^{(1)}$. Each of the quadratic potentials must satisfy the Laplace differential equation and the boundary conditions at the free surface sea bottom and far field mentioned in Section 5.2.2.

First order wave forces are then expressed by first order velocity potentials and first order motions, taken care of by the equation of motion. Second order wave forces can then be explicitly determined on the basis of the second order velocity potentials and first order potentials as well as hydrodynamic pressures and motions.

Eatock Taylor and Hung (1987) calculated numerically the complete second order forces on a cylinder.

Non-linear (second and higher order) wave forces generally are an order of magnitude less than the first order (linear) forces. However, if the period of the wave force coincides with a natural period, the effect of such forces can be large.

High-frequency horizontal forces on towers made up of slender members and vertical forces on tension leg platform hulls may be of importance. Low-frequency horizontal (and vertical) forces may be of importance to the motions of floating structures and tension-leg platforms.

Ringing loads

Steep, high waves encountering structural components extending above the still water level may cause non-linear transient loads and load effects. Figure 5.9 shows a measured irregular wave profile and the corresponding horizontal forces for a short time sample involving a steep wave. It is observed that a transient high frequency load occurs. Its amplitude is approximately 20 per cent of the steady state amplitude. Structural responses to these actions may be dynamically amplified and cause increased extreme response (ringing). Such transient non-linear actions may be important for structures consisting of large diameter shafts and having natural periods in the range of 2 to 8 sec and started to receive serious attention in connection with monotower, gravity base and tension leg platforms at the beginning of the 1990s. Ringing loads depend on the wave shape and particle kinematics close to the wave surface and are highly non-linear, and it is generally difficult to distinguish impact/slamming phenomena from higher order

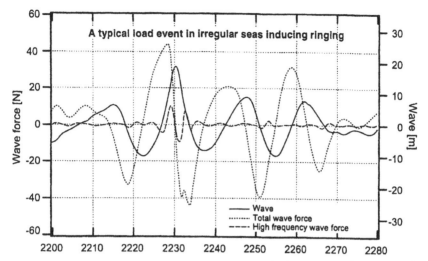

Figure 5.9 Measured horizontal force on a vertical cylinder piercing the wave surface (Krokstad *et al.*, 1996).

ones. However, it is agreed that the ringing load is an inertia-type loading that can be described by potential theory.

Various models for ringing loads have been proposed. They may be divided into: slender body and diffraction theories. The simplest slender body theory is based on Morison's equation and incident wave kinematics. Wave diffraction due to a relatively large diameter structure, may be accounted for by using the McCamy – Fuchs theory (see e.g. Farnes *et al.*, 1994). Rainey (1989) improved Morison's equation for the submerged part of the cylinder as well as a particular slamming term for the region where the free surface intersects the cylinder. This slamming term appears like the drag force term (eqn (5.27)); however, the coefficient C_D is replaced by a coefficient which depends on wave steepness. Kinematics has primarily been calculated by the Wheeler modification of Airy theory, but other theories, such as the second order irregular wave kinematics model, may be applied. Figure 5.10 shows how higher order wave components can affect the shape and especially the local steepness of the wave. While the second order component can increase crest height by 10–15 per cent, the effect of third order components seems to be less.

However, since contributions from the second order potential $\Phi^{(2)}$ are ignored, the accuracy of the slender body theory is limited.

For this reason efforts have been devoted to developing consistent ringing load models based on diffraction theory. Faltinsen *et al.* (1996) (FNV method) included non-linear contribution to the linear diffraction potential (MacCamy–Fuchs theory) and force components up to and including fifth order effects. A further development is reported by Krokstad *et al.* (1996) and Marthinsen *et al.* (1996). In this approach loads from a complete second order diffraction theory are combined

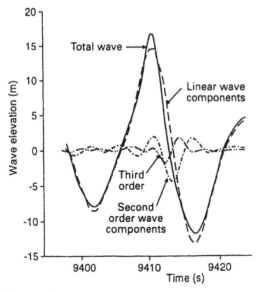

Figure 5.10 Contributions from linear, second and third order wave components to wave elevation of a steep wave (Stokka, 1994).

with third order loads from the FNV theory. The method yields an improved representation of second order forces. Although diffraction models yield estimates closer to model test results than Morison-type formulations, currently available methods are generally amenable to screening analysis of the ringing phenomenon. In this connection it is an advantage that diffraction theories seem to yield conservative load estimates. For platforms with multiple columns, the phenomenon is today best quantified by model tests.

5.4 CALCULATION OF WAVE LOAD EFFECTS

5.4.1 Dynamic models

Various dynamic models of marine structures, like those in Figure 5.1, are envisaged in this section, ranging from simple 'stick models' as shown in Figure 5.11 to sophisticated finite element models of the structure and foundation.

Excitation is due to wave loading and the structure, soil and water may contribute stiffness, mass and damping, depending on the support conditions of the structure.

Global models of, for example, platforms and buoyant bridges are commonly based on beam models. However, the caisson of gravity platforms is usually modelled as a rigid body. The P-Δ effect for platforms with 'large' motion displacement could be taken into account by linearized negative springs. Possible catenary

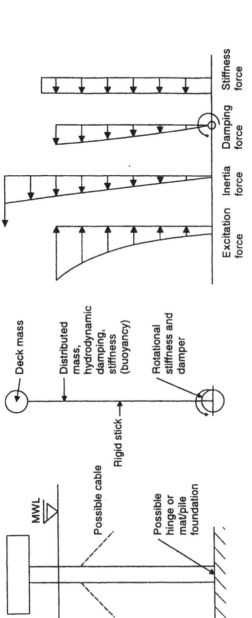

Figure 5.11 Simplified dynamic stick model of tower platforms.

mooring lines may be modelled by a simple spring-damper, or by a finite element model of the line. Particular attention to the modelling of the leg-deck connections (SNAME, 1994) in jack-ups is required. Pile foundations may be modelled by beam models and taking the interaction between pile and soil into account by a continuum model of the soil; or by representing the pile–soil behaviour by a simple spring-damper. A simplified multi-degree of freedom boundary element method of the pile and soil, referred to as disk-cone model, has proven to be computationally efficient (Wolf, 1994 and Emami Azadi, 1998). Mat or gravity foundations can normally be well represented by a spring-damper model.

While the structure is normally assumed to have linear elastic properties when load effects for component ultimate and fatigue limit states are determined, it may be necessary to account for non-linearity in soil behaviour. However, when dynamic behaviour up to system collapse is to be determined, non-linear material and geometric effects both of the structure, foundation and soil would normally be required.

Mass is contributed by structural and contained mass as well as the added hydro-dynamical mass. For a slender cylinder the latter mass is usually taken to be that of the displaced water. The added mass for large volume structures (e.g. caissons of floating gravity structures, floating bridges) has to be determined by potential theory for the relevant modes of behaviour. Particular attention should be paid to structural components which are close to the surface, relative to their size. Ogilvie (1963) and Vugts (1970) give data for an infinite cylinder moving horizontally at a certain distance below the water surface. Yeung (1989) determined added mass for a vertical cylinder, and an infinite cross-section shaped like a ship moving in the water surface. The added mass is frequency dependent.

Damping may be contributed by the structure, water and soil (rock) and is subject to significant uncertainties. Structural damping (Barltrop and Adams, 1991) in a welded steel structure may be of the order of 0.2–0.5 per cent of critical, and for concrete structures which are stressed so that microcracks occur, it may be of the order of 0.5–1.5 per cent (Langen *et al.*, 1997). Structural damping of platforms or submerged bridges may be about 1 per cent with pure structural modes of vibration.

Hydrodynamic damping stems from generation of waves (radiation damping) as well as from friction drag damping. The first source is determined from potential theory and is given for the special cases mentioned above in Ogilvie (1963), Vugts (1970) and Yeung (1989); it exhibits strong dependence on frequency and submergence. For significant drag damping to occur, vortex shedding must take place. The drag damping will be small if the KC number is below, say, 2. Hence, drag damping will be small for large diameter vertical columns in platforms and submerged bridges. The corresponding damping ratio may be less than 0.1 per cent. Similarly, potential (radiation) damping is found to be relatively small compared with drag damping for platform structures consisting of slender members. For floating bridges wave difference frequency (slow drift) excitation may be of importance. Both drag damping and second order (slow drift) potential damping are quite small at the excitation frequencies.

If the soil or rock is activated during the vibration, it will contribute radiation and hysteretic (material) damping. Soil damping for (embedded) plate and pile foundations is discussed, for example, by Moan *et al.* (1976), Barltrop and Adams (1991), Wolf (1994). Soil damping, especially in rocking motion, is frequency dependent. If a non-linear soil model is used, the hysteric damping will be implicitly included in the analysis.

If the damping of the structure or the soil is given with reference to a pure structural or foundation mode of vibration, the damping should be appropriately modified when it is included in an interaction mode. It is, for instance, shown by Moan *et al.* (1976) that the contribution from the structural damping ratio (ξ_s) to the damping ratio ξ for the first mode of a simple flexible tower rocking on soil

$$\xi = (\omega/\omega_s)^3 \xi_s \tag{5.39}$$

where ω and ω_s are the natural frequencies of a tower on flexible and rigid soil, respectively.

Similarly, the damping ratio (ξ_{wet}) referred to a wet system (including the effect of added mass) can be expressed by the damping ratio of the structure (ξ_{dry}) as follows

$$\xi_{wet} = \xi_{dry}(m^*_{dry}\omega^*_{dry}/m^*_{wet}\omega^*_{wet}) \tag{5.40}$$

where m^* and ω^* are the (generalized) mass and natural frequency of the relevant mode.

Stiffness is also contributed by the structure, water and the soil (rock). Linear elastic structural models are usually applied, except for possible catenary mooring lines. Water provides buoyancy that will influence the stiffness of a bridge supported by pontoons, but would be negligible for bottom supported platforms. The soil is of importance for bottom supported platforms and may be modelled by equivalent linear properties or by a more refined non-linear material model. Even if soil stiffness properties are frequency dependent the low frequency of water loading implies that the dynamic stiffness is close to the static values.

The mass, damping and stiffness properties presented above refer to ultimate and fatigue limit state criteria, based primarily on linear elastic global models. However, if prediction of the ultimate global capacity is required in connection to survival check in accidental limit states, models that account more properly for non-linear effects need to be applied.

Under such circumstances framed structures and possible piles are modelled with beam elements including strain hardening non-linear material and geometrical effects. Plasticity may efficiently be incorporated with plastic hinges. Pile–soil interaction may be modelled by non-linear spring elements along the piles with cyclic (hysteric) behaviour. Structural damping for elastic behaviour and radiation damping in the soil should be explicitly incorporated, while hysteric loss in the structure and soil are implicitly included by this model. Further details about this non-linear modelling may be found in Stewart (1992), Søreide and Amdahl (1994), Hellan (1995), Nadim and Dahlberg (1996) and Emami Azadi (1998).

In particular the assessment of damping and soil stiffness is susceptible to significant uncertainties. Hence, in-service measurements are useful to justify the assumptions made in design analyses. Hoen *et al.* (1991, 1993), for instance, show that the (total) modal damping ratios are about 2 per cent for the first three modes of gravity platforms. Karunakaran *et al.* (1997) found total damping ratios between 0.6 and 1.5 per cent for a jacket with natural periods around 1 sec. These references also provide information about assumed versus observed soil stiffness.

5.4.2 Equations of motion

Equations of motion may be formulated in the time or frequency domain (see e.g. Clough and Penzien (1993)). The choice of formulation depends especially on possible

- frequency dependence
- non-linearities

of the dynamic properties. A fairly general version of the dynamic equations of motion (in the time domain) can be written in matrix form in terms of the displacements $\mathbf{r}$ and their time derivatives $\dot{\mathbf{r}}$ and $\ddot{\mathbf{r}}$, as both the mass and damping matrices $\mathbf{M}$ and $\mathbf{C}$ are functions of time:

$$\int_{-\infty}^{\infty} \mathbf{M}(t - \tau)\ddot{\mathbf{r}}(\tau)\,d\tau + \int_{-\infty}^{\infty} \mathbf{C}(t - \tau)\dot{\mathbf{r}}(\tau)\,d\tau + \mathbf{K}\mathbf{r}(t) = \mathbf{Q}(\mathbf{r}, \dot{\mathbf{r}}, \ddot{\mathbf{r}}, t) \qquad (5.41)$$

Non-linearities in $\mathbf{r}$, $\dot{\mathbf{r}}$ and $\ddot{\mathbf{r}}$ are assumed to be small and are treated in the excitation load, $\mathbf{Q}(\cdot)$. The convolution integrals are due to the possible frequency dependence of mass and damping properties. For the case when $\mathbf{Q} = \mathbf{Q}(t)$, eqn (5.41) follows as an inverse Fourier transform of the frequency domain equation (5.45) given later.

The mass, damping and stiffness matrices are made up by contributions from the structure (st), water (w) and soil(s). In the frequency domain, $\mathbf{M}$, $\mathbf{C}$ and $\mathbf{K}$ are:

$$\mathbf{M} = \mathbf{M}^{(\text{st})} + \mathbf{M}^{(\text{w})}(\omega) \qquad (5.42\text{a})$$

$$\mathbf{C} = \mathbf{C}^{(\text{st})} + \mathbf{C}^{(\text{w})}(\omega) + \mathbf{C}^{(\text{so})}(\omega) \qquad (5.42\text{b})$$

$$\mathbf{K} = \mathbf{K}^{(\text{st})} + \mathbf{K}^{(\text{w})} + \mathbf{K}^{(\text{so})} \qquad (5.42\text{c})$$

The contribution to the stiffness by water is due to hydrostatic effects.

Since $\mathbf{M}(t)$ and $\mathbf{C}(t)$ in eqn (5.41) for $\tau > t$ and $\tau < 0$ are zero, the integration limit $(-\infty, \infty)$ in that equation could be changed to $(0, t)$.

If the properties are frequency independent, eqn (5.41) takes on the well known form

$$\mathbf{M}\ddot{\mathbf{r}} + \mathbf{C}\dot{\mathbf{r}} + \mathbf{K}\mathbf{r} = \mathbf{Q}(t) \qquad (5.43)$$

which is much simpler to solve. Equation (5.43) will be a reasonable approximation if one of two conditions are fulfilled:

(1) the retardation time for mass and for damping are so short that the time dependent properties are Dirac delta functions;
(2) the response is narrow-banded.

In practice the frequency dependent mass and damping properties are chosen to be the values corresponding to the peak frequency ω_p of the wave spectral density. Langen (1981) found that the error in the response of floating bridges by approximating mass and damping by their values at the wave spectral peak was less than 5 to 6 per cent.

The resulting equilibrium equation, eqn (5.43), is commonly written in incremental form for computational purposes:

$$\mathbf{M}\Delta\ddot{\mathbf{r}}(t) + \mathbf{C}\Delta\dot{\mathbf{r}}(t) + \mathbf{K}_I\Delta\mathbf{r}(t) = \Delta\mathbf{Q}(t) \qquad (5.44)$$

where $\mathbf{M}$, $\mathbf{C}$ and $\mathbf{K}_I$ are the incremental mass, damping and stiffness matrices valid within each time step, $\Delta\ddot{\mathbf{r}}(t)$, $\Delta\dot{\mathbf{r}}(t)$, $\Delta\mathbf{r}(t)$ and $\Delta\mathbf{Q}(t)$ are the corresponding increments of response acceleration, velocity, displacement and load vector.

It has been found convenient to cast the dynamic equilibrium equations in a form such that the coefficients on the left hand side are kept constant and the non-linearities are transferred to the right hand side.

Non-linearities in load processes (e.g. due to the relative motion term of Morison's equation and variations in added mass) are conveniently handled on the right hand side, and calculated by using the structural velocity in the previous time step. This approach is acceptable when Δt is less than 0.25 sec, but may not be so if larger time steps are applied. Also, the effect of non-linear springs due to a catenary mooring system may be handled on the right hand side. However, the added mass term up to the MWL, should be treated on the left hand side of the equation. Otherwise, many iterations may be required to have convergence, or no convergence at all may be experienced.

The Newmark $\beta = \frac{1}{4}$ method and time steps $\Delta t = 0.2$–0.25 are commonly used to determine load effects involving loads with periods with 3 sec or more (e.g. Langen, 1981; Mo, 1983; Farnes, 1990; Karunakaran, 1993). Alternatively, an improved Newmark method, the so-called α-HHT method (Hilber et al. 1978), is applied. Equilibrium iterations may be necessary to prevent drift-off in the solution.

If non-linear structural or pile–soil interactions are included, the relevant parts of the system matrices should be updated. A predictor–corrector approach, based on the α-HHT method, can be adopted to prevent large drift-off from the yield surface in elastoplastic problems.

An alternative approach for systems with linear and linearized system matrices is the frequency domain approach, which is very efficient for representation of the frequency dependent mass and damping terms. The transformed equilibrium equation then becomes

$$(\mathbf{K} + i\omega\mathbf{C}(\omega) - \omega^2\mathbf{M}(\omega))\mathbf{r}(\omega, \theta) = \mathbf{Q}(\omega, \theta) \qquad (5.45)$$

where $\mathbf{Q}(\omega, \theta)$, $\mathbf{C}(\omega)$ and $\mathbf{M}(\omega)$ are the Fourier transforms of the linearized version of the time domain counterparts in eqn (5.41). The first term may be regarded as a complex transfer function relating force amplitude to wave amplitude $\zeta(\omega, \theta)$ for a harmonic wave with frequency ω and a main wave direction θ.

5.4.3 Time domain analyses

Time Domain Analysis (TDA) is only of interest when, for example, non-linearities make a linearized frequency domain approach inaccurate or when a Frequency Domain Approach (FDA) which incorporates the non-linear features is very time consuming. The TDA is not attractive compared with the FDA when the behaviour is linear. This is because it implies a sampling uncertainty which will not be present in the FDA. Moreover, TDA is more time consuming than the FDA especially when frequency dependent dynamic properties need to be accounted for according to eqn (5.41). TDA may be performed with deterministic or probabilistic models of wave loading, as further discussed in Section 5.4.5.

The present discussion refers to TDA of systems with non-linear behaviour subjected to stochastic loading. In general, the load effects need to be calculated for all or representative sea states over a long term period for each sea state described by a wave spectrum (eqns (5.19) and (5.20)). Equation (5.43) is then solved by applying a number of load process samples which are generated by Monte Carlo simulation.

For short-crested seas, the sea-elevation process at a location $\mathbf{x} = [x_1, x_2]^T$ can be approximated by a discrete sum as

$$\zeta(\mathbf{x}, t) = \sum_{k=1}^{N_1} \sum_{i=1}^{N_2} a_{ik} \cos[\omega_i t - \kappa_i(x_1 \cos \theta_k + x_2 \sin \theta_k) + \varepsilon_{ik}] \tag{5.46}$$

where a_{ik} is the amplitude of frequency component i with direction θ_k; κ_i is the wave number corresponding to frequency ω_i. This amplitude is here taken as a deterministic value from the autospectral density and spreading function of a given state. The frequencies and directions are equidistant between specified upper and lower limits, while the phase angles ε_{ik} are independent and random with a uniform distribution between 0 and 2π. Expression (5.46) is effectively evaluated by the FFT technique (see e.g. Newland, 1984). An improved simulation procedure, especially for problems where subharmonics are of concern, may be obtained by taking a_{ik} as a Rayleigh distributed variable in ω, with a standard deviation of $\sqrt{2S_\zeta(\omega_i, \theta_k) \Delta\omega \Delta\theta}$.

Expressions similar to eqn (5.46) also hold for water particle kinematics (i.e. velocity and acceleration). The modifications required are introduction of a proper depth attenuation factor pertaining to a specific wave theory. Furthermore, phase shifts of the $\cos(\cdot)$ argument must be introduced to account for differentiation with respect to time. The hydrodynamic force time series are then obtained (e.g. by application of Morison's equation). The response is computed in the following manner:

- generate time series for the wave kinematics at discrete points along the structure by FFT;
- calculate corresponding loads by the Morison equation, and equivalent nodal forces;
- perform step by step integration of dynamic equations of motion, using methods mentioned in Section 5.4.2.

Finally, having obtained the response histories, and properly eliminated spurious start transients, statistical inference and estimation of relevant response quantities (variances, extremes) can be carried out. Filtering of the response may be considered especially in cases where the wave loading causes a combination of steady-state and transient response, and the transient part is deterministic when it has been initiated.

Limited sampling size introduces uncertainties, which may be quite significant for extreme values. It is particularly necessary to extrapolate from a limited sample size (say, half an hour) to extreme values in a short-term period of, say, 3–6 hours. Theoretical results are available for the distribution of individual response maxima of single cylinders with static response to non-linear loads associated with drag forces and surface elevation (e.g. Brouwers and Verbeek, 1983). For multi-DOF systems with dynamic behaviour only empirical studies of best fit can be made. A three-parameter Weibull distribution or a Weibull tail method (see e.g. Farnes, 1990) or Hermite models (see e.g. Winterstein, 1988) are frequently applied for this purpose.

5.4.4 Frequency domain analysis

FDAs outlined in the following are based on linearization of the system model. Like the surface elevation variation, load effect processes are then implicitly assumed to be Gaussian. The distribution of individual peaks or stress ranges, and expected maxima in a given short term period may then be achieved for single response quantities according to well known theory, see Chapter 10. Parameters, such as variance and spectral width, can be obtained from the response spectrum $S_r(\omega)$ for the response r in eqn (5.45).

The response spectral density matrix S_r of the response vector r may be obtained as

$$S_r(\omega) = H(\omega)S_Q(\omega)H^{*T}(\omega) \tag{5.47}$$

where $H(\omega) = (K + i\omega C - \omega^2 M)^{-1}$ and H^{*T} is the transpose of the complex conjugate of $H(\omega)$ and $S_Q(\omega)$ is the load spectral density matrix S_Q

$$S_Q(\omega) = F(\omega)S_\zeta(\omega) \tag{5.48}$$

where the hydrodynamic 'quasi'-transfer function $F(\omega)$ is:

$$F(\omega) = \int_{-\pi/2}^{\pi/2} Q(\omega, \theta)Q^{*T}(\omega, \theta)D(\theta)\,d\theta \tag{5.49}$$

and $S_\zeta(\omega)$ is the autospectral density of the sea elevation. $\mathbf{F}(\omega)$ is obtained by integrating the hydrodynamic load intensity over pairs of finite elements to produce pairs of nodal forces and adding them to the global spectrum matrix.

Having established response spectral densities, the corresponding variances (σ_{ri}^2) are computed by integration over the frequency range. The variance of a load effect s which is a linear combination of r_i and r_j in the $\mathbf{r}$ vector is

$$s = \alpha_i r_i + \alpha_j r_j \tag{5.50}$$

$$\sigma_s^2 = \alpha_i^2 \sigma_{ri}^2 + 2\alpha_i \alpha_j \rho \sigma_{r_i} \sigma_{r_i} + \alpha_j^2 \sigma_{r_j}^2 \tag{5.51}$$

where

$$\sigma_{r_i}^2 = \int_{-\infty}^{\infty} S_{r_i r_i}(\omega)\,d\omega; \qquad \rho = \int_{-\infty}^{\infty} S_{r_i r_j}(\omega)\,d\omega / \sigma_{r_i} \sigma_{r_j} \tag{5.52a, b}$$

5.4.5 Environmental load models for design calculations

A complete description of the dynamic load effect x under wave loading may be obtained by accounting both for short term variation of wave loading based on the Gaussian representation of the wave process with a mean direction θ as well as for long term variability (e.g. in terms of the joint density function $f_{H_{m0}, T_p, \theta}(b, t, \theta)$ of H_{m0}, T_p, θ) and possibly other sea-state parameters.

The long term distribution function $F_x(x)$ of x may be obtained by the total probability theorem as

$$F_x(x) = \int_{H_{m0}} \int_{T_p} \int_\theta w(b, t) F_{x|H_{m0}, T_p, \theta}(x|b, t, \theta) f_{H_{m0}, T_p, \theta}(b, t, \theta)\,db\,dt\,d\theta \tag{5.53}$$

where $F_{x|H_{m0}, T_p, \theta}(x|b, t, \theta)$ is the conditional distribution function for individual maxima for a given sea state; $w(b, t)$ is a weight function that accounts for the fact that the number of maxima per time unit vary and may be approximated by the exact formula for narrowband response:

$$w(b, t, \theta) = v_0^+(b, t, \theta) / E(v_0^+) \tag{5.54}$$

where $v^+(b, t, \theta)$ is the zero upcrossing frequency in a given sea state. By introducing the weight function, the probability distribution function $F_x(x)$ is defined as the number of maxima less than x divided by the total number of maxima.

In practice eqn (5.54) is calculated by using a discrete set of sea states and directions. Extreme value theory can then be applied to determine the characteristics of the maximum in a 100 year (with, say, $N = 5 \times 10^8$ maxima) or any other reference period.

In some cases we would need to know the joint probabilistic characteristics of several random response variables $\mathbf{x}$. For instance, the strength check of a steel member subjected to an axial load $N = x_1$ and a bending moment $M = x_2$ is accomplished with a non-linear interaction formula, for instance, of the type:

$$I(N, M) = N/N_u + M/[M_u(1 - N/N_E)] \leq 1 \tag{5.55}$$

where N_u and M_u are the ultimate strength under pure axial force and bending moment, respectively. N_E is the Euler buckling load.

The extreme values required for a design check can then most conveniently be based on the short-term statistics of individual maxima of the process $I(x_1, x_2)$ obtained by simulation (see Videiro and Moan, 1999).

Obviously, the long-term approach described by eqn (5.53) involves substantial effort when significant non-linearities need to be considered and alternative probabilistic load models therefore need to be used. Extreme load effects with, say, an annual exceedance probability of 10^{-2} may be estimated based on a limited set of sea states.

The overall aim of the design sea state concept is to estimate load (effects) corresponding to a prescribed annual exceedance probability (e.g. 10^{-2} or 10^{-4}) without having to carry out a full long-term response analysis.

An appropriate formulation of the design sea state concept is to use combination of significant wave height and spectral peak period located along an iso-probability density curve of $f_{H_{m0}, T_p}(h, t)$, denoted a contour line in the H_{m0} and T_p plane. Such contour lines can be established in different ways. The simplest way to establish the 10^{-2} contour line, is first, to estimate the 10^{-2} value of H_{m0} together with the conditional mean of T_p. The contour line is then estimated from the joint model of H_{m0} and T_p as the contour of constant probability density going through the abovementioned parameter combination. Alternative approaches to obtain the contour line are described by Haver *et al.* (1998). An estimate of the 10^{-2} action effect is then obtained by determining a proper extreme value for all sea states along the contour line and taking the maximum of these values.

If contour lines are used, the variability of the short-term extreme value needs to be artificially accounted for to obtain a proper long-term extreme value. This may be achieved in alternative ways, for example, by multiplying the expected maximum load effect calculated for a given sea state with a predetermined factor, typically in the range of 1.1 to 1.3, or by calculating the load effect as a predetermined, high fractile value, typically 90 per cent (see NORSOK N-003, 1999). Contour line methods, therefore, would have to be calibrated.

Alternatively, linearized analyses may first be applied to identify the range of sea states that contribute to the extreme value. Then, the complete non-linear short-term approach is used to determine the expected maximum for relevant sea states to obtain the largest one, which is taken to be the desired extreme value.

Instead of using a design sea state, a design wave specified by the wave height H, the wave period T and direction may be used to determine the extreme load effect. Load effects with, for example, annual exceedance probability of 10^{-2} can be determined in a simplified, conservative manner by the design wave approach for preliminary design of fixed platforms (NORSOK N-003, 1999). For fixed platforms with static behaviour, maximum action effects occur for the highest waves. The relevant wave height H_{100} is then taken to be that with the 10^{-2} exceedance probability. H_{100} may be taken to be 1.9 times the significant wave height H_{m0}, corresponding to an annual exceedance probability of 10^{-2}, as obtained from long-term statistics, when the duration of the sea state is 3 hours.

The period T used in conjunction with H_{100} should be varied in the following range:

$$\sqrt{6.5H_{100}} \leq T \leq \sqrt{11H_{100}}$$

The design wave to be used in detailed design for platforms in a relevant area should be established by special studies. If dynamic effects are moderate, they can be taken into account by applying equivalent inertia loads calibrated by stochastic analyses, as discussed in Section 5.5.2.

5.4.6 Stress ranges for fatigue design check

The repetitive load effects for fatigue limit states of welded structures are described by the distribution of stress ranges, S (see e.g. Almar-Næss, 1985). For basic (rolled, cast) material the joint distribution of mean stress and stress range is also required. The stress may be expressed by a nominal hot spot or hot spot notch value. The latter stress includes the notch effect of weld geometry. The fatigue strength is described by the number, N, of stress ranges, S, to failure (SN). It is crucial that the SN-curves applied are based on stresses that are defined in a consistent manner.

Fatigue design requires a description of the long term variation of local stresses due to wave – as well as possible sum-frequency wave actions, variable buoyancy, slamming – or current-induced vortex shedding. The effect of local (pressure) and global actions must be properly accounted for.

A simple expression for cumulative damage can be obtained by assuming that the SN-curve is defined by $NS^m = K$ and the number $n(s)$ of stress ranges is given by a Weibull distribution

$$n(S) = n_0 f_S(s) = n_0[(\gamma/\lambda)(s/\lambda)^{\gamma-1} \exp\{-(s/\lambda)^\gamma\}] \tag{5.56}$$

where n_0 = number of cycles as defined in relation to the stress range s_0, $\lambda = s_0/(\ln n_0)^{1/\gamma}$ is the scale parameter ($P[S \geq s_0] = 1/n_0$) and γ is the shape parameter

The damage D in a period τ with n_τ cycles is then

$$D = \sum_i \frac{n(S_i)}{N(S_i)} = \int_0^\infty n(S)S^m \, dS = \frac{n_\tau}{K}\left[\frac{s_0^\gamma}{\ln n_0}\right]^{m/\gamma} \Gamma(m/\gamma + 1) \tag{5.57}$$

Equations (5.56), (5.57) can be used to express the cumulative damage in a long-term period τ in two ways, namely by applying eqn (5.57) in conjunction with the stress range distribution for

- each sea state separately and summing up the contributions to the long-term D;
- the long-term period and determining D directly.

The narrow-band response in a single sea state (i) can be described by a Rayleigh distribution when the stress is taken to be twice the amplitude. This corresponds to a

Weibull distribution with $\gamma = 2$ and $\lambda^2 = 2\sigma_i^2$ (σ_i^2 is the variance of the response). In a long-term period τ, the number of (narrow-band) cycles associated with sea state i, is $n_i = \tau p_i \nu_i$, where p_i is the long-term probability and the number of cycles per time unit, respectively, of this sea state. Hence the cumulative damage in τ is

$$D = \sum_i \frac{\tau}{K} p_i \nu_i (\sqrt{2\sigma_i})^m \Gamma(m/2 + 1) \rho_i \qquad (5.58)$$

where ρ_i is a correction factor to account for wide-band and/or non-Gaussian load effects.

Stress ranges due to wide-band Gaussian or non-Gaussian response processes should be determined by an appropriate method of cycle counting (e.g. the rainflow method, see chapter 4 of Almar-Næss, 1985). Simple, conservative methods for combining high and low frequency responses may be applied. Fatigue damage may be calculated by assuming that the number of cycles is determined by the zero-upcrossing frequency and that the distribution of stress ranges follow a Rayleigh distribution. Wirsching and Light (1980) established an empirical correction to the fatigue damage determined by the narrow-band assumption. Extensive evaluations of various empirical, closed form methods for correcting the fatigue damage obtained by the narrow-band approach, show that Dirlik's (1985) method yields the best estimates. Jiao and Moan (1990) analytically derived a correction factor which yields reasonable estimates.

Leira et al. (1990) demonstrate that accurate fatigue estimates can be obtained for cases with non-linear effects by establishing a quasi-transfer function $H(\omega)$ that is used to calculate the response for all sea states, and is defined by $S_x(\omega) = |H(\omega)|^2 S_\zeta(\omega)$. $H(\omega)$ is obtained by calculating $S_x(\omega)$ using time domain samples of response for a sea state with spectral density $S_\zeta(\omega)$. The significant wave height of this sea state is given by

$$(H_{m0})^m_{eq} = \left(\sum_{i=1}^N p_i w_i^m H_{m0(i)}^m \right) \Big/ \left(\sum_{i=1}^N p_i w_i^m \right) \qquad (5.59)$$

where m is the exponent in the SN-curve, p_i is the relative frequency of a sea state number i and w_i is a weight function $w_i = C_D H_{m0(i)}/D_{av} + \pi C_M/\sqrt{2}$, D_{av} being the average diameter of loaded structural members. Even if the statistical uncertainty is less for the response relevant to fatigue than for extreme response, a sufficient sample to limit this uncertainty should be used. The load effects are described by a Weibull fit to the stress range distribution. The location and scale parameters are expressed by the standard deviation (Farnes and Moan, 1994). Hence, the relevant location and scale parameters for other sea states can be obtained when the variance is determined from the frequency domain results.

Equation (5.57) applied for a period τ with $n_\tau = n_0$ cycles is convenient as a basis for discussing the sensitivity of fatigue damage to various parameters. The shape factor γ of the Weibull distribution then depends on environmental conditions, relative magnitude of drag and inertia forces and possible dynamic amplification. For a quasi-static response in an extratropical climate, like the North Sea with 'continuous' storms, γ may be around 1.0 while γ may be as low as 0.4–0.6 for Gulf of Mexico platforms subject to infrequent hurricanes (Marshall and Luyties, 1982). For structures with predominantly drag forces, γ will be smaller than for pre-dominantly inertia forces. Note, for instance, that if u is Rayleigh distributed, $F_1 = c_1 u^2$ will follow an exponential distribution ($\gamma = 1$), while for $F_2 = C_2 u$, u will be Rayleigh distributed ($\gamma = 2$).

Dynamic effects may start to affect load effects relevant for fatigue when the natural period exceeds 2.0 sec. As illustrated by Marshall and Luyties (1982), in-creasing the natural period from 2 sec to 4 sec, may, for example, increase γ from 0.7 to 1.1 and from 0.9 to 1.3 for Gulf of Mexico and North Sea structures, respec-tively. The implication is a factor of the order of 10 on fatigue damage. Odland (1982) indicated similar results for jack-up platforms.

The stress range level that contributes most to D corresponds to the value that yields the maximum fatigue damage dD that is proportional to $f_s(s)s^m$. This stress range is found to be $s = s_0[(m + \gamma - 1)/(\gamma \ln n_0)]^{1/\gamma}$, implying that fatigue damage is primarily caused by stress ranges which typically are of the order of 10 to 20 per cent of s_0.

5.5 DYNAMIC ANALYSIS FOR DESIGN

5.5.1 Dynamic features of offshore platforms

The dynamic behaviour of platforms may be illustrated by considering two SDOF models with reference to Figure 5.11. In both models the loading is assumed to be proportional to the wave particle acceleration and hence written as:

$$q(z', t) = q_0(z') \sin(\bar{\omega}t) \tag{5.60}$$

where the co-ordinate z' refers to the seabed level and the mass consists of a deck mass M and a uniformly distributed mass m.

Otherwise, the two models have the following properties:

	Platform A (fixed platform)	Platform B (compliant tower)
Stiffness	Soil k_ψ	Uniform buoyancy
Damping	Soil c_ψ and uniform damper c	Uniform damper c

Model A will typically have a natural frequency above $\bar{\omega}$ while the natural frequency for Model B is below $\bar{\omega}$. The dynamic equation of equilibrium for this stick model is established by assuming that the motion is a rotation ψ about the support on the seabed. The horizontal displacement is then $r_x = z'\psi$. The equation of dynamic equilibrium is obtained by moment consideration and results in a SDOF version of eqn (5.43).

Based on the solution of eqn (5.43), with the integrated excitation force $Q = Q_0 \sin(\bar{\omega}t)$, the relative magnitude of generalized forces: inertia, damping and elastic/restoring force compared with the excitation force can be calculated.

Excitation force: $\quad Q = Q_0 \sin(\bar{\omega}t)$

Inertia force: $\quad Q_I = m\ddot{r}_x = -Q_0\beta^2\,\mathrm{DAF}\,\sin(\bar{\omega}t - \varepsilon)$

Damping force: $\quad Q_D = c\dot{r}_x = Q_0(2\xi\beta)\,\mathrm{DAF}\,\cos(\bar{\omega}t - \varepsilon)$

Elastic force: $\quad Q_S = kr_x = Q_0\,\mathrm{DAF}\,\sin(\bar{\omega}t - \varepsilon)$

where ε is the phase angle and

Dynamic Amplification Factor: $\quad \mathrm{DAF} = [(1 - \beta^2)^2 + (2\xi\beta)^2]^{-1/2}$

Frequency ratio: $\quad \beta = \bar{\omega}/\omega$ (where ω is the natural frequency)

The ratio of the maximum values of each force component is then:

$$|Q_0| : |Q_{I0}| : |Q_{D0}| : Q_{S0}| = 1 : (\beta^2\,\mathrm{DAF}) : (2\xi\beta\,\mathrm{DAF}) : \mathrm{DAF}$$

where Q_0, Q_{I0}, Q_{D0} and Q_{S0} denote the amplitude of the forces Q (excitation force), Q_I, Q_D and Q_S, respectively.

By assuming a frequency ratio of, say, $\beta = 0.3$–0.5 and 2.0–3.0 for platforms A and B, respectively and a damping ratio of $\xi = 0.02$ and 0.05, it is evident that elastic forces are predominant and balance excitation and inertia forces in platform A, while the dynamic equilibrium for platform B is achieved by inertia forces that balance excitation forces and elastic forces as illustrated in Figure 5.12.

The excitation and reaction forces in platform B yield a significantly smaller shear force and bending moment in the column than they do in platform A. It is noted in this connection that if the excitation force for platform B is balanced by the inertia force in the deck only, the bending action due to excitation forces will essentially be as for a column simply supported at both ends; this behaviour is illustrated in Blazy et al. (1971). On the other hand, the motions of platform B are much greater than those for platform A.

Various layouts of offshore platforms are envisaged. In Figure 5.13 the basic types of offshore platform are displayed. Typical natural periods for the structures are indicated in Figure 5.2. As a cantilever beam, the fixed tower will experience a significant overturning moment and shear force due to waves. Also, the fundamental natural period of vibration increases with increasing water depth and approaches the range of wave periods associated with significant energy. This fact implies that the response will be dynamically amplified to an increasing extent with increasing water depth. A better platform design for deep water is, therefore, to stiffen the tower as shown in Figure 5.13 by 'rigid' inclined members (which form a triangular truss). The bending moment in the central column will then be reduced, as the tower essentially becomes a beam supported at both ends. However, the inclined members also need to be sized adequately. Since these members are subject to significant lateral loads, the design may not be very cost-effective after all. A modified

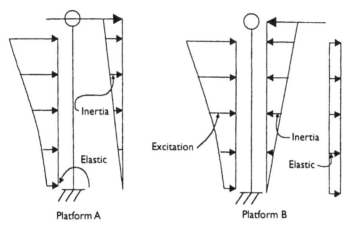

Platform A   Platform B

Figure 5.12 Schematic illustration of dynamic equilibrium.

tower may then be a compromise (see Figure 5.13). The maximum moment in the central tower is significantly reduced, and the 'truss' is less exposed to lateral loads, when it is located at a larger water depth.

An alternative approach would be to support the tower by catenary mooring (e.g. as in the guyed tower). The main tower is then let free to rotate on the seafloor and the low restoring force provided by catenary mooring makes the tower compliant (i.e. it follows the wave motion). This is a structure where excitation forces are balanced primarily by inertia forces as β would be larger than 1.0 (similar to platform B in Figure 5.12). The shear forces (and moments) along the tower become small because the dominant forces q_w and q_I (see Figure 5.13) counteract each other.

Catenary mooring may be partially or fully replaced by buoyancy, which typically is located in the upper part of the platform. Buoyancy contributes stiffness, mass and added mass and excitation forces. The buoyancy tank will commonly result in an increased fundamental natural period. The location of the buoyancy tank should be chosen so that the natural period of the second (flexural) mode (Figure 5.14) is not increased and that excitation forces for this mode are not increased.

The global flexibility of guyed and articulated towers are achieved by pivoting the base of the structure. In large water depths it may be possible to design a tower structure to be piled to the seabed and yet with sufficient bending flexibility to have the fundamental natural period, say, above 30 sec. Such platforms are called flexible towers (see e.g. Maus *et al.*, 1996).

Yet another alternative would be to use a TLP, which behaves like a pendulum where gravity is replaced by buoyancy. Their vertical mooring elements (tethers) are kept pretensioned by providing excessive buoyancy in the hull. The linearized stiffness for horizontal and vertical motion of a TLP is T/l and $EA/l + \rho g A_w$, respectively. T and EA are the total pretension and axial stiffness of the tethers, respectively and A_w is the water plane area. The corresponding natural periods

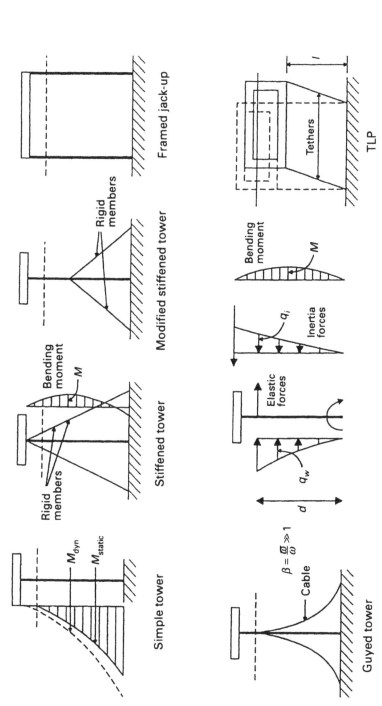

Figure 5.13 Platform concepts.

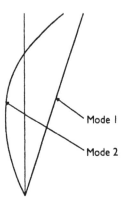

Figure 5.14 First two modes of vibration of compliant tower.

are above and below those of wave periods. Hence, the forces are inertia dominated in the horizontal direction and stiffness dominated in the vertical direction.

From the above discussion, it follows that wave-induced forces are smaller in compliant structures than structures 'rigidly' connected to the seafloor. On the other hand, the displacements/motions are larger in compliant platforms. While maximum displacements in extreme seas for 'fixed' platforms may be 0.5–1.0 m, they are of the order of the wave amplitude for compliant structures. This fact implies that the pipes (risers) from the deck of compliant platforms to the seafloor and subsoil reservoir must be carefully designed to avoid excessive stresses imposed by deformations.

It should be noted that wind forces may contribute significantly to the motions of (compliant) platforms with fundamental natural periods of 30 sec or more. Since the wind velocity spectrum contains energy in this range of periods, dynamic wind effects would also normally be of importance for such platforms. For compliant towers wind loads may also affect structural forces.

Among the dynamic features discussed earlier in this section, the natural period is particularly important. It is clearly desirable that natural periods for fixed platforms are as small as possible, while the natural period associated with 'rigid body' modes and flexural modes of the compliant towers (guyed tower, articulated towers, buoyant tower, flexible tower, etc.) should be as high and low as possible, respectively. During design the aim is normally to keep natural periods outside the range of 5 to 30 sec. This may be difficult, especially for flexural modes. If natural periods then exceed 5 sec, it is particularly important to reduce global wave loads in this range of periods.

Since the intensity of wave loads (e.g. according to Morison's equation) is largest in the surface zone where particle accelerations and velocities are largest, the loads and load effect may be minimized by making the structure in the 'splash zone' as wave transparent as possible. Moreover, the phase lag, for example, between the wave loads on various vertical members can be utilized to achieve cancellation of the total wave loads. For instance, for a regular wave of length λ cancellation occurs for members with a distance $\lambda/2$. For identical vertical members with a distance of 40 m, complete wave force cancellation in deep water then would occur for a wavelength of 80 m, or a wave period of 7.2 sec. This means that it is possible to obtain a very beneficial cancellation of forces for waves with a period close to the natural period of flexural vibration.

5.5.2 Calculation of extreme load effects for ULS check

Modern design codes require environmental design load effects to be determined based on characteristic loads which correspond to an annual exceedance probability of, say, 10^{-2} and appropriate load factors (API, 1993/1997; NORSOK N-003, 1999), using appropriate models of sea loading, structure and soil. Models of different refinement are used at different design stages – conceptual, pre-engineering and detailed engineering – with a balance of probabilistic and mechanics features.

The simple global behaviour (like 'stick' models of platforms) used in early design phases are refined towards detailed design. At this stage a detailed finite element model (Figure 5.15) of the structure is required to determine the relevant load effects for each structural component.

Design analyses for fixed platforms, like jackets, gravity platforms and jack-ups are commonly based on a regular (design) wave. When dynamic effects are of concern, an improved model – recognizing the stochastic features of waves – is necessary. It is then important to ensure that the refined model is properly based on current design practice. This means, for instance, that a stochastic analysis approach should be consistent with the design wave approach for structures with quasi-static behaviour. Moreover, dynamic effects should preferably be considered by their additional forces as compared with their quasi-static ones. To illustrate these two issues consider wave load effects obtained for a three-legged jack-up platform (Karunakaran *et al.*, 1994). With typical member diameters in the range of 0.15 to 0.8 m, drag forces predominate in extreme sea states. (This fact is observed in Table 5.1 which shows that load effects are proportional to C_D.)

The structural damping was taken to be 2 per cent and hydrodynamic drag damping was included by the relative velocity term. A non-linear soil-structure model for the spud can foundation was used. The first natural period is 5.7 sec at extreme load levels. In stochastic analyses C_D and C_M were taken to be 1.0 and 2.0, respectively. C_D for the design wave is 0.7. A Gaussian and non-Gaussian model for surface elevation are considered. The non-Gaussian model is based on a second order Stokes expansion. The kinematics is based on the Wheeler modification. The regular design wave is modelled by a Stokes fifth order theory.

The results in Table 5.1 show that the quasi-static and dynamic load effects increase by introducing second order (non-Gaussian) waves.

The comparison between quasi-static load effects obtained by the stochastic time domain and a design wave approach (in terms of the factor R_{QS}) shows the importance of consistent definition of the total procedure for calculating load effects. In

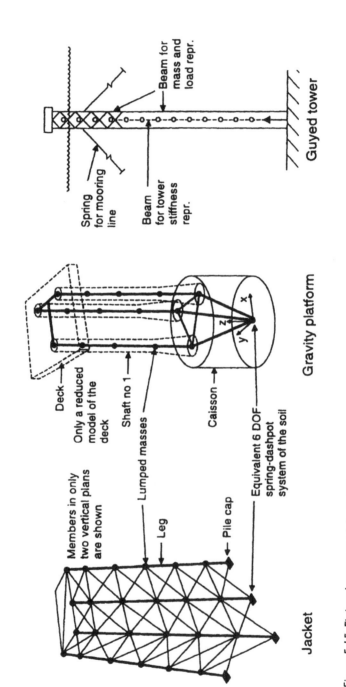

Figure 5.15 Finite element models of offshore structures.

Table 5.1 Extreme load effects in three-legged North Sea jack-up in a sea state with $H_{m0} = 14.8$ m and $T_p = 16$ sec, and design wave of $H = 27$ m and $T = 14.5$ sec (Karunakaren et al., 1994).

	Load effect a					
	Base shear		Overtuning moment		Deck displacement	
	R_{QS}	DAF	R_{QS}	DAF	R_{QS}	DAF
$C_D = 1.0$ in time domain analysis:						
Gaussian waves	1.07	1.14	1.04	1.29	1.04	1.25
Non-Gaussian waves	1.26	1.13	1.33	1.24	1.33	1.20
$C_D = 0.7$ in time domain analysis:						
Gaussian waves	0.76	1.14	0.74	1.29	0.74	1.25
Non-Gaussian waves	0.92	1.13	0.96	1.24	0.95	1.20

[a] For each load effect two characteristics are given:
  (1) the ratio R_{QS} of the expected maximum load effect obtained by stochastic analysis and the load effect obtained by design wave approach with no dynamics accounted for;
  (2) DAF obtained as the ratio of the expected maximum load effect obtained in stochastic analysis based on a dynamic and quasi-static model, respectively.

particular, a time domain stochastic approach based on Wheeler kinematics and $C_D = 1.0$ is seen to yield slightly larger load effects than the design wave approach that has been commonly used. A C_D of 0.8 used in conjunction with the second order theory would yield similar results.

Dynamic effects are measured by dynamic amplification factors. Dynamic effects can then be accommodated in the load effects used for design by:

• a stochastic dynamic analysis based on a refined dynamic model;
• a stochastic analysis based on a simplified dynamic model to calibrate inertia loading to be used with a refined structural model

The direct calculation of extreme dynamic load effects is based on the methods outlined in Section 5.4. To determine design values of load effects, load factors γ_Q are generally applied on loads while 'expected value' of mass, damping and stiffness properties are applied (e.g. ISO 2394, 1998). This approach causes a problem when the Morison equation with the relative velocity formulation (eqn 5.31) is used. This term implies both an excitation and a damping term. Application of load factors greater than 1.0 on the relative velocity term will then implicitly increase the damping beyond its 'expected value'. This problem can obviously be resolved by applying the load factor γ_Q on load effects rather than on loads.

As an alternative to this direct determination of stochastic dynamic load effects using the relevant refined dynamic model, a simplified dynamic model may be used to express the dynamic effects by equivalent inertia forces. A relevant model for a tower-type platform may then be a simple stick model to represent the mass, stiffness and damping properties. However, it is important to determine the loads by properly including the phase lag on different components. For this reason it is convenient to include elements in the model which are only used to introduce loads properly. This kind of model is indicated for the guyed tower in Figure 5.15. The first step is to determine the DAF, as the ratio between the dynamic extreme response and the quasi-static extreme response. It is important to determine these responses for representative sea states, and to perform time domain simulations such that statistical uncertainties do not affect the results too much (Karunakaran *et al.*, 1993). Dynamic amplification will vary along the structure. For a jacket with a fundamental natural period of about 4 sec, the DAF for the overall bending moment may vary between 1.2 and 2.5 from the seabed up to the mean water level. In particular, the quasi-static bending moment induced by wave loads in the structure above the sea surface is zero. The dynamic amplification factor $DAF = M_{dyn}/M_{stat}$, for that part of the platform will actually be infinitely large (see moment diagram indicated for the sample tower in Figure 5.13).

The dynamic effects are therefore, in general, most conveniently simulated by applying inertia loads (mass × accelerations) on the deck and tower structure masses. Since the masses are given, the acceleration field is tuned such that the DAF for the base shear and overturning moment are fairly accurately represented for the extreme wave condition.

Obviously, the method outlined in this section is expected to yield accurate estimates when the dynamic response is dominated by a single mode, the response is narrow-banded and the dynamic response is associated with wave periods well separated from those that cause quasi-static response. This approach is, for instance, adopted in design approaches for jack-ups (SNAME, 1994).

The behaviour of compliant towers is more complex since dynamic contributions stem from two modes, with natural periods on either side of the dominant wave excitation period. This means that the inertia forces in the first mode balance excitation forces while the inertia forces in the second mode add to the excitation. However, Vugts *et al.* (1997) show that fairly accurate results can be obtained by calibrating a quasi-static approach with inertia loads for this kind of platforms as well.

The magnitude of the load factor should reflect uncertainties involved in the determination of load effects (see e.g. Moan, 1995). It is noted that steady-state wave-induced drag loads normally are subject to more uncertainty than inertia loads. This is because the drag force is more empirical in nature and also because it is more critically dependent on the kinematics model for the splash zone. No design code currently reflects this difference in uncertainty level by load factors dependent on the relative magnitude of drag and inertia forces.

Ringing and other higher order wave loads are subject to even larger uncertainties. Uncertainties associated with lack of knowledge are often compensated by using conservative approaches. Actually load model uncertainties may be so large that experiments are required to determine the characteristic load effects, as discussed in Section 5.5.5.

When the inertia and damping forces are induced by the loading, uncertainties associated with these reaction forces add to those in the excitation forces. When dynamic effects are represented by an equivalent inertial load pattern as mentioned above, uncertainties may add to those in the wave loads themselves. The main source of uncertainty is associated with damping. API (RP2A-LRFD API, 1993) specifies an additional load factor γ_D on inertia forces, before the load factor $\gamma = 1.35$ for wave loads is applied on the excitation forces and factored inertial forces. The total load factor on dynamic load contributions is therefore about 1.7. This factor was determined (Moses, 1985) based on an estimate of the additional uncertainty associated with dynamic loads. This approach is limited to jackets.

No other codes for jackets, jack-ups and other fixed platforms include this kind of additional load factor γ_D. It is important to consider the load factor γ_D in view of the possible conservatism built into the procedure used to estimate load effects, and especially the damping model assumed. Extreme dynamic load effects in fixed platforms are sensitive to equivalent damping values below 1.0 per cent of the critical value (Karunakaran, 1993). By conservative estimate of the damping in that range, no additional load factor would be required. If the equivalent damping is more than 1 per cent, the sensitivity to damping is so small that no γ_D is required despite the large uncertainty in estimating the damping ratio. This is often the case in practice.

5.5.3 Calculation of stress ranges for FLS check

Fatigue design is commonly based on resistance data specified by SN-curves. In special cases, fracture mechanics approaches may be applied. Stress ranges are based on expected long-term distributions of stress ranges, without any load factor. Moreover, the design criterion is based on linear cumulative damage, such as the Miner–Palmgren law, typically allowing damage in the range of 0.1 to 1.0. The significant uncertainties in fatigue loads and resistance imply a high failure probability. Acceptable safety is hence ensured by a proper inspection, maintenance and repair strategy. For this reason simplified design analyses may also be justified.

Fatigue estimates may be based on alternative approaches – in a hierarchy of procedures with increasing accuracy and complexity. Here, three main alternatives are considered:

- Assume that stress ranges follow a two-parameter Weibull distribution, obtained by estimating s_0 corresponding to an exceedance probability of $1/n_0$; and assume γ according to guidance – including the effect of dynamics – mentioned in Section 5.4.6. Calculation of s_0 and selection of γ obviously need to be conservative.
- FDA for each sea state (i) to determine response variance and assume narrow-band response, implying Rayleigh distribution of stress ranges. Moderate non-linearities may be accounted for by determining a quasi-transfer function based on time domain analysis, or another linearization approach. Factors may be introduced to correct for wideband or non-Gaussian response.
- TDA combined with rainflow counting of cycles for a representative set of sea states that are found (e.g. by frequency domain analysis) to contribute most to the fatigue damage.

Screening in order to identify joints with high dynamic stresses and stress concentration, which require more detailed fatigue analyses, may be undertaken by using the first approach. s_0 may be based on the nominal member stress for the extreme event and an appropriate stress concentration factor, and the shape parameter γ could be obtained by general guidance.

Detailed fatigue analyses should be performed using conservative deterministic methods or frequency domain techniques and, in particular situations, by TDA. Stochastic approaches should be applied for dynamic sensitive structures. For linear systems, frequency domain techniques are efficient.

More complete time domain approaches may especially be necessary in case of strong non-linearities (e.g. associated with local splash zone behaviour), at least to calibrate simpler methods.

5.5.4 Non-linear system assessment for ultimate or accidental limit states

Current ultimate strength code checks of marine structures are commonly based on load effects (member and joint forces) that are obtained by a linear global analysis while resistances of the members and joints are obtained by experiments or theory which account for plasticity and large deflection. This methodology then focuses on the first failure of a structural component and not the overall collapse of the structure, which is of main concern in view of the failure consequences. The advent of computer technology and the finite element method have made it possible to develop analysis tools that include second order geometrical and plasticity effects and to account for possible redistribution of the forces and subsequent component failures until the system's collapse.

Ultimate strength analysis aims at providing a more realistic measure of the overall strength of a platform, by using methods to account for global and inelastic features (e.g. to represent redistribution of loads to alternative paths).

Initially such methods were developed for seismic analysis and for calculating the residual strength of systems with damage (e.g. according to the accidental limit state checks). More recently, such methods have also been applied to reassessment of ageing structures to determine the ultimate capacity of the intact system as well as the global strength after fatigue-induced fracture of members in connection with inspection planning.

Models which have been used to idealize structural members include phenomenological models and various finite element-type models (see e.g. Hellan *et al.*, 1994; Hellan, 1995; Nichols *et al.*, 1997). Cost-effective solutions are obtained by using large deformation theory for beam elements and special displacement functions (e.g. Livesley 'stability' functions) and concentrating the material non-linearities in yield hinges at predefined locations or at locations where maximum stress occurs. Yield hinge models are developed with different refinements, from yield hinges with zero extension along the element to models that account for the extension of the yield hinge; with elastic–perfectly plastic or gradual plastification

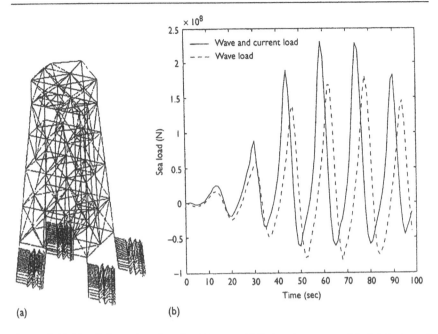

Figure 5.16 North Sea jack and sea load history. (a) Finite element model of eight-legged North Sea jacket; (b) sample of wave and current load history for cyclic analysis.

of the cross-section, strain hardening and the Bauschinger effect. The joint behaviour may be modelled by a plastic potential, with interaction between the axial force, in-plane bending and out of plane bending. Formulations have also been published that account for brace to brace interaction by adding 'beam' elements between the brace ends.

Fixed platform analyses are carried out by modelling the pile–soil behaviour by equivalent linear or non-linear concentrated springs or distributed springs along the piles, or by the continuum (finite element) model (Horsnell and Toolan, 1996; Lacasse and Nadim, 1996). As demonstrated, for example, by Moan *et al.* (1997) the choice of pile–soil model can affect the load distribution in the structure and, hence, the failure mode and corresponding ultimate strength. The most important issue is, of course, that a pure linear pile–soil model would not represent a possible soil failure and hence overestimate the system strength if the pile–soil is the critical part of the system. For the jacket in Figure 5.16(a) with plugged piles the pile–foundation is not critical. Yet the difference in jacket failure mode when using a linear instead of a non-linear model results in an ultimate load which is about 15 per cent smaller for the former case (Figure 5.17).

Determination of the global ultimate capacity by monotonically increasing wave loading has become a well established approach (see e.g. API RP2A (API, 1993/1997)).

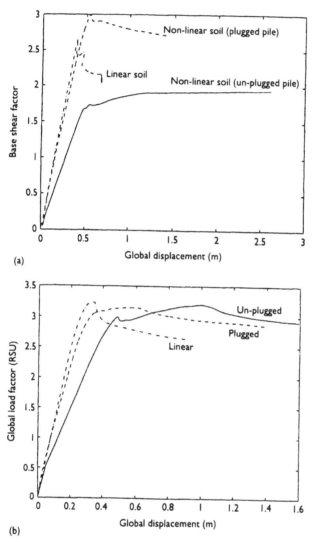

(a)

(b)

Figure 5.17 Static load–deformation characteristic of jacket for different pile–soil models. (a) Broad side loading; (b) end on loading (Moan et al., 1997).

Utilization of the true ultimate limit of the structural system may imply inelastic deformations. Cyclic wave or earthquake loading may cause degradation of the strength and lead to failure at load amplitudes which are less than for monotonically increasing loading (Hellan et al., 1991; Stewart et al., 1993).

The dynamic behaviour of fixed platforms under load levels that ensure linear elastic behaviour is stiffness dominated and inertia forces amplify the response as discussed in Section 5.5.1. However, as the ultimate strength of the structure as a

Table 5.2 Ultimate capacity of an eight-legged North Sea jacket with plugged piles (Figure 5.16a) using a non-linear pile–soil model (Moan *et al.*, 1997).

Limit state	Static load capacity factor: (100-year load)		Cyclic dynamic load capacity factor: (static capacity)	
	End on loading	Broad side loading	End on loading	Broad side loading
First member failure	1.94	1.79	–	–
Ultimate limit	2.89	2.73	1.12	0.96

whole is approached, the stiffness decreases and the system becomes inertia dominated. In this situation the external forces are partly balanced by inertia forces and the ultimate strength of the system increases (Stewart, 1992; Bea and Young, 1993; Schmucker, 1996; Emami, 1995). The ultimate strength of the platform in Figure 5.16(a) is calculated using the typical load history shown in Figure 5.16(b). It is seen from Table 5.2 that the ultimate capacity under dynamic cyclic loads is larger than the pushover capacity for end on loading, while the opposite occurs for broad side loading. This is because the inertial resistance effect near ultimate failure is larger for end on loading than for the broad side loading case due to a more ductile load–deformation characteristic (Figure 5.17).

Various simplified methods, based on the monotonic (static) load–deformation characteristics, to estimate the dynamic relative to the static global capacity are assessed and compared with results obtained from analyses of complete jacket–pile–soil systems by Emami Azadi (1998).

5.5.5 Ringing load effects for ULS design check

To illustrate the ringing phenomenon, consider the dynamic response of the monotower platform shown in Figure 5.18 which was analysed by Farnes *et al.* (1994). The platform/soil system has a fundamental natural period of 5 sec. Calculation of the higher order loading associated with surface elevation is very complex. Farnes *et al.* (1994) used a very simple, Morison-type approach, which included the MacCamy and Fuchs diffraction effects, using the Wheeler modification of wave kinematics. The contribution from the drag term was found to be negligible. The dynamic response was obtained by a time domain approach.

The wave profile, linear quasi-static overturning moment and the additional moment from non-linear wave loads are shown for an extreme, steep wave in Figure 5.18. The additional non-linear wave load has the shape of a double triangular impulse with the top close to the top of the wave where the linear wave loads are zero. The minimum of the impulse appears before the instant wave surface is equal to the MWL. The impulse has diminished when the linear wave load achieves its extreme minimum. The impulse is too small to give an extreme maximum load in the wave crest or increase the extreme minimum of the total

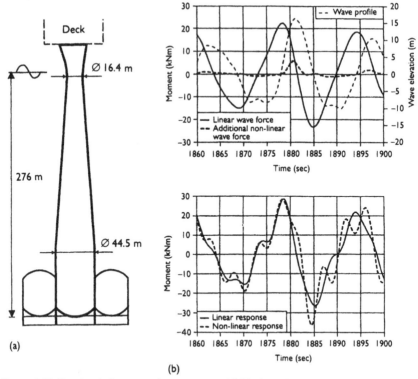

Figure 5.18 Ringing behaviour of monotower. (a) Platform layout; (b) wave profile and overturning moment (Farnes *et al.*, 1994).

load when it is superimposed on the linear wave load. Hence, small non-linear wave loads have no effect on the extreme value distribution of a quasi-static system and the response may be considered Gaussian.

The inertia of a dynamically responding system delays and amplifies the response. The delay and amplification are dependent on the ratio of the load period and the natural period. With a β of about 0.3 for the steady-state wave load, the DAF is about 1.1. The duration of the non-linear wave load impulse (which is about 5.5 sec in Figure 5.18(b)) is close to the natural period and implies a DAF of about 1.4 according to elementary results for a SDOF system. The response from the impulse is, hence, delayed and amplified more than the response from the linear load. The relative delay of the response from the impulse compared with the linear part shifts the minima responses from the two components closer together in time and the non-linear response contributes considerably to the total extreme minima. This is shown in Figure 5.18b. The response from the impulse will continue to oscillate with the natural period and it is rather unlikely, except for some particular wave periods, that a maximum in proceeding oscillation will increase the following maximum in the linear response. The distribution will, hence, be skewed upwave.

Significant impulses from the non-linear wave loading will, by their nature, only be induced by large waves. The response from the impulse will usually be damped out well in advance before arrival of the next wave that is large enough to induce a new impulse. The damping and scattered appearance of large waves indicate that the non-linear amplification is larger for the extremes than for the standard deviation of the response.

Farnes *et al.* (1994) also compared calculations with test results, based on a tuned dynamic model and wave surface elevation according to test samples. The accuracy was quite good. However, later work on other structures (gravity platforms, TLPs with more complex geometry) were not as encouraging. As mentioned in Section 5.3.5 there is no satisfactory theoretical method for calculating ringing loads. Design loads would therefore have to be obtained primarily by model tests, however, supported as much as possible by analyses.

Loads or load effects for final design should be established by recognizing that combined steady state and (transient) ringing effects with an annual exceedance probability of 10^{-2} and 10^{-4} for ultimate and accidental collapse limit states checks, respectively, are aimed at. Some guidance on the determination of ringing loads by theoretical and experimental methods are presented by DNV (1995). Ringing is known to be caused by random long crested waves in sea states corresponding to a steepness $s_p = 2\pi H_{m0}/g(T_p)^2$ of approximately 0.03–0.05 and $H_{m0} \geq 10$ m. It is important to have a sufficient number of ringing events within the time series of observations in order to establish estimates of extreme values. This fact implies long time series.

It is important to separate the steady state and ringing response by filtering and to compare the observations with predictions. Close agreement between experimental and theoretical values for steady state loads is expected. The analysis method for the ringing contribution may be applied to tune the analysis model, which can be used for other predictions and, hence, provide an additional reference for judging the uncertainties involved.

Load factors are applied to cover uncertainties in the environmental condition used, and load estimation procedure, and would, hence, depend on whether loads are obtained by tests or analyses or a combination. An uncertainty associated with selecting sea states from the long term data basis is that critical conditions could be omitted. To limit this uncertainty, analyses – despite their uncertainty – can be used without too much effort to screen important conditions. Uncertainties in model tests may, for instance, be concerned with scaling effects, model simplifications, non-uniformity of wave elevation across the basin, finite dimensions causing wave reflections the and data acquisition system. The main uncertainty in the theoretical model is concerned with the kinematics and hydrodynamic model, which in general have been calibrated against experimental results to some extent. Statistical uncertainties are present in both experimental and theoretical analyses.

Under these circumstances it is clearly not possible to set a general level of safety factors for the ringing component. It must rather be set on the basis of the possible conservatism built into the procedure and the resulting random uncertainties for the relevant case.

5.6 REFERENCES

Almar-Næss, A. (ed.) (1985) *Fatigue Handbook*, Tapir Publishers, Trondheim.

API (1993/1997). 'Recommended Practice for Planning, Designing and Constructing Fixed Offshore Platforms', API RP2A-WSD July 1993 with Supplement 1 with Section, 17.0, 'Assessment of Existing Platform', February 1997.

Barltrop, N. D. P and Adams, A. J. (1991) *Dynamics of Fixed Marine Structures*, Butterworth–Heinemann, London.

Bea, R. G. and Young, C. N. (1993) 'Loading and Capacity Effects on Platform Performance in Extreme Storm Waves and Earthquakes', Paper No. 7140 given at 25th Offshore Technology Conference, Houston.

Blazy, J. P. (1971) 'Full scale tests and mathematical model', Paper No. 1401 given at Offshore Technical Conference, Houston.

Borgman, L. E. (1972) 'Statistical models for ocean waves and wave forces', *Advances in Hydroscience*, Vol. 8, Academic Press.

Brouwers, J. J. H. and Verbeek, P. H. J. (1983) 'Expected fatigue damage and extreme response for Morison-type wave loading', *Applied Ocean Research* 5(3): 129–33.

Clauss, G., Lehmann, E. and Østergaard, C. (1991) *Offshore Structures*, Vol. 1, Springer-Verlag, Berlin.

Clough, R. W. and Penzien, J. (1993) *Dynamics of Structures*, McGraw Hill, New York.

Dirlik, T. (1985). 'Application of computers in fatigue analysis', Ph.D. thesis, University of Warwick.

DNV (1995) 'Preliminary Guidelines for Analysis of Ringing in Design of Offshore Platforms', Report No. 94-3536, Det Norske Veritas, Oslo.

Eatock Taylor, R. and Hung, S. M. (1987) 'Second order diffraction forces on vertical cylinder in regular waves', *Applied Ocean Research* 9(1): 19–30.

Emami Azadi, M. (1998) 'Analysis of static and dynamic pile-soil-jacket behaviour', Dr.Ing. thesis, MTA-rapport 1998: 121, Department of Marine Structures, NTNU, Trondheim.

Emani Azadi, M. R., Moan, T. and Amdahl, J. (1995) 'Dynamic effects on the performance of steel offshore platforms in extreme waves', *Proc. First Eurosteel Conf., Athens*, A. A. Balkema, Rotterdam, pp. 411–420.

Faltinsen, O. M. (1990) *Sea Loads on Ships and Offshore Structures*, Cambridge University Press, Cambridge.

Faltinsen, O.M., Newman, J. N. and Vinje, T. (1996) 'Nonlinear wave loads on a slender vertical cylinder', *J. Fluid Mech.*, **289**: 179–98.

Farnes, K. A. (1990). 'Long-term statistics of response in non-linear marine structures' Dr.ing.thesis, MTA-rapport 1990: 74, Division of Marine Structures, NTH (NTNU), Trondheim.

Farnes, K. A. and Moan, T. (1994) 'Extreme dynamic, non-linear response of fixed platforms using a complete long-term approach', *J. Applied Ocean Research* 15: 317–326.

Farnes, K. A., Skjåstad, O. and Hoen, C. (1994) 'Time domain analysis of transient resonant response of a monotower platform', paper given at BOSS '94 Proc. Pergamon Press, Oxford.

Gran, S. (1992) '*A Course in Ocean Engineering*', Elsevier, Amsterdam.

Gudmestad, O. T. (1993) 'Measured and predicted deepwater wave kinematics in regular and irregular seas', *J. Marine Structures* **6**: 1–73.

Hasselman, K. *et al.* (1973) 'Measurements of the wind-wave growth and swell decay during the Joint North Sea Wave Project (JONSWAP)', *Erganzungsheft zur Deutschen Hydrographischen Zeitschrift*, Reihe A, No. 12, Hamburg.

Haver, S. (1980) 'Analysis of uncertainties related to the stochastic modelling of ocean waves', Dr.ing.thesis, Report UR-80-09, Division of Marine Structures, NTH (NTNU), Trondheim.

Haver, S. and Moan, T. (1983) 'On some uncertainties related to the short-term stochastic modelling of ocean waves', *J. Applied Ocean Research* **5**(2): 93–108.

Haver, S., Gran, T. M. and Sagli, G. (1998) 'Long-term response analysis of fixed and floating structures', Statoil Report No. 97-4715, Stavanger.

Heideman, J. C. and Weaver, T. O. (1992) 'Static wave force procedure for platform design', paper givent at Civil Engineering in the Oceans Conference, Vol. V, pp. 496–517, College station, Texas, ASCE.

Hellan, Ø. (1995) 'Nonlinear pushover and cyclic analyses in ultimate limit state design and reassessments of tubular steel offshore structures', Dr.ing. thesis, MTA-report 1995: 108, Division of Marine Structures, NTH(NTNU), Trondheim.

Hellan, Ø., Skallerud, B., Amdahl, J. and Moan, T. (1991). 'Reassessment of offshore steel structures: shakedown and cyclic non-linear FEM analysis', paper given at 1st ISOPE Conference, International Society of Offshore and Polar Engineers (ISOPE), pp. 34–42.

Hellan, Ø., Drange, S. O. and Moan, T. (1994). 'Use of non-linear pushover analyses in ultimate limit state design and integrity assessment of jacket structures', paper given at 7th BOSS Conference. Pergamon Press, Oxford, Vol. 3, pp. 323–45.

Hilber, H. M., Hughes, T. J. R. and Taylor, R. L. (1978) 'Collocation, dissipation and 'overshoot' for time integration schemes in structural dynamics', *Earthq. Engng. and Structural Dynamics* **6**: 99–117.

Hoen, C. (1991). 'System identification of structures excited by stochastic load processes', Dr.ing. thesis, MTA-report 1991: 79, Division of Marine Structures, NTH (NTNU), Trondheim.

Hoen, C., Moan, T. and Remseth, S. (1993). 'System identification of structures exposed to environmental loads', paper given at EURODYN '93, Vol. 2, pp. 835–844. A. A. Balkema, Rotterdam.

Horsnell, M. R. and Toolan, F. E. (1996) 'Risk of foundation failure of offshore jacket piles', paper given at 28th Offshore Technology Conference, Houston, pp. 381–92.

ISO 13819 (1994/1997) 'Petroleum and Natural Gas Industries – Offshore Structures – Part 1: General Requirements', (1994), 'Part 2: Fixed Steel Structures', (1997), Draft C, International Organization for Standardization, London.

ISO 2394 (1998). 'General Principles on Reliability for Structures', Draft for approval, International Standards Organizationk London.

Jiao, G. and Moan, T. (1990) 'Probabilistic analysis of fatigue due to Gaussian load processes', *J. Prob. Engng. Mechanics* **5**(2): 76–83.

Jahns, H. O. and Wheeler, J. D. (1972) 'Long term wave probabilities based on hindcasting of severe storms', Paper No. 1590 given at Offshore Technology Conference, Houston.

Karunakaran, D. N. (1993) 'Nonlinear dynamic response and reliability analysis of drag-dominated offshore platforms', Dr.ing.thesis, MTA-report 1993: 96, Division of Marine Structures, NTH (NTNU), Trondheim.

Karunakaran, D., Spidsøe, N., Gudmestad, O. and Moan, T (1993). 'Stochastic dynamic time-domain analysis of drag-dominated offshore platforms', paper given at EURODYN '93, A.A. Balkema, Rotterdam.

Karunakaran, D. N., Spidsøe, N. and Haver, S. (1994) 'Nonlinear dynamic response of jack-up platforms due to non-Gaussian waves', paper given at 13th OMAE Conference ASME, Houston, TX, Vol. II, 85–92.

Karunakaran, D., Bærheim, M. and Leira, B. J. (1997) 'Measured and simulated dynamic response of a jacket platform', paper given at 16th OMAE Conference, ASME, Houston, TX, Vol. II, 157–64.

Kinsman, B. (1965) *Wind Waves*. Englewood Cliffs, N.J. Prentice-Hall Inc.

Krokstad, J. R., Stansberg, C. T., Nestegård, A. and Marthinsen, T. (1996) 'A new non-slender ringing load approach verified against experiments', paper given at 15th OMAE Conference, Florence.

Lacasse, S. and Nadim, F. (1996). 'Model uncertainty in pile axial capacity calculation's, Paper No. 7996 biven at Offshore Technology Conference, Houston.

Langen, I. (1981). On Dynamic Analysis of Floating Bridges', Report No. 81-1, Division of Structural Mechanics, NTH (NTNU), Trondheim.

Langen, I., Skjåstad, O. and Haver, S. (1997) 'Measured and predicted response of an offshore gravity platform', paper given at BOSS '97, Vol. 3, 213–27, Pergamon Press, Oxford.

Leira, B. J. (1987). 'Gaussian vector-processes for reliability analysis involving wave-induced load effects', Dr.ing.thesis, Report UR-87-57, Division of Marine Structures, NTH (NTNU), Trondheim.

Leira, B. J., Karunakaran, D. and Nordal, H. (1990). 'Estimation of fatigue damage and extreme response for a jack-up platform', *J. Marine Structures* 3: 461–93.

Longuet-Higgins, M. S. (1963) 'The effects of nonlinearities on statistical distributions in the theory of see waves', *J. Fluid Mechanics* 17.

MacCamy, R. C. and Fuchs, R. A. (1954) 'Wave Forces on Piles: A Diffraction Theory', U.S. Army Force of Engineers, Beach Erosion Board, Tech. Memo No. 69.

Marshall, P. W. and Luyties, W. H. (1982) 'Allowable stresses for fatigue design', paper given at BOSS '82, McGraw Hill, New York.

Marthinsen, T., Stansberg, C. T. and Krokstad, J. R. (1996) 'On the ringing excitation of circular cylinders', paper given at 6th ISOPE, Los Angeles, Vol. 1, 196–204.

Maus, L. D., Finn, L. D. and Danaczko, M. A. (1996). 'Exxon study shows Compliant Piled Tower cost benefits', *Ocean Industry*, 20–25.

McKay, M. D. et al. (1979). 'A comparison of three methods for selection values of input variables in the analysis of output from a computer code', *Technometrics* 21: 239–45.

Mo, O. (1983) 'Stochastic time domain analysis of slender offshore structures', Dr.ing.thesis, Report UR-83-33, Division of Marine Structures, NTH (NTNU), Trondheim.

Mo, O. and Moan, T. (1984). 'Environmental load effect analysis of guyed towers', paper given at 3rd OMAE Conference, Houston.

Moan, T. (1995) 'Safety Level Across Different Types of Structural Forms and Material – Implicit in Codes for Offshore Structures', SINTEF Report STF70 A95210, Trondheim (prepared for ISO/TC250/SC7).

Moan, T., Syvertsen, K. and Haver, S. (1976). 'Stochastic Dynamic Response Analysis of Gravity Platforms', Report SK/M33, Division of Marine Structures, NTH (NTNU), Trondheim.

Moan, T., Leira, B. J. and Olufsen, A. (1990) 'Stochastic dynamic analysis of marine structures subjected to sea loads', paper given at EURODYN '90, Trondheim, Norway, A. A. Balkema, Rotterdam, 661–9.

Moan, T., Emami Azadi, M. and Hellan, Ø. (1997). 'Nonlinear dynamic vs. static analysis of jacket systems for ultimate limit state check', paper given at International Conference on Advances in Marine Structures, DERA, Dunfirmline.

Moses, F. (1985) 'Implementation of a Reliability-based API -RP2 A Format', Report API-PRAC 83-22 (project report).

Nadim, F. and Dahlberg, R. (1996). 'Numerical modelling of cyclic pile capacity in clay', paper given at OTC Conference, Vol. 1, pp. 347–356.

Newland, D. E. (1984) *An Introduction to Random Vibrations and Spectral Analysis* (2nd ed.), Longman, London.

Nichols, N. W. Birkenshaw, M. and Bolt, H. M., (1997) 'Systems strength measures of offshore structures', paper given at 8th BOSS Conference, Pergamon Press, Oxford, Vol. 3, 343–57.

NORSOK N-003 (1999) *Actions and Action effects*, NTS, Oslo.

Odland, J. (1982). 'Response and strength analysis of jack-up platforms', *Norwegian Maritime Research* 10: 4.

Ogilvie, T. H. (1963) 'First- and second-order forces on a cylinder submerged under a free surface', *J. Fluid Mech.* 16: 451–72.

Olufsen, A., Karunakaran, D. and Moan, T. (1989) 'Uncertainty and sensitivity analysis of wave and current induced extreme load effects in offshore structures', paper given at 8th OMAE Conference.

Price, W. G. and Bishop, R. E. D. (1974) *Probabilistic Theory of Ship Dynamics*, Chapman and Hall, London.

Rainey, R. C. T. (1989) 'A new equation for calculating wave loads on offshore structures', *J. Fluid Mech.*, 204.

Sarpkaya, T. and Isaacson, M. (1981) *Mechanics of Wave Forces on Offshore Structures*, Van Nostrand Reinhold, New York.

Schmucker, D. G. (1996) 'Near failure behaviour of jacket-type offshore platforms in the extreme wave environment', Ph.D thesis, Report No. RMS-21, June 1996, Stanford University, CA, USA.

Sigbjørnsson, R. (1979) 'Stochastic theory of wave loading processes', *Engineering Structures* 1(2): 58–64.

SNAME (1994). 'Site specific assessment of mobile jack-up units: (a) Guideline (August 1991); (b) Recommended Practice (May 1994); (c) Commentaries to the Recommended Practice (May 1994)', *Technical & Research Bulletin*, 5-5A (Society of Naval Architects and Marine Engineers, New York).

Stewart, G. (1992) 'Non-linear structural dynamics by pseudo-force influence method, Part II: Application to offshore platform collapse', paper given at 1st ISOPE Conference, San Francisco, CA.

Stewart, G., Moan, T., Amdahl, J. and Eide, O. (1993) 'Nonlinear reassessment of jacket structures under extreme storm cyclic loading, Part I: Philosophy and acceptance criteria', paper given at 12th OMAE Conference, ASME, New York, 492–502.

Stokka, T. (1994). 'A Third Order Wave Model', DNV Research Report No. 93-2071, Det Norske Veritas, Høvik.

Søreide, T. H. and Amdahl, J. (1994). 'USOFS – A Computer Program for Ultimate Strength Analysis of Framed Offshore Structures', Theory Manual, Report No. STF71 A86049, SINTEF Structural Engineering, Trondheim.

Torsethaugen, K. (1996). 'Model for a Double Peaked Wave Spectrum', SINTEF Civil and Environmental Engineering, Report No. STF22 A96204, SINTEF Structural Engineering, Trondheim.

Videiro, P. M. and Moan, T. (1999) 'Efficient evaluation of long-term distributions', paper given at 18th OMAE Conference, St Johns, Newfoundland.

Vinje, T. and Haver, S. (1994) 'On the non-Gaussian structure of ocean waves', paper given at BOSS '94 (MIT, Cambridge, MA), Pergamon Press, Oxford.

Vugts, J. H. (1970). 'The hydrodynamic forces and ship motions in waves', Dr. thesis, TH Delft, Uitgeverij Waltman, Delft.

Vugts, J. H., Dob, S. L. and Harland, L. A. (1997) 'Strength design of compliant towers including dynamic effects using an equivalent quasi-static design wave procedure', paper (Delft, The Netherlands) given at BOSS '97 Proc, Elsevier, Amsterdam.

Wheeler, J. D. (1970). 'Method for calculating forces produced by irregular waves', *Journal of Petroleum Engineering* 359–67.

Winterstein, S. (1988) *Nonlinear Vibration Models for Extremes and Fatigue*, J. Engng. Mech., ASCE, **114**(10): 1772–1790.

Wirsching, P. H. and Light, M. C. (1980) 'Fatigue under wide band random loading', *J. Str. Div. ASCE*, No. 7, 1593–607.

Wolf, J. P. (1994) *Foundation Vibration Analysis Using Simple Physical Models*, Prentice Hall, Englewood Cliffs, NJ.

Yeung, R. W. (1989) 'Added mass and damping of a vertical cylinder in finite-depth water', *Appl. Ocean Research* **3**: 3.

Chapter 6

Loading from explosions and impact

Alan J. Watson

6.1 INTRODUCTION

Commonly, blast and impact loads are of subsecond duration and magnitude tens of times larger than any other loads in the design life of the structure. The maximum positive or rebound negative peaks of stress or displacement are critical for the structure's survival and subsequent vibrations will only be important if the loads are repetitive. For some industrial structures blast and impact forces are repeated in-service loads and the response must be checked as a serviceability limit state including cracking, vibration and fatigue.

The design and construction of structures against accidental or deliberate impact or explosions is now often considered a part of normal design in the ever increasing importance of safety against industrial and transportation accidents or terrorism. Ronan Point (1968), Flixborough (1974), Chernobyl (1986), Piper Alpha (1988), Peterborough (1989), Oklahoma City (1995), and Eschede (1998) all had a profound effect on design philosophy. These accidents highlight the fact that safety is a multi-disciplinary activity and have shown that structural design changes would be beneficial without enormously increasing the cost. If solid abutments had been used instead of columns in the design of the Eschede bridge, or if the rail lines had been given a greater clearance, the bridge would have been more robust. If compartmentalized construction and moment frames had been used in the Oklahoma Federal Building, increasing the total building cost by 2 per cent, the extent of the progressive collapse which followed the explosion would have been reduced. The public inquiry into the collapse of Ronan Point (Griffiths *et al.*, 1968), revealed that the gas explosion produced a peak lateral pressure on the walls of about $42 \, kN/m^2$ for a few milliseconds which, aided by the upward explosive pressure on the slab above, displaced the top of the wall removing all support from the floor slab of the flat above. Collapse progressed upwards and impact from the collapsing floor slabs then caused collapse to progress downward. Ronan Point had little restraint against rotational or translational displacements between floor and wall slabs and the blast pressure had been enough to fail the joints designed only for modest wind pressures. A subsequent risk assessment showed that Ronan Point, with 110 flats and a design life of 60 years, had a 2 per cent risk of one of the flats having a structurally damaging explosion in 60 years.

The material and human consequences of such incidents are so severe that low risk is not an adequate reason for ignoring the danger. Secondary consequences such as loss of public or business confidence can be equally costly.

6.1.1 Philosophy of design

Since Ronan Point collapsed, the stability of all buildings over four storeys in the UK must be checked with key elements designed for $35 \, kN/m^2$ static loading in the critical direction, or with continuity to limit the area of collapse if a key element fails. The $35 \, kN/m^2$ static load has no statistical significance as an impulsive load, either from blast or impact.

The limit state philosophy for structural design uses elastic response for service loads, plastic response for ultimate loads and prevention of overall collapse disproportionate to a local failure. Many buildings have brittle or non-structural elements such as windows and suspended ceilings that are extremely vulnerable to blast pressure and produce hazardous debris, both within and outside a building.

The resistance of the fixings and supports of external cladding on a building as well as of the panel itself determines the blast or impact resistance and the interaction with the characteristics of the blast loading function, but there is little blast design guidance available for cladding fixings.

Cladding fixings are often hard to inspect but some indication of damage can be obtained from the residual deformations in frames and cladding panels (EPSRC, 1997; Pan and Watson, 1996).

Structures must have safe and serviceable paths for all loads, including extreme loads. British Standard Codes of Practice since 1972 have recommended that by using nominal peripheral and horizontal ties buildings would be more robust to resist extreme loads. Explosions and impact loads may differ in both magnitude and direction from static design loads and produce local damage such as cratering of concrete elements or local buckling of steel elements that would reduce the moment or shear capacity locally. Deflections are very similar for structures under distributed static or dynamic loads but not when the load is concentrated (Watson and Ang, 1984).

6.1.2 Diagnosis of extreme loads

The damage to structural elements from extreme loads can be back analysed to find the load parameters, such as the $35 \, kN/m^2$ equivalent static loading from Ronan Point (1968). From an analysis of damaged lamp posts at Flixborough (1974), Roberts and Pritchard (1982) estimated the peak dynamic pressure produced by the explosion. Sadee et al. (1976) estimated overpressure–distance curves from observations of damage to brickwork and concrete structures. The case study in Section 6.4 uses the damage to buildings from an explosion to evaluate the dynamic loads and so assess the cause of damage to other buildings on the site.

A compression wave from an explosion in air expands as a three-dimensional blast wave propagating at maximum velocities well above that of low amplitude sound waves. It reflects and refracts from solid surfaces and from atmospheric discontinuities.

Explosions also produce high temperatures, which are more locally concentrated than the pressure and decay rapidly, and also produce high velocity fragments from any confining structure, which may impact with a surface before the blast wave arrives. The synergistic effects of blast and fragment impact are not well understood.

If an explosion occurs in contact with a solid it produces stress of the same order of magnitude as the elastic modulus of the solid. The air pressure produced at close range has an initial peak, which is orders of magnitude larger than normal atmospheric pressure, but decreasing with distance travelled. Behind the peak the pressure is still above atmospheric but decreasing with time and falls below atmospheric. The potential of this underpressure to produce structural damage is not certain and in part depends on synchronization with the rebound of the structure.

An impact produces a localized application of pressure on the surface of the structure, which can only spread into the structure from the point of application. This is in contrast to explosions where the blast pressures rapidly engulf the entire surface of the structure. The important parameters of an impact, for diagnostic or forensic purposes, are the shape, velocity and mass of the impactor, and whether or not the impactor deformed.

When pressure is applied very rapidly to the surface of a structure then strain waves are generated which transfer the local dynamic surface deformations into overall structural deformations. An analysis of the transient stress state is necessary when the applied pressure changes more rapidly than the time taken for the strain waves to travel between the boundaries of the structure and establish a state of equilibrium between overall structural resistance and applied pressure effects. During this transition period the transient strain and stress conditions may produce local failures that are decoupled and of different shape from the failures that can occur due to overall structural deformation.

The strain waves propagate at characteristic velocities for the material and transfer momentum into the structure by dynamic displacements of the boundary surface. The rates of strain and stress that are produced locally in the material are orders of magnitude greater than those produced in the overall structural deformation, which are again orders of magnitude greater than under slowly applied loading. Most construction materials have enhanced properties at these high rates of strain.

6.2 BLAST PHENOMENA

6.2.1 Explosive sources

A detonation wave travels through high explosives at 5,000–10,000 m/s. At a free air boundary the gaseous products expand at high velocity, pressure and temperature

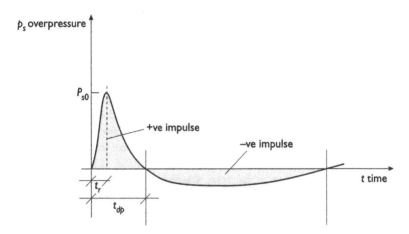

Figure 6.1 Typical overpressure (above atmospheric p_0) at a fixed distance from the explosion.

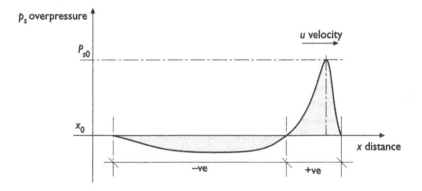

Figure 6.2 Typical overpressure at a fixed time along a radial line from the explosion.

to produce a shock wave with an infinitesimal rise time, producing rapid fluctuations in air pressure and a dynamic wind as it travels from the explosion (Figures 6.1 and 6.2).

Air–gas mixtures, dust and vapour clouds release energy by a process of rapid burning known as deflagration. Air shock from a deflagration propagates more slowly and has a longer rise time.

For vapour clouds the degree of confinement is critical in determining whether or not there is a detonation or a deflagration. Various forms of organic dust can also produce an explosive reaction. Propane, butane and similar gases in stoichiometric concentrations will explode if there is a source of initiation.

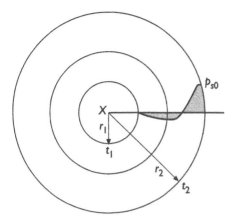

Figure 6.3 Spherical shock wave.

6.2.2 Shock wave parameters

Most explosions, after propagating a short distance, produce a spherical shock wave of surface area $A = 4\pi r^2$ (Figure 6.3).

The characteristics of the spherical air shock are as follows:

(a) The energy of the shock front/unit surface area decreases with r^2 (inverse square law).

(b) The peak overpressure p_{s0} decreases with distance r from the explosion and eventually reduces to a sound wave (Figure 6.4).

(c) The velocity of the shock front u is given by:

$$\frac{dr}{dt} = u = u_s \left(1 + \frac{6p_{s0}}{7p_0}\right)^{1/2}$$

where $u_s = 340 \, \text{m/s}$ is the sound velocity in air for normal conditions at sea level and atmospheric pressure $p_0 = 0.1 \, \text{N/mm}^2$. The different units used for pressure are $1 \, \text{N/mm}^2 = 1 \, \text{MPa} = 145 \, \text{p.s.i.} = 10 \, \text{bar}$.

(d) t_{dp} is the positive duration of the shock wave which increases with distance from the explosion because higher pressures travel faster (Figure 6.5).

(e) p_s is the overpressure which decays with time at a fixed location depending on p_{s0} and t_{dp} (Figure 6.6).

(f) v is the velocity of the air particles behind the shock front as they move radially away from the explosion during the positive phase (e.g. for $p_{s0} = 3p_0$ then $v = 300 \, \text{m/s}$) and towards the explosion in the negative phase.

(g) p_d is the dynamic pressure $= \frac{1}{2}\rho v^2$ where $\rho = $ air density.

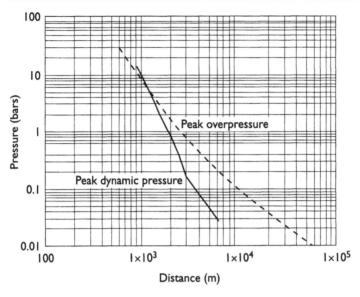

Figure 6.4 Overpressure and dynamic pressure versus range from a 1-MT explosion (Biggs, 1964).

6.2.3 Comparing explosives

The TNT equivalence of an explosive is the weight of TNT which produces a pressure wave in air with one of its characteristics equal to that of the shock wave produced by the explosive at the same distance. The peak pressure or impulse define the shock wave but its shape is also distinguished by the rise time, decay time, positive phase duration or negative phase duration. All of these characteristics vary with the distance the shock wave has travelled in air.

The equivalent weight of TNT is based on peak pressure or impulse and is larger for peak pressure than for impulse.

Explosives differ in the rate at which they detonate and the heat produced and these influence the characteristics of the shock wave in air.

6.2.4 Shock wave scaling

The parameters of the shock wave from one explosive charge can be related to the parameters of the shock wave from a similar shaped charge of the same explosive, but of a different size.

(1) Principle of similitude

If for two charges of the same shape and the same explosive all the dimensions of the

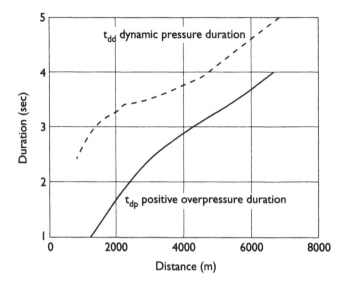

Figure 6.5 Overpressure and dynamic pressure positive phase duration versus range (Biggs, 1964).

first are *k* times those of the second, then the peak pressure p_{s0} measured at any distance R from the centre of the first charge will be equal to those measured at distance kR from the centre of the second charge. The +ve impulse, energy and duration of the second will be *k* times the corresponding quantities for the first at these related distances.

The characteristics vary with the size of the explosive charge and it is experimentally observed that if two spherical charges are made from the same explosive, then the peak pressures in the air blast waves produced by these charges will be equal at distances that are in the same ratio as the cube root of the weight of each charge when the atmospheric pressures are the same in the two cases. This cube root scaling allows empirical charts to be published from the results of experiments using a wide variety of charge sizes.

(2) Cube root scaling

Since densities are presumed to be equal for the two charges of the same explosive, if one is *k* times larger in its linear dimensions then its mass will be k^3 times greater and the principle of similitude can be stated using the cube root of the mass as the scaling factor.

If the masses of two geometrically similar charges of the same explosive are M_1 and M_2 then the peak pressures at distances proportional to $(M_1)^{1/3}$ and $(M_2)^{1/3}$, respectively will be equal and are said to occur at homologous times (i.e. corresponding but not necessarily equal times).

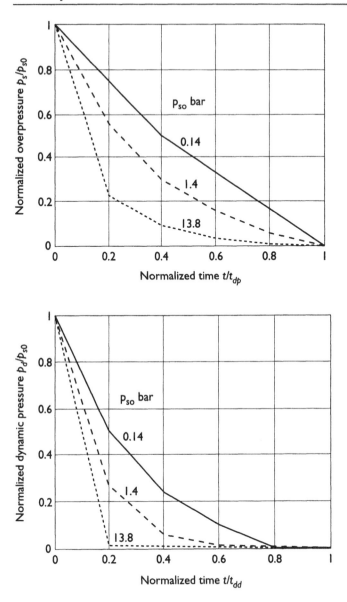

Figure 6.6 Overpressure and dynamic decay curves (Biggs, 1964).

The positive impulses, durations and energy will be proportional to $(M_1)^{1/3}$, $(M_2)^{1/3}$, respectively at those distances and times. That is:

$$p_{s0} = f\left(\frac{R}{M^{1/3}}\right); \qquad \frac{I}{M^{1/3}} = F\left(\frac{R}{M^{1/3}}\right); \qquad \frac{t_d}{M^{1/3}} = \phi\left(\frac{R}{M^{1/3}}\right)$$

where p, I, t_d are peak pressure, +ve impulse and duration, respectively measured at R from W kg of explosive and f, F, ϕ are unspecified functions. Cube root scaling has been verified by experiment but does not describe the decay of peak pressure with distance.

(3) Application of scaling laws

In practice scaling laws are used:

1. to obtain shock parameters for any size of explosive charge from those of a standard of the explosive;
2. to produce a standard charge using experimental methods.

Example I:

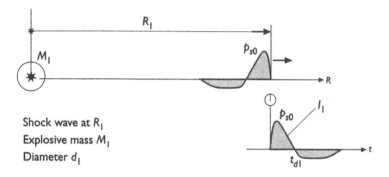

Shock wave at R_1
Explosive mass M_1
Diameter d_1

Cube root scaling indicates that if a similar charge of mass M_2 has a diameter $d_2 = kd_1$ then p_{s0} occurs at $R_2 = kR_1$; that is:

$$p_{s0} = f\left(\frac{R}{M^{1/3}}\right)$$

then

$$(p_{s0})_1 = f\left(\frac{R}{d_1}\right) = (p_{s0})_2 = f\left(\frac{R_2}{kd_1}\right), \qquad R_2 = kR_1$$

and

$$\frac{t_{d_2}}{M_2^{1/3}} = \phi\left(\frac{R_2}{M_2^{1/3}}\right) = \phi\left(\frac{R_1}{M_1^{1/3}}\right) = \frac{t_{d_1}}{M_1^{1/3}} \frac{t_{d_2}}{kd_1} = \frac{t_{d_1}}{d_1} \quad \text{and} \quad t_{d_2} = kt_{d_1}$$

$$\frac{I_2}{M_2^{1/3}} = F\left(\frac{R_2}{M_2^{1/3}}\right) = F\left(\frac{R_1}{M_1^{1/3}}\right) = \frac{I_1}{M_1^{1/3}} \frac{I_2}{kd_1} = \frac{I_1}{d_1} \quad \text{i.e.} \quad I_2 = kI_1$$

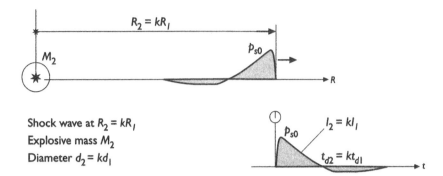

Shock wave at $R_2 = kR_1$
Explosive mass M_2
Diameter $d_2 = kd_1$

That is, the scaled parameters for cube root scaling are:

$$\text{Scale distance}\quad \bar{R} = \frac{R}{M^{1/3}}\left(\text{i.e. } \frac{R_1}{M_1^{1/3}} = \frac{R_2}{M_2^{1/3}} \quad \therefore \frac{R_1}{R_2} = \left(\frac{M_1}{M_2}\right)^{1/3}\right.$$

$$\left. = \frac{d}{kd} = \frac{1}{k}\right)$$

$$\text{Scale time}\quad \bar{t} = \frac{t}{M^{1/3}}\left(\text{i.e. } \frac{t_1}{t_2} = \frac{1}{k}\right)$$

$$\text{Scale impulse}\quad \bar{I} = \frac{I}{M^{1/3}}\left(\text{i.e. } \frac{I_1}{I_2} = \frac{1}{k}\right)$$

$$\text{Scale pressure}\quad \frac{\bar{I}}{\bar{t}} = \frac{I}{M^{1/3}}\frac{M^{1/3}}{t} = \frac{I}{t}$$

$$\text{Scale velocity}\quad \frac{\bar{R}}{\bar{t}} = \frac{R}{t}$$

that is, pressure and velocity are the same for the prototype at homologous times.

Example 2

Use cube root scaling to compare the shock wave from a 300 kg explosive charge with the shock wave from a 300 g charge of the same explosive type and shape.

$$\text{Scale distance}\quad \bar{R} = \frac{R_1}{(300)^{1/3}} = \frac{R_2}{(0.3)^{1/3}}$$

$$\text{Scale factor}\quad \frac{R_1}{R_2} = \left(\frac{300}{0.3}\right)^{1/3} = 10$$

that is, the same peak overpressure and shock wave velocity occurs at 100 m from the 300 kg charge as occurs at 10 m from the 300 g charge, but the +ve duration t_d and impulse I are ten times greater.

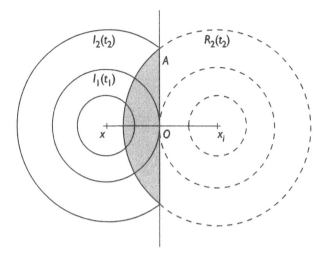

Figure 6.7 Weak blast wave reflection.

6.2.5 Interaction of shock waves with plane surfaces

(a) Reflection of weak shocks

Spherical shock waves of low overpressure reflect from a plane surface as if the reflected shock waves (Figure 6.7) came from an imaginary source equidistant, and on the same perpendicular, from the surface as the real source but on the opposite side of the surface. The reflected waves propagate with the same velocity as the incident waves.

Influence of surface properties:

1. If the plane surface is a rigid protective wall, then at $(0, t_1)$, the particle velocity $v = 0$ and the peak pressure p_r is larger than p_{s0}. At t_2 the real shock covers a circular area of the surface, radius OA. Peak pressure $p_{s0}(t_2)$ is increased around the circumference of the circle of effect by reflection. Providing $A\hat{X}O \leq 35°$, p_{s0} has the same magnification by reflection as when $A\hat{X}O = 0°$.
2. If the plane surface is the external wall of a normal building, it is less than rigid and at $(0, t_1)$ the surface is accelerated and has a velocity and a displacement. The surface continues to accelerate as long as an overpressure p_s exists on one side. The reflected pressure is of lower amplitude than for the rigid surface.

The surface may not exceed the limiting elastic deflection if the reflected overpressure is low or +ve duration is short. For greater overpressure or longer +ve duration, plastic deformation and possibly collapse may occur.

If the +ve duration of the shock wave is much longer than the natural period of the surface then surface response is similar to that of a spring instantaneously loaded with a constant load.

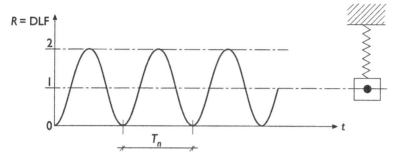

Figure 6.8 *Figure 6.8* Spring–mass model.

That is, the surface overshoots the equilibrium position, is restored by the spring force but once more overshoots and vibrates about equilibrium position at the natural frequency of the spring (Figure 6.8).

If the +ve duration of the shock wave is much shorter than T_n then overpressure reduces to zero before any significant deflection occurs and hardly any spring resistance is developed during the +ve phase.

Assuming constant force P and acceleration $\ddot{y}$:

$$\text{Surface velocity} \quad \dot{y} = \ddot{y} t_{d_p} = \frac{P}{M} t_{d_p} = \frac{I}{M} \quad \text{where } M = \text{Effective mass}$$

$$\text{Kinetic energy of shockwave} = \tfrac{1}{2} M \left(\frac{I}{M} \right)^2 = \frac{I^2}{2M}$$

Hence peak overpressure p_{s0} determines the response of a non-rigid surface barrier to shock waves with a relatively long +ve duration and +ve impulse I determines the response to shock waves with a relatively short +ve duration.

(b) Reflection of strong shocks

Spherical shock waves of high overpressure ($p_{s0} \gg p_0$) reflect from rigid or non-rigid plane surfaces in a more complicated way than weak shocks, because the reflected shocks are advancing into air with pressure, density and velocity very different from normal atmospheric conditions (Figure 6.9).

At time t_1, shock wave I_1 reaches the surface at O and reflects. Boundary conditions are $v = 0$ and peak pressure $> 2p_{s0}(t_1)$.

The velocity of the reflected shock front R is not constant and so R cannot be drawn on concentric spheres from an imaginary source.

At $t > t_1$, the intersection of the incident wave $I(t)$ and reflected wave $R(t)$ is no longer on the surface and a new shock surface M (Mach stem) connects the ring of intersection points of I, R, M (triple point) to the surface s.

The shock wave system depends on the distance OX (e.g. if $OX = 0$ no separate reflections are formed, and there is only the Mach wave).

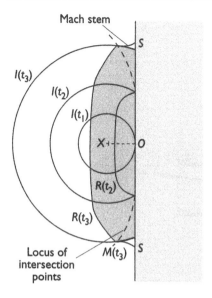

Figure 6.9 Strong wave reflection.

6.2.6 Blast loading effects on buildings

Consider the building $h \times b \times l$ with a plane shock wave normal to the wall F. The blast loading from the positive overpressure is (Figure 6.10).

1. Initial diffraction: the incident wave reaches F at t_0 and is reflected. Resultant pressure $> p_{s0}(t)$ over a clearing period $t_c = (3S_c)/u$ where $h \geq S_c \leq b/2$ (i.e. after t_c reflection effects no longer act).

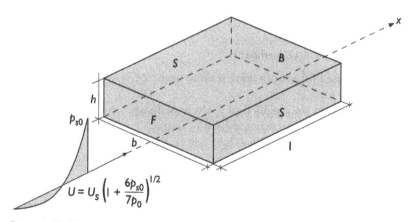

Figure 6.10 Blast on buildings.

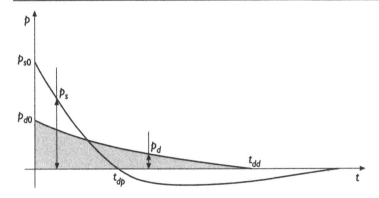

Figure 6.11 Pressure variations p_s, p_d on front wall F.

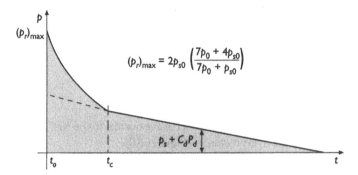

$$(p_r)_{max} = 2p_{s0}\left(\frac{7p_0 + 4p_{s0}}{7p_0 + p_{s0}}\right)$$

Figure 6.12 Total reflected pressure p_r on front wall F.

2. General overpressure: acts for as long as the front wall F and the back wall B are subjected to different overpressures.
3. Drag loading: The particle velocity v of the air behind the shock front produces a dynamic wind pressure $p_d = \frac{1}{2}\rho v^2$ and a drag pressure $C_d p_d$ where $C_d =$ appropriate drag coefficient.

The negative phase of the shock wave is often neglected in assessing blast effects (Figures 6.11 and 6.12).

Although p_d decays less rapidly than p_r at a fixed distance from the explosion, it decays more rapidly with distance and $t_{dd} > t_{dp}$ (Figures 6.4 and 6.5).

Side walls s, back wall B, roof R all have negative drag coefficients.

Back face B reaches a steady state pressure at $t = (4S_c)/u$ after the shock wave reaches the back face.

The external walls and roof of a building receive the shock wave first from an external explosion. There will be a leakage of pressure into the building through openings for as long as there is a positive difference between external and internal pressure $(P - P_j)$, depending on the area of the openings A_0 and the volume of the structure V_0. The internal pressure, P_i varies within the internal space and is highest close to the leak. For structures where A_0/V_0 is small and $P < 10$ bar, the average internal pressure increment ΔP_i in time Δt msec is:

$$\Delta P_i = C_L \left(\frac{A_0}{V_0} \right) \Delta t$$

where C_L is the leakage pressure coefficient given in TM5-855-1 (1986).

When an explosion occurs inside a building then it is the interior surface of the walls and ceiling of the building which are first loaded by the pressure of the shock wave that reflects and increases the pressure. If there are openings in the walls or ceiling then there will be venting of pressure out of the building for as long as there is a negative difference in the external and internal pressure $(P - P_i)$, and the internal pressure will decrease. Internal reflections become so complicated that for preliminary analysis re-reflected shocks are neglected. Arrival times of re-reflected shocks can be calculated if a more exact analysis of loading is required. Reflections are also simplified into normal incidence but slant distances are used in determining the reflected pressure. In addition to the reflected blast loading, internal explosions produce a quasi-static pressure which depends on the charge weight to room volume ratio for peak value and on the venting for the quasi-static decay characteristics.

With internal explosions the transmission of blast waves within the corridors and connected rooms must be analysed. In experimental work using tunnels and ducts, the following observations are given in TM5-855-1:

(a) An increase in overpressure occurs if the cross sectional area of the corridor decreases.

(b) A decrease in overpressure occurs if the corridor has sharp turns or bends. The peak pressure P_n after n bends of $90°$ when friction and pressure attenuation between bends is neglected, is given as $P_n = P_{s0} (0.94)^n$ where P_{s0} is the peak overpressure before the first bend.

(c) Overpressure also depends on friction losses along the tunnel walls, the viscosity and the rate of decay of the shock front.

(d) Overpressure attenuates with distance into a smooth corridor. It depends on the charge weight and the distance from the explosion to the tunnel entrance, but does not depend on the dimensions of the normal corridor.

(e) Overpressure attenuates as a long duration pulse goes from one corridor into another of larger area A_2, according to the relationships:

$$P_2 = \left[\sqrt{\frac{A_1}{A_2}} \right] P_1 \qquad \text{for } \frac{A_1}{A_2} \geq 0.$$

$$P_2 = \frac{A_1}{A_2} P_1 \qquad \text{for } \frac{A_1}{2} < 0.$$

6.3 IMPACT PHENOMENA

6.3.1 Introduction

The independent variables of impact loading include the mass, shape, velocity vector, structure and material properties of the impacting structure. In civil engineering the impacted structure is usually stationary and the magnitude of a very short duration impact load is less critical than the impulse or kinetic energy of the impactor which must be absorbed by deformation. Temperature effects of impact are often ignored but may alter the material properties. Impactors can be of low mass and high velocity, such as bullets with velocities up to 1,000 m/sec and fragments of damaged structures, or of large mass and much lower velocity such as vehicles with velocities nearer to 10 m/sec. The larger the mass the more likely it is that the impact will cover a large area of contact. The greater the impulse the more energy there is to absorb and the area of contact determines the distribution of surface pressure, overall structural displacements and local deformations.

Local damage includes penetration and perforation by the impactor, cratering or depression on the impact face, scabbing or bulging on the distal face, radial or circumferential cracking, punching and shear failure, Amde *et al.*, 1996. At high rates of loading, stress waves from the impact and the high strain rate properties of the material determine the location and type of damage (Watson and Chan, 1987). Figure 6.13 shows impact on a concrete beam with the cracks formed at 1 msec after impact.

Overall deflection from an impact or a static load at the same location, may be similar but initially the inertia of the structure produces higher modes of

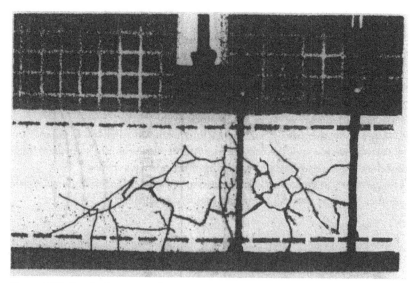

Figure 6.13 Cracks forming under an impact load on a concrete beam.

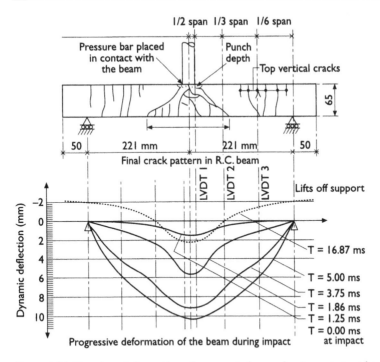

Figure 6.14 Transient deformation of a concrete beam after impact at midspan.

deformation causing impact damage such as top face cracking, which is not typical of static loading, Figure 6.14 (Watson and Ang, 1982).

The impact velocity determines the strain rate, mode of response and the type of damage (Zukas *et al.*, 1982). Velocities producing strain rates of about 10^0 do not enhance the properties of concrete and structural response is primarily elastic with some local plasticity. Structural response times are measured in msec in the concrete beam, Figure 6.14 where the rigid mass of 1.8 kg impacted at 16 m/sec, and the local and overall structural response did not occur coincidentally. Velocities over about 500 m/sec produce loading and structural response times measured in µsec. The local response depends on material properties around the impact area. The phenomena requires a stress wave analysis and the strain rate and material constitutive relations are significant influences on the plastic flow and failure criteria. In this velocity regime overall global response becomes secondary and is decoupled from the local response. Impact velocities above 2000 m/sec are characteristic of shape charge impact and produce pressures which exceed the material strengths by several orders of magnitude so that solids behave as fluids at the early stages of impact. Table 6.1 (CEB, 1988) shows the strain rates from different types of impact.

Alternatively the kinetic energy density of an impactor, defined as the kinetic energy per unit area of contact, can be used to determine the damaging capability

Table 6.1 Typical strain rates for types of impact loading.

Type of loading	Strain rate (s^{-1})
Traffic	10^{-6}–10^{-4}
Pile driving	10^{-2}–10^{0}
Aeroplane impact	$5(10^{-2})$–$2(10^{0})$
Hard impact	10^{0}–$5(10^{1})$
Hypervelocity impact	10^{2}–10^{6}

of a projectile by penetration of the target. This parameter is considered to be a measure of the impact shear stress produced and Smith and Hetherington (1994) list the transition zones. This is a difficult parameter to define when the projectile is irregularly shaped.

If the impactor has the higher dynamic yield, then much less plastic deformation will occur in the impactor and this is a hard impact. If the impactor has the lower dynamic yield then plastic deformation will be much greater in the impactor and this is a soft impact.

6.3.2 Modelling impact

Analysis of a mass dropped onto an undamped spring mass system of stiffness k shows that the ratio between the dropped mass m_1 with velocity v_1 and the target mass m_2 is significant in determining the response. If the dropped mass is small relative to the target mass and both are elastic, then the mass may rebound immediately after it impacts. If the dropped mass is relatively large then the two masses may move together after impact. The two masses may also move together, regardless of relative size, if the impact surfaces are inelastic. Using conservation of momentum at the instant of impact and conservation of energy for motion after impact, the maximum deflection of the spring when m_1 and m_2 move together is:

$$\Delta x = \left[m_1 g \pm \sqrt{(m_1 g)^2 + \frac{k(m_1 v_1)^2}{(m_1 + m_2)}} \right] \frac{1}{k}$$

In practice m_2 is the effective mass of an impacted structure, and k is determined by its boundary conditions, structural and material properties. The positive sign gives the maximum downward deflection of the structure, and negative gives the maximum upward or minimum downward deflection of the structure.

The impulse on the structure is given by the area under the impact force–time relationship and is equal to the rate of change of momentum of the impacting mass:

$$\Delta(mv) = \int_0^t f \, dt$$

At the instant of impact when the impactor rebounds:

$$v = \frac{\text{Impulse}}{m}$$

After impact if the mass is in harmonic free vibration at its natural undamped frequency $\omega = \sqrt{k/m}$ rad/sec:

$$x = X \sin \omega t \qquad \text{and} \qquad \dot{x} = X\omega \cos \omega t$$

where X is the amplitude of vibration.

At the instant of impact $t = 0$, and $\dot{x} = X\omega = \text{Impulse}/m$.

The equation of motion when the positive direction of displacement is the same as that of the impulse giving the deflection x at any time t is:

$$x = \frac{\text{Impulse}}{m\omega} \sin \omega t$$

6.3.3 Low velocity impact by low mass projectiles

Cladding of composite sandwich construction is commonly used on buildings and is exposed to impact by small mass projectiles at low velocity:

(1) Determination of the magnitude and distribution of surface pressure when an isotropic solid is impacted at normal incidence and low velocity by a spherical impactor.

Zukas et al., (1982) assume the impactor and the target are linear elastic and the duration of the impact is long relative to stress wave transit times. On impact, the target and impactor remain in contact and compress for a total distance α at a rate of compression $\dot{\alpha} = v_1 + v_2$ where v_1, v_2 are the approach velocities of the impactor and target respectively. Assuming that the Hertz law of contact $P = n\alpha^{3/2}$ applies during impact, the maximum deformation is:

$$\alpha_1 = \left(\frac{5v^2}{4Mn} \right)^{2/5} \qquad \text{when } \dot{\alpha} = 0 \qquad (6.1)$$

where:

$$n = \frac{4\sqrt{R}}{3\pi(k_1 + k_2)}, \qquad k_1 = \frac{1 - v_1^2}{\pi E_1},$$

$$k_2 = \frac{1 - v_2^2}{\pi E_2}, \qquad M = \left(\frac{1}{m_1} + \frac{1}{m_2} \right)$$

and v is Poisson's ratio, R is radius of spherical impactor, E is Young's modulus, m is mass and subscripts 1 and 2 denote impactor and target parameters, respectively.

The maximum contact force is then:

$$P_{\text{max}} = n \left[\left(\frac{5 \, v^2}{4 \, Mn} \right)^{2/5} \right]^{3/2} = n^{2/5} \left(\frac{5 \, v^2}{4 \, M} \right)^{3/5} \qquad (6.2)$$

Assuming that when a sphere impacts a flat surface, with a force P, the radius of the area of contact a, is given by the Hertz equation for the area of contact when the load P is static:

$$a = \left[\frac{3\pi P}{4}(k_1 + k_2)R_1\right]^{1/3}$$ (6.3)

$$a_{max} = (R_1)^{1/2}\left(\frac{5v^2}{4Mn}\right)^{1/5}$$ (6.4)

The Timoshenko pressure distribution over the area of contact is:

$$q_{x,y} = q_0\left[1 - \left(\frac{x^2}{a^2} + \frac{y^2}{a^2}\right)\right]^{1/2}$$ (6.5)

where q_0 = pressure at the centre of the contact area $x = y = 0$, at the boundary of the contact area $x^2/a^2 + y^2/a^2 = 1$, $q_{x,y} = 0$, and summing the pressure over the area of contact and equating this to P:

$$q_0 = \frac{3P}{2\pi a^2}$$ (6.6)

From eqns (6.2), (6.4)–(6.6) and using polar co-ordinate r, the magnitude and distribution of the surface pressure is obtained as:

$$q_r = \left(\frac{3n}{2\pi R_1}\right)\left(\frac{5v^2}{4nM}\right)^{1/5}\left[1 - \left(\frac{r}{a}\right)^2\right]^{1/2}$$ (6.7)

This pressure produces internal stresses to compare with the limiting stresses producing failure modes in the target:

(2) Determination of the dynamic force, area and duration of contact when a flexible cladding panel is impacted by a spherical impactor at normal incidence and low velocity.

When the panel is flexible then local and overall deformation without punching failure is likely at low velocity impact. Zukas et al., (1982) present an analytical method for determining the response of isotropic and anisotropic laminated panels impacted by a spherical impactor. The local deformation α is the Hertzian contact deformation determined by the force–deformation relationship: $P_c = n'\alpha^{3/2}$. Plate bending deflection δ_p is determined by the force–deflection relationship: $P_p = k_p\delta_p$, where k_p is the stiffness of the plate and is a function of the elastic constants and the boundary conditions. For a circular, isotropic plate of radius R, thickness h, Young's modulus E_r and Poisson's ratio $= v_r$, with simply supported boundaries,

$$k_p = \frac{4\pi E_r h^3}{3(1 - v_r)(3 + v_r)R^2}$$

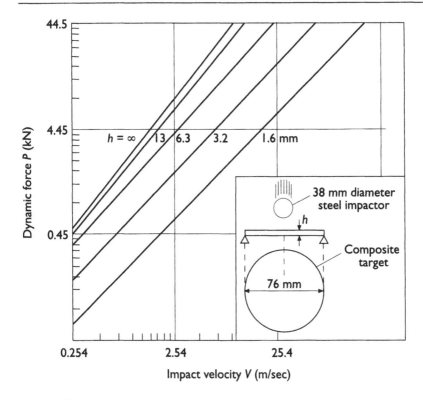

Figure 6.15 Free boundaries.

Assuming impact by a rigid impactor on a stationary plate at an approach velocity $v = v_1$, the kinetic energy of impact equals the work done on the plate in local and overall deformation.

$$\tfrac{1}{2} m_1 v^2 = \int_0^{\delta_{max}} P_p \, d\delta_p + \int_0^{\alpha_1} P_c \, d\alpha \tag{6.8}$$

Solving the equation for P at a given embed v with known properties of the impactor and composite plate, shows that P increases linearly with v but at a reducing rate with h, whether the plate has simply supported or fixed boundaries. With fixed boundaries the effective masses will be different and dynamic force is greater at a given impact velocity than with simply supported boundaries, Figures 6.15 and 6.16.

For a given impact velocity the dynamic force P and area of contact decrease but the contact duration increases as the target flexibility increases (i.e. plate thickness h decreases).

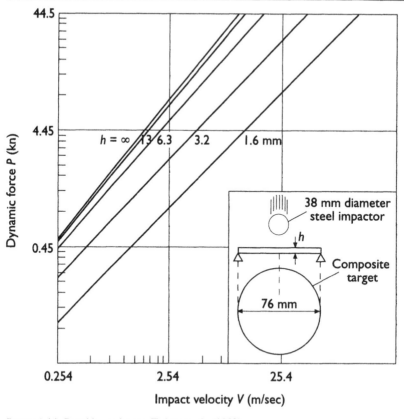

Figure 6.16 Fixed boundaries, Zukas et al., (1982).

6.3.4 Empirical formulae for low velocity impact on concrete

The penetration, perforation and scabbing of reinforced concrete impacted by flat faced cylindrical missiles was investigated by Barr *et al.* (1980) using experimental and analytical techniques. Comparisons were made with the modified NDRC (US National Defense Research Committee) empirical formula (Kennedy, 1976), and the CEA/EDF (French Atomic Energy Commission) empirical formula (Berriaud *et al.* 1978).

The modified NDRC formulae for a semi-infinite target are:

$$\left(\frac{x}{2d}\right)^2 = (2.74)(10)^{-5}\left(\frac{Dd^{0.2}}{(\sigma_c)^{1/2}}\right)v^{1.8} \quad \text{when } \frac{x}{d} \leq 2 \tag{6.9a}$$

and

$$\frac{x}{d} - 1 = (2.74)(10)^{-5}\left(\frac{Dd^{0.2}}{(\sigma_c)^{1/2}}\right)v^{1.8} \quad \text{when } \frac{x}{d} > 2 \tag{6.9b}$$

where x in metres, d in metres, M in kg and v in msec^{-1}, are missile penetration depth, diameter, mass and velocity, respectively, and $D = M/d^3$ is calibre density, and σ_c is concrete compressive strength, in Pa. The perforation thickness e can be obtained from:

$$\frac{e}{d} = 1.32 + 1.24 \left(\frac{x}{d} \right) \quad \text{when } 18 \geq \frac{e}{d} \geq 3$$

$$\frac{e}{d} = 3.19 \left(\frac{x}{d} \right) - 0.718 \left(\frac{x}{d} \right)^2 \quad \text{when } \frac{e}{d} \leq 3$$

The CEA/EDF formula gives the perforation velocity as:

$$V_p^2 = 1.7\sigma_c\rho^{1/3} \left(\frac{de^2}{M} \right)^{4/3} \tag{6.10}$$

where ρ is the concrete density and the range of applicability is:

$$20 < V < 200 \, \text{msec}^{-1}$$

$$0.3 < \frac{e}{d} < 4$$

$$30 < \sigma_c < 45 \, \text{MPa}$$

$$120 < \text{Reinforcement} < 300 \, \text{kg m}^{-3}$$

flat nosed cylindrical missile.

The NDRC formulae were within 30 per cent of the penetration velocity measured in the experiments for different mass and diameter missiles and thickness and strength of concrete. The CEA/EDF formulae were within 100 per cent of the penetration velocity but the reinforcement and concrete strengths used in the experiments were outside the range of validity.

6.4 DESIGN ACTIONS

Concept definition and resistance requirements specify the design parameters, and extreme loads set the upper bound actions for the ultimate accidental limit state.

Accident load cases for the design of a double skin concrete containment structure surrounding a nuclear reactor pressure vessel, took the upper bound as catastrophic failure requiring the evacuation of people living outside the reactor site (Eibl, 1993). The structure was designed to resist quasi-static internal pressures, fast dynamic internal pressures, fast dynamic external forces and high temperature from the heat generated in a core melt accident. The outer wall thickness of the containment was 1.8 m to resist a 20 tonne military aircraft crashing at 215 m/sec. The inner concrete barrier, 0.7 m thick, was designed as a fragment shield against high velocity missiles from bursting plant and equipment and from an assumed hydrogen detonation pressure wave. The inner concrete barrier has an outer metal plate, which also serves as the necessary concrete reinforcement of this composite structural member. The design method utilized the strain rate sensitivity of the concrete in a Hugoniot curve of hydrostatic pressure against volumetric concrete strain.

To obtain a statistical base for characteristic blast pressures the damage to structural elements from an explosion can be back analysed. Analysis of the Ronan Point gas explosion produced the requirement to analyse key elements for $35\,kN/m^2$ equivalent static loading. The 1974 Flixborough vapour cloud explosion damaged many structures and from an analysis of lamp posts Roberts and Pritchard (1982) estimated the dynamic pressure produced by the explosion. Sadee *et al.* (1976) estimated the overpressure–distance curve from observations of the damage to brickwork and concrete structures. In many cases, because of the unknowns, sophisticated analytical techniques are not justified.

After a gas explosion that destroyed a building in Peterborough, 1987, a survey of the surrounding damage to windows, traffic signs and lamp posts, and the distance travelled by debris, was used to estimate the characteristics of the explosion (Watson 1994). A survey was made of the frame dimensions and glass thickness for all the windows exposed to the direct blast wave, and whether or not the glass had been broken. Eyewitness accounts indicated that window panes might have broken either inward or outward.

The resistance of glass to blast pressure depends upon the edge conditions, dimensions, thickness and ultimate tensile strength of the glass. Dragosavic (1973) analysed a rectangular pane of glass, assuming simple supports on all four sides and uniform pressure on the pane, giving the ultimate resistance q (kN/m^2) as:

$$q = f_{kb}(10^3)d_1^2\beta^2/\beta^2 b^2 \qquad (6.11)$$

where f_{kb} = ultimate tensile strength of glass, assumed to be $84\,kN/mm^2$; d, b = thickness (mm) and short side length (mm), respectively; β = a function of the side lengths L, b.

Because of the variability in the strength of glass, and in the degree of fixity to the frame, the calculated results probably do not predict the actual ultimate resistance by better than ± 50 per cent (Mainstone, 1971).

The calculated resistance q (kN/m^2) for each window, is plotted against r, the distance from the explosion, Figure 6.17, showing whether or not the glass was broken. The resistance of broken and unbroken panes gives an estimate for the blast overpressure at various distances assuming normal incidence. Upper and lower limits for this blast overpressure are indicated as (UL) and (LL). Windows possibly broken by effects other than overpressure are identified but not used, for example, those that could have been broken by flying debris.

A lower bound estimate of peak overpressure is plotted in Figure 6.17. This smoothed curve has no broken windows above it if a 50 per cent reduction is made on the theoretical resistance of the broken windows and there are only 10 unbroken windows below it if the 50 per cent tolerance is used.

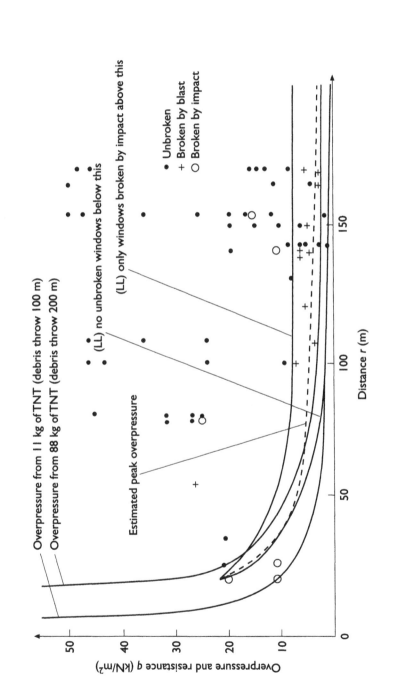

Figure 6.17 Explosive peak overpressure estimated from a window survey.

Metal posts close to the explosion provided simple elements for analysis. None had any damage that could be attributed to the explosion and the analysis would therefore be an upper bound estimate of the pressure.

The response to blast pressure depends upon the duration of the blast t_d relative to the fundamental natural period of vibration of the post, and by simple measurement, $T = 0.4$ sec. If $t_d \ll T$ it responds to impulse and to peak pressure if $t_d > T$. Between these limits it responds to both. The duration t_d was estimated by assuming a triangular pressure time curve with peak pressure p_m. The post had no visible damage, indicating that it had not exceeded the elastic limit. Using the analysis given by Biggs (1964), and assuming a linear resistance–deflection curve:

$$t_d = \frac{R_m T}{A p_m \pi} \tag{6.12}$$

where R_m is maximum elastic resistance (kN), T is natural period (sec), A is area subjected to the blast pressure (m^2), p_m is peak blast pressure (kN/m^2).

Analysing a post at 60 m from the explosion for first mode deformation, and using $p_m = 7.5$ kN/m^2 from Figure 6.17, gives $t_d = 0.18$ sec. When Biggs' analysis is applied to an undamaged post at 16 m using a peak pressure of 30 kN/m^2 extrapolated from Figure 6.17, the duration of the blast pulse t_d is 0.093 sec. As expected it is less than at 60 m.

The peak pressure predicted is sensitive to the assumed shape of the pressure pulse. If the pulse had a rise time of 16 per cent of the decay time then p_m is calculated to be 150 kN/m^2 which fits reasonably well with the extrapolated peak overpressure line from the window survey. The building at the centre of the explosion was completely destroyed.

Eyewitness accounts and press photographs indicated that debris from the exploded building was thrown up to 200 m from the centre of the explosion. The debris throw distance was compared to that of TNT using an analysis by Kinney and Graham (1985). This showed that 11 kg of TNT would have thrown debris 100 m and 88 kg TNT would have thrown it 200 m. The overpressures produced by these quantities of TNT at different ranges are plotted in Figure 6.17, and are very sensitive to range at less than 25 m.

6.4.1 Idealization of high rate dynamic loads

Design loads or actions are usually computationally manageable idealizations such as the pressure–time idealization for fragment impact, Figure 6.18, and for a hydrogen pressure wave, Figure 6.19 (Eibl, 1993). Chen and Chen (1996) give a load idealization, for impact on shallow buried plates. Yang and Yau (1997), have idealized impact loads from vehicles moving over simple and continuous beams using impact formulas. They claim that current codes specify impact factors that may significantly underestimate the beam response.

The main characteristics of an idealized impulsive load such as those produced by impact or blast loading, are the peak pressure, rise and decay functions and the

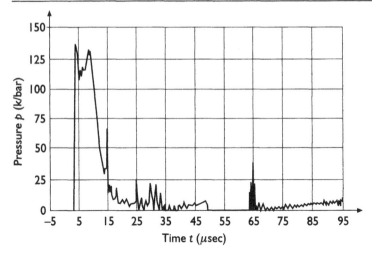

Figure 6.18 Fragment impact pressure.

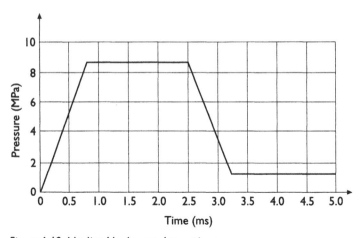

Figure 6.19 Idealized hydrogen detonation pressure.

total duration. The response of the structure then depends on how these load parameters relate to the parameters chosen to model the structure. For high rates of loading an undamped spring–mass system is frequently chosen to model the structure and several authors, for example Craig (1981) have used this as a useful model to show how load and structural parameters interact.

A potentially damaging condition imposed on a structure by any form of loading, is displacement, whether it is local strain or overall structural deflection. If that part of the structure, which will displace the most, can be identified, then a Single Degree Of Freedom (SDOF) model can be constructed as an equivalent structure where displacement at the critical point on the structure is given by the extension of the spring. To ease computation the model may be undamped because damping does not significantly alter the magnitude of the first peak of oscillation and it is this deflection which determines whether or not the structure survives extreme loads from explosions or impact.

6.4.2 Influence of load characteristics on the response of an elastic spring–mass SDOF system

The peak transient deflection of a structure under a specified loading condition is the same as the equivalent SDOF model that can be more easily analysed (Craig, 1981). The result gives a response function or dynamic load factor:

$$R(t) = \frac{U(t)}{U_{st}} = \frac{\text{Dynamic deflection}}{\text{Static deflection}} = \frac{u(t)}{P_0/k}$$

For instance the sudden release of the mass m in an SDOF system causes a force $P_0 = mg$ to act on the unstretched spring of stiffness k, producing an ideal step load with rise time $t_r \to 0$ and duration of load $t_d \to \infty$ (Figure 6.20).

This sets the SDOF system into elastic vibration and the dynamic deflection $u(t)$ varies with $R(t)$ as the mass overshoots the equilibrium position.

After several cycles of vibration where the maximum dynamic deflection has reached twice the static deflection (Figure 6.21), the damped SDOF system comes to rest at $R(t) = 1$. On the first cycle, the maximum value of $R(t)$ is similar for both the damped and undamped systems and so with only a small error, an undamped system can be used to determine $R(t)_{max}$ which occurs at $t = T_n/2$ where $T_n = (2\pi)/w_n$ is the undamped natural period.

A rectangular pulse load with $t_r \to 0$ but $t_d \ll \infty$ sets the SDOF system into forced vibration over the time t_d, and it then continues in free vibration (Figure 6.22).

The deflection of the mass during the forced vibration $0 \le t \le t_d$ is the same as that for an ideal step load and $0 \le R(t) \le 2$ (Figure 6.23).

If $t_d \ge T_n/2$ the maximum deflection occurs at $t = T_n/2$ when $R(t) = 2$. If $t_d < T_n/2$ then $R(t) < 2$ in the range $0 \le t \le t_d$. When $t > t_d$ then the mass is in

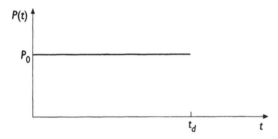

Figure 6.20 Ideal step load.

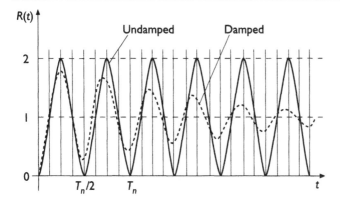

Figure 6.21 Response ratio for ideal step load.

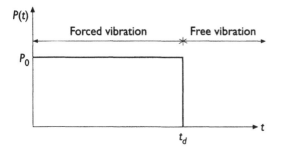

Figure 6.22 Rectangular pulse load.

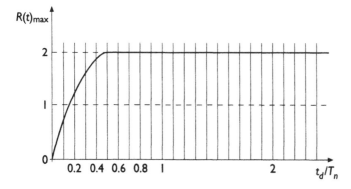

Figure 6.23 Maximum response ratio for a rectangular pulse load.

free vibration and Craig's analysis shows that the maximum deflection is given by $R(t)_{max}$ when the SDOF system is undamped.

$$R(t)_{max} \approx 2 \sin\left(\frac{\pi t_d}{T_n}\right)$$

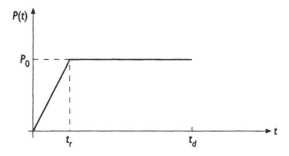

Figure 6.24 Ramp load.

$R(t)_{max} \geq 1$ during the free vibration when $\sin(\pi t_d/T_n) \geq \frac{1}{2}$ (i.e. $t_d/T_n \geq 0.16$), and during the forced vibration when $t_d/T_n \geq 0.25$ (i.e. the maximum dynamic deflection exceeds the static deflection).

$R(t)_{max} = 2$ in the free vibration era when $\sin(\pi t_d/T_n) = 1$ (i.e. $t_d/T_n = 0.5$), or in the forced vibration era when $t_d/T_n \geq 0.5$ maximum dynamic deflection reaches twice the static deflection).

A ramp load of finite rise time t_r and duration $t_d \to \infty$ is shown in Figure 6.24.

When this acts on an undamped SDOF system Craig's analysis shows that the deflection of the mass depends upon the duration of the rise time and the ratio of rise time to the natural period of the system T_n.

From Figure 6.25 $R(t)_{max} = 2$ when $t_r = 0$ (i.e. an ideal step input); as t_r/T_n increases, the overshoot reduces and small oscillations occur about $R(t) = 1$; and if $t_r > 3T_n$ then load can be treated as static and dynamic effects ignored.

6.4.3 Elastoplastic response of an SDOF system

In all the cases considered above, the spring in the SDOF system is taken to be linearly elastic. If, however, the load causes the structure to become plastic, then it can be modelled using an undamped SDOF system with an elastoplastic spring with the stiffness function shown in Figure 6.26.

This elastoplastic undamped SDOF system has been analysed by Biggs (1964) to obtain the elastic, plastic and recovery deflection for an ideal step load $t_r \to 0$; $t_d \to \infty$. The maximum deflection u_m will have to satisfy the limiting ductility factor u_m/u_1. If the system survives the plastic deflection then it will partially recover and vibrate about a residual deflected position. Design charts have been produced from Biggs' solution; one example is given in Figure 2.17 of this book, showing the maximum deflection of an elastoplastic undamped SDOF system for a step load $t_r \to 0$; $t_d \ll \infty$.

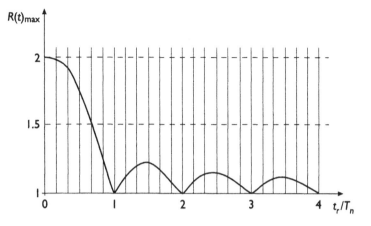

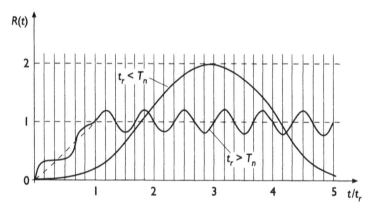

Figure 6.25 Response ratio for a ramp load.

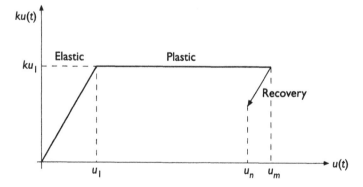

Figure 6.26 Elastoplastic spring stiffness function.

6.5 DESIGNED RESPONSE

6.5.1 Principles of design for high rate dynamic loads

Extreme actions due to high rate loading such as come from impact or blast, produce a response in the structure which depends on how the parameters of the load and the structure relate to each other (see Section 6.4). A variety of analytical methods have been proposed to predict this structural response and so allow comparisons to be made with the performance limits chosen by the designer for deflection or rotation.

These analytical methods can be independent of the rate of loading but high rate loading accelerates the mass of the structure producing inertia forces, alters the constitutive stress–strain relationship for many construction materials and produces strain waves causing local concentrations of stress. For analysis under dynamic load the real continuous structure is often converted into an equivalent spring–mass system with lumped masses supported by elastic or elastoplastic springs that model the resistance–displacement function of the real structure. The number of springs equals the number of degrees of freedom the designer has to consider to accurately define the modes of deformation. Vibrations of this system are damped by an energy absorbing function that relates to the damping in the real structure.

For blast and impact design a SDOF is often assumed for the structural element which is then modelled as an undamped system of a single lumped mass on a spring. This only gives deflection at the centre of loading on the structure and so implies a deformed shape, and assumes that only the first peak amplitude of vibration is significant and that damping does not essentially alter this value. Since mass relates acceleration to inertia force, the single mass of the model must be equivalent to the distribution of acceleration on the full mass of the real structure and the resistance of the spring in the model depends upon the stiffness of the real structure for the particular load arrangement.

The following section reviews some of the analytical methods available in the literature of dynamic design, including some continuous mass models.

6.5.2 Methods of design for extreme dynamic loads

6.5.2.1 Elastic impact factor method

This method for concept design assumes structural resistance is that of a massless linear elastic spring of stiffness k, there are no inertia forces, the force–deflection relationship is linear for both static and dynamic loads and energy is conserved.

A static force W produces deflection $u_s = W/k$ and impact force F produces a deflection

$$u_d = \frac{F}{k} = \frac{F}{W}u_s \quad \text{so that} \quad F = W\frac{u_d}{u_s}$$

(i) If force F is produced by mass M falling from height h and static force $W = Mg$ then by conservation of energy of the falling mass and assuming that all KE is transferred into strain energy:

$$\text{KE}_{\text{impact}} = \text{Loss of K.E.} = W(h + u_d) = \tfrac{1}{2}Fu_d$$

by substituting for F and solving the quadratic equation:

$$\therefore \quad u_d = u_s\left(1 + \sqrt{1 + \frac{2h}{u_s}}\right) \quad \text{and} \quad F = W\left(1 + \sqrt{1 + \frac{2h}{u_s}}\right)$$

where $(1 + \sqrt{1 + (2h)/u_s}) = C$ is an impact factor on deflection and force.

(ii) If W moves horizontally and impacts at velocity v then $\text{KE} = \tfrac{1}{2}(W/g)v^2 =$ strain energy $= \tfrac{1}{2}Fu_d = \tfrac{1}{2}W(u_d^2/u_s)$ and by substituting for F:

$$u_d = u_s\sqrt{\frac{v^2}{gu_s}} \quad \text{and} \quad F = W\sqrt{\frac{v^2}{gu_s}}$$

where the impact factor $C = \sqrt{v^2/(gu_s)}$.

In concept design F is used as an equivalent static load and u_d is checked against the ductility ratio.

6.5.2.2 Equivalent systems

A more rigorous design converts the structure into an equivalent spring–mass system with lumped masses supported by elastic or elastoplastic springs that model the resistance–displacement function of the real structure. The number of springs equals the number of degrees of freedom the designer has to consider and although most structural elements have a large number of degrees of freedom their response to dynamic loading can be approximated by a single degree of freedom equivalent system. When checking whether the structure can survive the first cycle of response to extreme blast or impact loading, damping of the SDOF system is often neglected.

Consider the response of a simply supported beam under a time varying load/unit length of beam $w(t)$ and assume that the deflected shape of the beam is the same as produced by the static application of the load (Baker et al., 1983) (Figure 6.27).

$$\text{Elastic deflection} \quad y_x = \frac{16y_0}{5L^4}(xL^3 - 2Lx^3 + x^4)$$

for $y_0 = $ maximum displacement at midspan and $y_{\text{max}} = (5wL^4)/(384EI)$:

$$wL = \frac{384EI}{5L^3} = \text{Stiffness } k_b$$

$$\text{External WD} = \int_0^L w(t)y\,dx = w(t)y_0\int_0^L \frac{16}{5L^4}(xL^3 - 2Lx^3 + x^4)\,dx$$

$$= \frac{16}{25}Lwy_0$$

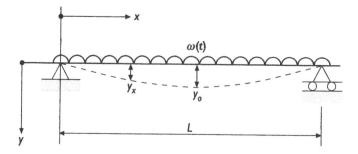

Figure 6.27 Elastic deformation.

$$\text{Internal strain energy } U = \int_0^L \frac{M^2}{2EI} dx = \frac{EI}{2} \int_0^L \left(\frac{d^2 y}{dx^2}\right)^2 dx$$

$$= \frac{EI}{2} \left(\frac{16\, y_0}{5\, L^4}\right)^2 \int_0^L 144(x^2 - Lx)^2 dx$$

$$= \frac{3,072\, EIy_0^2}{125\, L^3}$$

$$\text{Kinetic energy } \quad KE = \int_0^L \tfrac{1}{2}\rho A \left(\frac{dy}{dt}\right)^2 dx$$

$$= \tfrac{1}{2}\rho A \left(\frac{16}{5L^4}\right)^2 \dot{y}_0^2 \int_0^L (xL^3 - 2Lx^3 + x^4)^2 dx$$

$$= \frac{1984}{7875} \rho A L \dot{y}_0^2$$

If the beam is to be represented by an undamped SDOF model with a massless linear spring, Figure 6.28, then for equivalence y_0 is equal in the beam and the model.

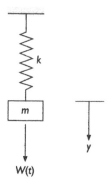

Figure 6.28 Equivalent SDOF model.

External WD $= Wy_0$; Internal strain energy $U = \frac{1}{2}ky_0^2$; KE $= \frac{1}{2}m\dot{y}_0^2$ equating WD, U and KE for the beam and model gives:

$$y_0 = \frac{16}{25}Lwy_0 \therefore \text{ Load factor } K_L = \frac{W}{wL} = \frac{16}{25} = 0.64$$

$$\frac{1}{2}ky_0^2 = \frac{3072}{125}\frac{EI}{L^3}y_0^2$$

$$\text{Spring stiffness } k = \frac{6144}{125}\frac{EI}{L^3} = 0.64\,k_b$$

$$\text{That is, stiffness factor } K_e = \frac{k}{k_b} = 0.64 = K_L$$

$$\frac{1}{2}m\dot{y}_0^2 = \frac{1984}{7875}\rho A L\dot{y}_0^2$$

$$\text{Mass factor } K_m = \frac{m}{\rho A L} = \frac{2(1984)}{7875} = 0.50$$

These factors transform an elastic structure into an equivalent SDOF model and are derived and listed by Biggs (1964), for several beams and slabs with different support and loading conditions. Baker *et al.* (1983) have shown that these transformation factors do not change significantly if the beam deforms to the first mode shape thus the higher modes contained in the static deformed shape can be neglected.

In resisting extreme actions from impact and blast it is uneconomic to design the structure using only the elastic resistance. A considerable resistance is obtained from the plastic behaviour. If the simply supported beam under a time varying uniformly distributed load $w(t)$ has a plastic hinge formed at midspan then the elastic deformation can be neglected by assuming that the beam has a rigid plastic resistance deformation function (Baker *et al.*, 1983) (Figure 6.29):

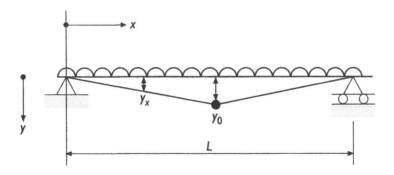

Figure 6.29 Plastic deformation.

$$\text{Since } y_x = y_0 \frac{2x}{L} \left(0 \le x \le \frac{L}{2} \right) \text{ and } y_x = y_0 \frac{(L-x)2}{L} \left(\frac{L}{2} \le x \le L \right)$$

$$\text{External } WD = \int_0^L wy_x \, dx$$

$$= wy_0 \frac{2}{L} \left[\int_0^{L/2} x \, dx + \int_{L/2}^L (L-x) \, dx \right] = \frac{wy_0 L}{2}$$

$$\text{Strain energy } U = M_p 2\theta = M_p 2\frac{2y_0}{L}$$

$$KE = \int_0^L \frac{1}{2} \rho A \left(\frac{dy}{dt} \right)^2 dx$$

$$= \frac{1}{2} \rho A \int_0^{L/2} \left(\frac{2x}{L} \dot{y}_0 \right)^2 dx + \int_{L/2}^L \left(2 \left(\frac{L-x}{L} \right) \dot{y}_0 \right)^2 dx$$

$$= \rho A \dot{y}_0^2 L/6$$

For the equivalent system with plastic behaviour, only strain energy differs from the equivalent elastic SDOF. Strain energy $U = Ry_0$ for the equivalent plastic SDOF system where $R = $ the plastic resistance of the spring (i.e. yield force).

$$Ry_0 = 4M_p \frac{y_0}{L}$$

$$R = \frac{4M_p}{L} = 0.5R_b \text{ where } R_b = wL = \frac{8M_p}{L} = \text{Beam resistance}$$

$$\text{Mass factor } K_m = \frac{m}{\rho AL} = \frac{1}{3}$$

$$\text{Load factor } K_L = \frac{W}{wL} = 0.5$$

Note that K_L, K_m change when yielding occurs.

Equivalence between the structure and the SDOF system is based on deflection, not force or stress and dynamic reactions are not given by the spring force. An analysis by Baker *et al.* (1983), for a simply supported beam under a time varying UDL $= w(t)$, uses the model shown in Figure 6.30.

Using a free body diagram, for the elastic deflection where $a = $ distance from LH support to the point of action of the resultant of the inertia force:

$$Va - \left(\frac{wL}{2} \right) \left(a - \frac{L}{4} \right) = M_{x=L/2}$$

to find the value of a, the moment of elemental inertia forces about the left-hand

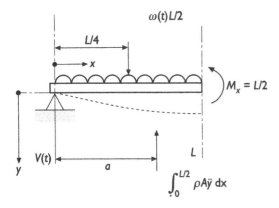

Figure 6.30 Dynamic reaction model.

support is equated to the moment of the resultant inertia force

$$\int_0^{1/2} \rho A \frac{d^2 y}{dt^2} x \, dx = a \int_0^{1/2} \rho A \frac{d^2 y}{dt^2} \, dx$$

For the elastic deformed shape under static load

$$\frac{d^2 y}{dt^2} = \frac{16}{5L^4}(xL^3 - 2Lx^3 + x^4)\ddot{y}_0$$

$$\therefore \quad a = \frac{61}{192}L$$

$$\therefore \quad V\left(\frac{61}{192}L\right) - \left(\frac{wL}{2}\right)\left(\frac{61}{192}L - \frac{L}{4}\right) = y_0\left(\frac{48\,EI}{5\,L^2}\right)$$

and the dynamic reaction when the beam remains elastic up to the maximum load is:

$$\therefore \quad V(t) = \frac{30.216\,EI}{L^3}(y_0) + 0.1066(wL)$$

When yielding occurs at the maximum load then assuming the beam still has its elastic shape curve and substituting M_p for $M_{x=L/2}$

$$V(t) = \frac{192 M_p}{61L} + 0.1066(wL)$$

that is, V_{max} occurs when wL is max.

6.5.2.3 Structural response diagrams

The maximum deflection of an equivalent SDOF system loaded by blast or impact can be obtained from the solution to the equation of motion. Such response

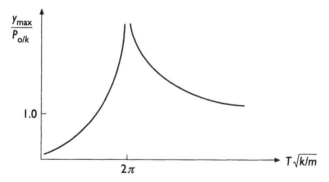

Figure 6.31 Response ratio for a sinusoidal forcing function.

diagrams can be replotted as pressure–impulse diagrams and used when it is a limit state and not the time history of the structure under transient loads, that is of interest.

For example, the response of an undamped, linear elastic SDOF system, stiffness k, mass m, to a sinusoidal forcing function $P(t) = P_0 \sin(2\pi/T)t$ is plotted in Figure 6.31 on axes:

$$\frac{y_{max}}{p_0/k} = \frac{\text{Max dynamic deflection}}{\text{Max static deflection}} = R_{max}$$

$$T\sqrt{\frac{k}{m}} = T\bigg/\sqrt{\frac{m}{k}} = \frac{\text{Duration of loading}}{\text{Natural period of system}} = T\omega$$

where $R =$ Response ratio $=$ dynamic load factor and circular frequency $\omega = \sqrt{k/m}$ rad/sec.

Using Baker's mathematical approximation Figure 6.32 for an air blast wave, $P(t) = P_0 e^{-t/T}$, where since $e^{-t/T}$ never reaches zero, T is used as an equivalent duration of loading to solve the equation of motion $m\ddot{y} + ky = P_0 e^{-t/T}$ for boundary conditions $y = 0, \dot{y} = 0, t = 0$,

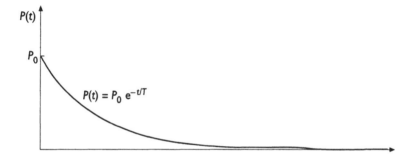

Figure 6.32 Approximate air blast wave.

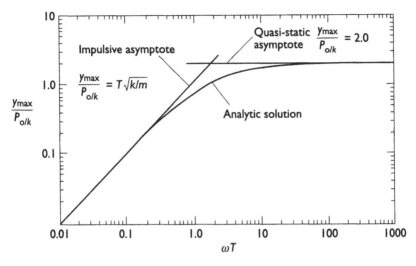

Figure 6.33 Response ratio for approximate air blast wave (Baker et al., 1983).

$$y(t) \Big/ \frac{P_0}{k} = \frac{(\omega T)^2}{(1 + (\omega T)^2)} \left[\frac{\sin \omega t}{\omega t} - \cos \omega t + e \frac{-\omega t}{\omega T} \right]$$

using $dy/dt = 0$, the time ωt_{max} at which $y(t) = y_{max}$ is found by trial and error for specific values of ωT and the analytical solution is plotted giving $y_{max}/(P_0/k)$ as a function of ωT (Figure 6.33).

Since two straight line asymptotes can be used to approximate the analytical solution, Baker identifies three different loading regions:

1. Quasi-static region when $\sqrt{(k/m)}T > 40$ and the structural response $y_{max} = \Psi(P_0/k)$ depends on the peak load P_0 and stiffness k, but not on the mass or duration T. Since the duration $T > 40(\sqrt{m/k})$, the applied load $P(t)$ dissipates very little before y_{max} is reached and the displacement is given by the quasi-static asymptote

$$\frac{y_{max}}{P_0/k} = 2.0$$

2. Impulsive region when $\sqrt{(k/m)}\,T < 0.4$ and the displacement is given by the impulsive asymptote:

$$\frac{y_{max}}{P_0/k} = \sqrt{\frac{k}{m}}T \quad \text{that is} \quad \frac{\sqrt{km}}{P_0 T}y_{max} = 1.0 \quad \text{where} \quad P_0 T = \text{Impulse } I$$

therefore $y_{max} = (I/\sqrt{km})$ is directly proportional to $I = P_0 T$ and also depends on the stiffness k and the mass m. The applied load drops to 0 before y_{max} is reached since the duration of load $T < 0.4(\sqrt{m/k})$.

3. Transition region when $(40 > \sqrt{(k/m)}\,T > 0.4)$ and the displacement

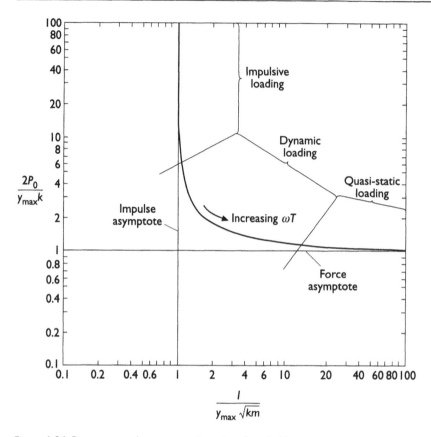

Figure 6.34 Response to the pressure–impulse of an air blast wave.

$y_{max}/(P_0/k)$ depends on P_0, k, I, m. The load and response time is of the same order of magnitude and y_{max} depends on the entire loading history since the duration of load $40\sqrt{m/k} > T > 0.4\sqrt{m/k}$.

Figure 6.33 is converted into the pressure–impulse (P–I) diagram of Figure 6.34 by manipulation of the asymptotes. For an undamped linear elastic SDOF system the ordinate and abscissa are respectively:

$$\frac{2P_0}{y_{max}k} \quad \text{and} \quad \frac{I}{y_{max}\sqrt{km}}$$

The rectangular hyperbolic curve of the P–I diagram is an isodamage curve defining critical combinations of P_0, I which produce the damaging deformation limit y_{max} in an undamped linear elastic SDOF specified by k and m, equivalent to a specific structure:

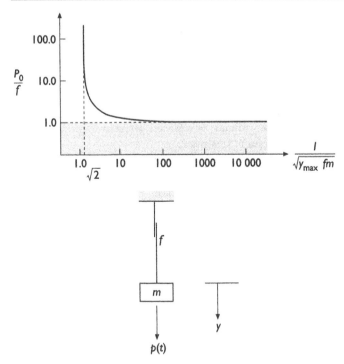

Figure 6.35 Response of a rigid–plastic system.

- deformation will be larger than the threshold damage if (P_0, I) moves to the region above and to the right of the curve;
- changes in I, but not changes in P_0, will move (P_0, I) combinations off the impulsive asymptote;
- only changes in P_0 will move (P_0, I) combinations off the quasi-static asymptote.

The rigid–plastic SDOF system shown in Figure 6.35 uses a Coulomb friction element to model the structural retarding force f where the deformation $y = 0$ if $P_0 \leq f$.

An energy balance gives the equation to the quasi-static and impulse asymptotes, respectively, as:

$$\frac{P_0}{f} = 1.0 \quad \text{and} \quad \frac{I}{\sqrt{y_{max} fm}} = \sqrt{2}$$

6.5.2.4 Isodamage curves

Structural response to transient load can be impulsive or quasi-static, and P–I diagrams can be used for different levels of damage. Damage that occurs at a specific displacement y_{max} can be caused by impulse or by peak pressure and the rectangular hyperbolic curve is then an isodamage contour.

Using the following procedure a P–I diagram can be constructed for an equivalent elastic SDOF system where k and m are known, for example:

$$k = 0.29 \frac{\text{kPa}}{\text{mm}} \quad \text{and} \quad m = 11.36 \, \text{kg/m}^2$$

(1) On log–log paper mark the ordinate as reflected pressure $P_0(kPa)$ and the abscissa as the reflected impulse I (kPa-msec).

(2) Plot curves of distance v explosive charge weight using pressure and impulse values from a table of air blast parameters such as the curves for TNT hemispherical surface blast from Kingery and Bulmash (1984).

(3) Draw the impulsive asymptote of the isodamage curve for the mode of damage occurring at y_{max}; for example, for $y_{max} = 200$ mm the impulsive asymptote:

$$I = y_{max}\sqrt{km} = 200\sqrt{0.29 \times 11.36} = 363 \, \text{kPa} - \text{msec} \quad \text{since} \frac{I}{y_{max}\sqrt{km}} = 1.0$$

(4) Draw the quasi-static asymptote of the isodamage curve for the mode of damage occurring at $y_{max} = 200$ mm:

Since $\dfrac{y_{max}}{P_0/k} = 2.0,$ the quasi-static asymptote

$$P_0 = \frac{y_{max} \, k}{2} = \frac{200}{2} 0.29 = 29 \, \text{kPa}$$

(5) If the transition curve is omitted, the asymptotes then give a conservative estimate of the (I, P_0) combination, which produces the specified mode of damage.

6.6 DAMAGE MITIGATION

6.6.1 Energy absorbing crush-up materials

Shock attenuating materials are used to reduce both the peak pressure and the impulse transmitted to structures from air or ground borne shock waves. When the materials are applied to the external surfaces it is known as 'backpacking' and stress waves are attenuated before reaching the structure. The strength and thickness of the structure can then be reduced, so reducing its cost.

External shock mitigating materials must have a low compressive strength with a high compressibility and energy absorption. Materials with an elastoplastic stress–strain curve such as rigid polystyrene, polyurethane foams and cellular concrete are used as shock mitigators. So also are materials with a plasto-elastic stress–strain curve such as foamed rubber, expanded clay, shale and slag.

An external Shock Mitigating (ESM) system described by Muszynski and Rochefort (1993) uses empty plastic bottles to confine and entrap air in a low-density cementitious matrix. Another system uses epoxied hollow ceramic beads to form foam. Static compression tests on these materials demonstrate the importance of confinement on the stress–strain curve. A dynamic test was performed using ground shock effects of close-in detonations against a concrete basement wall. Both ESM systems reduced the transmitted stress on the wall by about 90 per cent of the peak free field stress. The impulse was reduced by at least 12.5 per cent.

Hulton and MacKenzie (1998) use a qualitative energy analysis to explain why some ESM systems apparently increase the vulnerability of a wall in some circumstances. It was concluded from experiments using high explosive air blast against wall panels of lightly reinforced concrete, that when response was in the impulsive range, damage to the wall was increased when an ESM cellular steel panel was used that absorbed energy through plastic crushing of the cells. Experimentally it was observed that the ESM reduced the reflected pressure on the wall from the blast wave. If this corresponds with a reduction in the impulse on the wall, there will be a reduction in the kinetic energy transferred to the wall and damage should be reduced. Damage, however, was increased in the impulsive range (i.e. when the positive duration of the blast was much less than the natural period of the wall). Considering the overall energy it was argued that a reduction in the reflected impulse implies a reduction in the energy reflected from the wall which implies an increase in the energy absorbed by the wall and an increase in damage as observed.

Zhao and Gary (1998) tested two types of aluminium honeycombs in the three orthogonal directions X_1, X_2 and X_3 under static loading and at impact velocities of 2, 10, 28 m/sec. The minor cell diameters were 4.7 mm and 6.2 mm and densities were 130 kg/m^3 and 100 kg/m^3, respectively, before testing.

There were no visible differences between the static and impact loaded honeycombs for X_1 and X_2 in-plane, lateral loading, but for each direction the failure mode was different. Under X_1 loading the mean pressure was 0.09 MPa at all rates of loading and the Presssure–Crush per cent curves were ideal plasto-elastic curves. Under X_2 loading the Pressure–Crush per cent curve is ideal elastoplastic, and the mean pressure is close to the static value. In the X_3 out of plane, axial loading tests there was a 40 per cent difference in the mean crushing pressure between static and impact loading. The mean impact crushing pressure was constant at 5.4 MPa and the Pressure–Crush per cent curve was ideal elastoplastic for all impact loads. Wierzbicki (1983) gives the mean crushing pressure p_m as a function of $\mathcal{J}_0$ the flow stress of honeycomb foils, h the cell wall thickness, and S the minor cell diameter:

$$p_m = 16.56 \mathcal{J}_0 \left(\frac{h}{S}\right)^{5/3} \tag{6.13}$$

The observed enhancement of the crushing strength is likely to depend more on structural inertia and less on the strain rate sensitivity of aluminium foils.

The specific energy absorbed (J/cm^3) vs. normalized deformation of high density metal honeycombs at initial strain rates from quasi-static to 2,000/sec has been obtained experimentally by Baker *et al.* (1998). The honeycombs were aluminium with a density 32 per cent that of solid aluminium and stainless steel with a density 37 per cent that of solid steel. The stainless steel honeycomb absorbed almost double the energy absorbed by the aluminium honeycomb at similar deformation. Strain rate effects were observed on stress in both metals. Aluminium type 5052, is not expected to be strain rate dependent and the rate effect observed in these tests was considered to be solely caused by a change in collapse mode as the strain rate increased. Post-test inspection of the aluminium honeycomb, however, did not show a distinct difference in the permanent deformation between the quasi-static and impact tests. The stainless steel honeycomb is likely to have combined material and collapse rate effects. It was observed that deformation was distributed more uniformly along the length of the quasi-static specimen but propagated from the impact end in the dynamic test.

Sierakowski and Ross (1993) demonstrated that the properties of novel thermo-plastic honeycomb structures manufactured from high impact polystyrene, poly-carbonate and surlyn, were strain rate sensitive. The compressive dynamic/static strength ratio for the materials tested in a split Hopkinson bar, at strain rates of approximately 230/sec, was between 1.40 and 1.47 for the polycarbonate and poly-styrene and was 3.72 for the surlyn. Strain at peak stress decreased with increasing strain rate and there was no increase in energy absorption for any of these materials at high strain rate but the honeycomb structure has considerable potential for energy absorption. The longitudinal wave speed in the polycarbonate was measured to be 500 m/sec.

Harrigan *et al.* (1998) demonstrated the inertia effects in the performance of energy absorbing materials and structures, both experimentally and computation-ally. During the dynamic internal inversion of metal tubes of uniform thickness, inertia produced an initial peak force in excess of the steady state force and reduced the steady state force compared with its quasi-static value. Peaks in the crushing force of ESM systems increase the shock on the structure. For cellular aluminium honeycombs, inertia makes the crushing stress sensitive to impact velocity and modifies the crushing mechanisms at the cell wall, increasing the initial crushing stress and the plateau stress. The static load displacement character-istics of the aluminium honeycomb specimens are given in Figure 6.36. Some specimens were pre-crushed to initiate inelastic deformation and some were uncrushed.

To assess the dynamic properties of aluminium honeycombs, cylindrical specimens on the end of an instrumented Hopkinson pressure bar were impacted at velocities of 20 m/sec up to 300 m/sec. The dynamic force pulse was measured, giving the initial peak stress and the energy absorption characteristics.

The quasi-static uniaxial load–displacement curves are elastoplastic with an initial peak stress for the pre-crushed specimens typically 60 per cent of that for the initially uncrushed specimens. The plateau stress was the same for crushed and uncrushed specimens at approximately the initial peak stress of the pre-crushed specimens. All specimens had a locking displacement and the specific energy

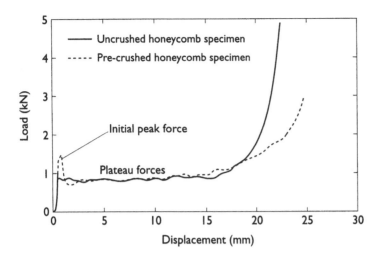

Figure 6.36 The static load–displacement characteristics of an aluminium honeycomb.

absorption capacity was determined from the area under the load-displacement curves. If lockup occurs before sufficient energy has been absorbed, then the ESM system may act to increase the shock loading on the structure. The ratio of dynamic to static initial peak crushing stress and plateau stress, increase significantly with impact velocity and would need to be evaluated when designing an ESM system. It is beneficial that the energy absorption increases significantly with increasing impact velocity.

Density and strain rate dependence influence the yield or peak initial stress, the plateau stress and the lock-up or compaction strain in polyurethane foams and can have important consequences when they are used for ESM systems. Beneficial effects occur when the changes increase the area under the stress strain curve so increasing the energy absorption capacity. The effects are detrimental if lock-up occurs, so increasing the transmitted load. The strain rate sensitivity of polyurethane foams of different density has been reported by Kuennen and Ross (1991). These experiments have shown that there is a tendency for the lock-up strain to decrease as strain rate increases. A reduction of 10 to 30 per cent was discerned at strain rates above 10^3/sec. Polyurethane with density between 0.16 and 0.48 g/cc tested at strain rates from 2×10^3 to 3×10^3/sec had, respectively, dynamic yield and plateau stress from 1.5 to 2.0 times greater than the static stress at 15 per cent strain. The static and dynamic compressive stress–strain curves had the characteristic elastoplastic compaction shape.

Fujimoto *et al.* (1991) tested ESM systems for underground structures by impacting the surface of sand or bentonite clay covering a beam. This was protected from the subsurface shock by an ESM system of either a paper honeycomb of three different strengths, or polyethylene foam. The ground shock was determined by measuring the flexural stress in the beam at midspan. The static compressive strength–displacement curves for the paper honeycomb was elastoplastic with no compaction curve, but was elastic with a compaction curve, for the polyethylene foam. The paper honeycomb was the most successful ESM and reduced the stress at midspan to about 30 per cent of the value with no ESM. The Polyethylene gave hardly any stress reduction in the beam, probably because it compacted early in the deformation. Without an effective ESM the beam would need much greater hardness or depth of earth cover to limit the stress in the beam.

Shock mitigating systems do not have to be applied to the external surface of an underground structure to reduce the damaging effects of ground shock. Krauthammer *et al.* (1995) reported that a trench filled with soft material reduced the peak stress from 5.7 MPa to 2.75 MPa from an underground explosion.

If a subsurface explosion produces fragments, soft materials that mitigate the ground shock will be ineffective in stopping the fragments. Anderson *et al.* (1995) tested concrete slabs by impact from a 7.62 mm armour piercing bullet. A surface layer with polystyrene beads replacing some of the aggregate, was easily perforated although it did reduce the concrete crater. Most successful was the use of slurry infiltrated concrete (SIFCON) cast as a surface layer on a concrete slab with conventional aggregates, Anderson *et al.* (1992).

6.7 DESIGN CODES

6.7.1 General principles of design codes for dynamic loading

Codes of Practice have an important function of reassurance to the general public. They are the interface between the designer, constructor, researcher and manufacturer and give legislative control over the standards of the building and construction industry. It is through the codes of practice and building regulations that Governments bring about changes. After the Ronan Point collapse in 1968, changes to improve the robustness of buildings were introduced by an amendment to the Building Regulations 1970 and then to Codes of Practice from 1972.

Damage resulting from the failure of an element should not spread disproportionately, a robustness that Ronan Point clearly did not possess. Vulnerable key elements and subframes, must be designed to resist extreme loads and if this is impractical, then the structure must be provided with ways to limit the collapse following the failure of the element.

This section deals only with the recommendations provided in standard codes of practice to improve the robustness of constructed facilities. They can be regarded as a minimum requirement and it is possible that no other action needs to be taken against extreme loads when there is only a remote possibility of them occurring. Facilities which are at risk, or would have to provide essential recovery services such as hospitals, bridges and public utilities, should be designed for the extreme loading.

6.7.2 Accidental and extreme loading

BS8110: Part 1:1997, Part 2: 1985

These Codes of Practice for the structural use of concrete recommend ways to improve robustness and structural integrity. Dynamic effects include wind loads and vibration, special hazards for flourmills and chemical plants which could include explosions, and shock loading on prestressed concrete beams which could include impact. A structure is robust if it is not disproportionately susceptible to the effects of accidental loading including unexpected and extreme dynamic loading.

To make it more robust, reinforcing bars tie the structure together to resist a notional horizontal load. This load is applied simultaneously at each floor or roof level and at each level has the value of 1.5 per cent of the characteristic dead weight of the structure to mid-height of the storey above and below. This notional load acts like wind loading or an equivalent horizontal force as a result of frame sway, and so does not alter the deformed shape of the structure from that under normal design loads. Dynamic loading from impact and explosions produces a different response to that from the normal design loads of gravity and wind. These dynamic effects can include:

- support reactions in the same direction as the shock load;
- shock loads opposite in direction to gravity or wind loads;
- a change in the material and structural properties with rate of loading;
- a change in the mode of flexural and longitudinal deformation;
- rapid changes in the distribution of the load across the structural element.

Structural integrity depends on having a load path for all loads including accidental, to transfer them from the point of application to a foundation and the ground. Extreme accidental loads might sever the load path and so redundancy is needed to provide alternative load paths.

The codes do not relate robustness or structural integrity to any particular loading and a concrete structure of more than four storeys is assumed to have adequate robustness against a general array of accidental actions.

A flow chart of the design procedure for ensuring robustness and the empirical design of ties using partial load factors on loads and material properties is given in Part 2 of the code. The load factor takes into account possible increases in load, and the effects of exceptional loads caused by misuse or accident, combined with dead load and a proportion of the wind and live loads. For these accidental load cases the partial safety factors for loads $\gamma_f = 1.05$ and materials $\gamma_m = 1.3$ for concrete in flexure and 1.0 for steel reinforcement, are less than those used in design for the standard static loads, although the uncertainty of the load and of the material strength is unlikely to be less for accidental load cases. If the loads are applied at a rate of straining above 10/sec, the steel and concrete yield strength will be enhanced and possibly offset uncertainty in the design load. The designer could change the γ factors on load to achieve the same effect.

Internal ties are reinforcing bars spread evenly within the floor slabs or grouped in walls or beams, and must be continuous and well anchored. The tie cross sectional area is determined from a prescribed tensile force, calculated from the total characteristic loads applied to the floor, the number of storeys in the structure and the effective span of the slab in the direction of the tie. Flexural reinforcement may also be used to complement the tie steel. The flexural steel is assumed to be acting at its design strength and the tie steel at its characteristic strength. Curtailment of flexural longitudinal reinforcement to match the bending moment diagram must not reduce the tie steel at any section.

Peripheral ties are designed to resist a nominal tensile force F_t that depends only upon the number of storeys for buildings up to ten storeys, and is constant at 60 kN for buildings above ten storeys and continuous around the edge of the building. They resist the force in the internal ties to provide them with adequate anchorage.

Columns and walls around the perimeter of a building are to be anchored to the structure at each floor and roof level. The horizontal tie must have a tensile resistance based on the number of storeys, the floor to ceiling height or the total design load in the column or wall. The benefit of having this reinforcement is clear but it can be omitted if the peripheral tie is within an external wall and horizontal tying anchors the internal ties to the peripheral ties. Corner columns must be tied in two approximately orthogonal directions at each storey. Vertical ties in each column or wall must be continuous from the lowest to the highest level.

Key elements in buildings over four storeys are designed for an ultimate load of $34 \, kN/m^2$ applied from any direction on the projected area of the member, and on any horizontal member providing essential lateral support to the key member, and on any attached element such as a cladding panel, supported by the key element. An allowance can be made for the strength of the attached element and its connection.

In buildings over four storeys, beam or slab elements supporting the maximum design ultimate dead and imposed load, must be designed to bridge the increased span when a supporting vertical element on the storey below, is removed. Catenary action may be utilized for this when the necessary horizontal reactions are provided at the adjacent supports.

A wall is able to provide lateral restraint if it is capable of resisting the peripheral tie force F_t kN applied horizontally on each metre height of wall.

Under extreme loading from impact or explosions the serviceability limit states would be exceeded and the feasibility of carrying out repairs would have to be assessed.

During construction the risk of imposing extreme loads is present when lifting elements by crane and in collapse of the partially constructed frame.

To provide continuous ties in precast concrete construction, part of the length of each bar in an element, is lapped or spliced with anchored bars from the supports. Such connections are topped with *in situ* concrete and the static load in the tie gives the bond or the bearing stress. The overall stability of the building during construction or after accidental damage is likely to depend on these connections.

ENV 1993–1–1: 1992 Eurocode 3: Design of Steel Structures
Part 1.1: General rules and rules for buildings

This recent code is more explicit than many of the BS codes in requiring that a structure shall resist explosions, impact, or the consequence of human errors, and avoid disproportionate damage. To limit the potential damage, designers could:

- avoid, eliminate or reduce the hazards to which the structure is exposed;
- select a structural form which has low sensitivity to the hazards;
- select a structural form and design to survive the accidental removal of an individual element;
- tie the structure together.

These requirements are met by choosing suitable materials, by appropriate design and detailing and by specifying control procedures for their production, construction and use. The *National Application Document* for use in the UK with ENV 1993–1–1: 1992, states that substantial permanent deformation of members and connections is acceptable in achieving these requirements.

The general recommendations for structural integrity are very similar to those in BS8110: 1997 for reinforced concrete buildings. Internal ties, column ties, edge ties and peripheral ties are all required at each principal floor and roof level to localize accidental damage. These ties must have a design tensile resistance not less than 75 kN at floors and 40 kN at the roof. Ties in the floors of multi-storey buildings depend upon w_f the total design load on the slab, s_t the mean transverse spacing of the ties, and L_a the greatest distance in the direction of the tie between centres of adjacent lines of support. The force must be:

- for internal ties, $0.5\,w_f\,s_t\,L_a$;
- for edge ties, $0.25\,w_f\,s_r\,L_a$;
- for peripheral ties, the greater of these or 75 kN at floors and 40 kN at the roof, or 1 per cent of the design vertical load in the column at that level;
- column lengths must have splices with a design tensile resistance not less than $\frac{2}{3}$ of the design vertical load applied to the column from the floor below the splice.

EC3 differs from BS8110 in specifying that if a single column is removed, there must be not more than $70\,\text{m}^2$ or 15 per cent of the area of the storey that collapses under persistent floor loads factored by the ψ factors.

Key elements are those supporting more than $70\,\text{m}^2$ or 15 per cent of the area of the storey, and their essential lateral restraining elements. These elements must not fail when loaded by an accidental load A_K not less than $34\,\text{kN/m}^2$ factored by $\gamma_A = 1.05$, applied in the appropriate directions. The reactions from other building components attached to the key element and subjected to A_K must also be included but limited by the ultimate strength of the components or connections. The value of A_K is not limited to $34\,\text{kN/m}^2$ but depends upon the importance of the key element and the consequences of failure. This accidental load is assumed to act in combination with the dead and imposed loads using a combination factor.

EC3 is a limit state code and structural elements must be designed for ultimate limit states when subjected to impact or blast defined as accidental actions. Structures under smaller impact loads applied cyclically, for instance by some industrial machinery, must be checked from first principles for the serviceability limit state of vibration. Actions are classified by their variation in time and impact and blast, although short relative to other forms of loading, can often be treated as an equivalent static load if the duration is long relative to the natural period of the structural element. Actions are also classified by their spatial variation and this may change rapidly when confined blast pressures are applied, or from a close range.

Characteristic values for impact and blast are most likely to be specified by the client and the designer but need to satisfy the minimum $A_K = 34\,\text{kN/m}^2$ specified by the code. The design value of the accidental action is $A_d = \gamma_A A_K$ where γ_A is the partial safety factor for accidental actions.

Material properties are characteristic values, which allow for variability in a sample from variations in manufacture. The use of two characteristic values is useful when an accidental action is dynamic with a variable rate of loading. Many materials used in construction have an increase in strength and stiffness at high strain rates. To neglect this increase may be conservative for checking the response of the element but not for checking the response of its connections. The design material property is $X_d = X_k / \gamma_M$, where γ_M is the partial safety factor.

The design requirement for safety of a structure is that the ultimate limit state design capacity is at least equal to the accidental actions:

$$\sum_j \gamma_{GA,j} G_{k,j} + A_d + \psi_{1,1} Q_{k,1} + \sum_{i>1} \psi_{2,i} Q_{k,i}$$

where $G_{k,j}, Q_{k,1}$ and $Q_{k,I}$ are the characteristic values of the permanent actions, the main variable action and any accompanying variable actions respectively. The values $\gamma_{G,j}, \gamma_{GA,j}$ and $\gamma_{Q,I}$ are the partial safety factors for the permanent actions, the permanent actions in accidental design situations and the variable action respectively and are taken as equal to 1.0. The factors $\psi_{1,1}$ and $\psi_{2,I}$ give the quasi-permanent fraction of the variable actions $Q_{k,I}$ and $Q_{k,I}$. This design value can be used after an accidental event for checking the remaining design capacity, in which case $A_d = 0$. When considering an accidental situation, the direction and position of the accidental actions may be different from the permanent and variable actions arising from the normal use of the structure.

ENV 1991-2-7: 1998 Eurocode 1: Basis of design and actions on structures. Part 2–7: Actions on structures – Accidental actions due to impact and explosions

The three-year period of experimental application for this code began in August 1998, during which period it is only approved for provisional application and open for comment.

Accidental impact forces from road or rail traffic under bridges or other structures and from vehicles on the bridge are given for the design of structural elements or their protection systems. Forces for ship impact are given, and for helicopter emergency landing impact when the landing pad is on the roof of a building.

Accidental explosions are considered to be of the deflagration type from air–gas or air–dust mixtures. The pressure rise is slow relative to detonations and there is a constant gas pressure phase following the peak. The duration is usually longer than that of detonations.

Impact and explosive actions are categorized in terms of injury and death to people, unacceptable change to the environment, and large economic loss to the community. When the consequences are low to medium the static equivalent forces, or prescriptive design and detailing, can be adopted for design of the structural elements and protective systems. A more advanced analysis is indicated for the most serious consequences. The acceptable risk level and seriousness of the consequences has to be determined case by case and by public reaction to the cost and disturbance of installing safety measures and reaction after an accident. A structure is considered to be at risk from an accidental action when the probability of the action exceeds 10^{-4} per year. The necessary statistics are not often available and nominal design values are given for use in practice.

Horizontal static equivalent design forces due to vehicle impact on the supporting substructure of a bridge, such as walls and columns, are tabulated for the type of road and vehicle. The maximum force is 1,000 kN in the direction of normal travel for a truck on a motorway and the minimum is 40 kN for a car in a parking garage. The nominal forces perpendicular to the direction of normal travel vary from 500 kN to 25 kN for these cases. These loads are spread over prescribed impact zones. No horizontal forces need to be considered on overhead elements unless the clearance is less than 6 m. If it is less, then a prescribed horizontal force determined by the clearance is applied to vertical surfaces and a force inclined at 10° to the horizontal acts on the underside of the bridge over the traffic lane.

Horizontal static equivalent design forces due to impact of rail traffic on overhead bridges or nearby structures, are also specified parallel and perpendicular to the track direction. The maxima are 10,000 kN and 3,500 kN, respectively, depending on the speed of the train and act on a specified area. Impact forces on the superstructure of the bridge are not specified.

Static equivalent design forces from 22,000 kN to 4,000 kN are specified for accidental actions caused by ship impact. The bow impact zone is dimensioned above and below the water line but could be altered by the lifting of the bow on impact with a column foundation block. Modification factors are used on the static equivalent forces for stern and side impacts and bow impact by a ship off course.

Equivalent static loads, or prescriptive design and detailing rules are also specified for the design of structural elements and protective systems under accidental actions produced by an explosion. A full dynamic analysis is only recommended for category three consequences.

Typical pressures for air–gas and air–dust deflagrations are given as $1,500 \text{ kN/m}^2$ and $1,000 \text{ kN/m}^2$ but will depend on the size, shape and venting of the enclosure. The notional accidental static pressure for the design of a key element in category 2 is 20 kN/m^2 from any direction with an added reaction from an attached building component subjected to the same pressure. This contrasts with the 34 kN/m^2 used in British Standard codes.

6.8 REFERENCES

Amde, A. M., Mirmiran, A. and Walter, T. A. (1996) 'Local damage assessment of metal barriers under turbine missile impact', *ASCE J. Struct. Engineering*, **122**(1), p. 99.

Anderson, W. F., Watson, A. J. and Kaminskyj, A. (1992) 'The resistance of SIFCON to high velocity impact', paper given at 2nd International Conference on Structures Under Shock and Impact, Southampton, pp. 89–98.

Anderson, W. F, Watson, A. J. and Kaminskyj, A. (1995) 'Construction to defeat small arms attack', paper given at 7th International Symposium on Interaction of the Effects of Munitions with Structures, Mannheim, Germany, April 1995, pp. 181–8.

Baker, W. E, Togami, T. C. and Weydert, J. C. (1998) 'Static and dynamic properties of high density metal honeycombs', *Int. J. Impact Engng*, **21**(3): 149–63.

Baker, W. E., Cox, P. A., Westine, P. S., Kulesz, J. J. and Strehlow, R. A. (1983) *Explosion Hazards and Evaluation*, Elsevier Scientific Publishing.

Barr, P., Brown, M. L., Carter, P. G., Howe, W. D., Jowett, J., Neilson, A. J. and Young, R. L. D. (1980) 'Studies of missile impact with reinforced concrete structures', paper given at BNES Symposium on Design of Chemical and Nuclear Installations against Impact from Plant Generated Missiles, Leicestershire.

Berriaud, C, *et al.* (1978)Local behaviour of reinforced concrete walls under missile impact, *Nuclear Engineering and Design*, **45**: 457.

Biggs, J. M. (1964) *Introduction to Structural Dynamics*, McGraw-Hill, New York.

BS 8110 (1997) *The Structural Use of Concrete: Part 1, Code of Practice for Design and Construction*.

BS 8110 (1985) *The Structural Use of Concrete: Part 2, Code of Practice for Special Circumstances (incorporating Amendment 1)*.

Burton, R. (1958) *Vibration and impact*, Dover Publications, Inc., New York. Library of Congress Catalog Card No. 68–17,404.

CEB (1988) *Concrete Structures Under Impact and Impulsive Loading – Synthesis Report*, CEB Bulletin d'Information No. 187, August.

Chen, H-Liang and Chen, S-En. (1996) Dynamic responses of shallow-buried flexible plates subjected to impact loading. *ASCE J. Struct. Engineering* **122**(1), 55–60.

Craig, R R. (1981) *Structural Dynamics*, John Wiley, New York.

DD ENV 1993–1–1 (1992) Eurocode 3, *Design of Steel Structures: Part 1.1, General Rules and Rules for Buildings* (together with United Kingdom National Application Document).

Dragosavic, M. (1973) 'Structural measures against natural gas explosions in high rise blocks of flats', *Heron* **19**(4).

Eibl, J. (1985) 'The design of impact endangered concrete structures', paper given at 8th SMiRT Conference, Vol. H, Amsterdam. pp. 265–70.

Eibl, J. (1987) 'Soft and hard impact', paper given at 1st International Conference on Concrete for Hazard Protection, Edinburgh.

Eibl, J. (1993) *Civil Engineering Aspects of New Nuclear Reactor Containments*, CEB Bulletin d'Information No. 219, Safety and Performance Concepts, August, pp. 113–29.

ENV 1991-2-7 (1998) Eurocode 1, *Basis of Design and Actions on Structures: Part 2–7, Actions on Structures – Accidental Actions due to Impact and Explosions*, CEN, Brussels.

EPSRC (1997) *Engineering and Physical Research Council*, Research Focus No. 31, November.

Fujimoto, K. Yamaguchi, H. Sayama, M. and Kuzuha, Y. (1991) 'Effects of cushion layer for underground structure subjected to impact loading', paper given at 5th International Symposium on Interaction of Conventional Munitions with Protective Structures, Mannheim, Germany, April, pp. 126–31.

Griffiths, H., Pugsley, A. and Saunders, O. (1968) *Report of the Inquiry into the Collapse of Flats at Ronan Point, Canning Town*, Ministry of Housing and Local Government, HMSO, London.

Harrigan, J. J., Peng, C. and Reid, S. R. (1998) 'Inertia effects in impact energy absorbing materials and structures', paper given at International Symposium on Transient loading and response of structures, Trondheim, Norway, pp. 447–74.

Hulton, F. and MacKenzie, J. F. (1998) 'The mitigation of blast loading on structures: energy and failure', paper given at International Symposium on Transient Loading and Response of Structures, Trondheim, Norway, pp. 649–57.

ISO 2394: *General Principles for the Verification of the Safety of Structures*.

Kennedy, R. P. (1976) A review of procedures for the analysis and design of concrete structures to resist missile impact effects, *Nuclear Engineering and Design* **37**: 183.

Kingery, C. N. and Bulmash, G. (1984) *Airblast Parameters from TNT Spherical Air Burst and Hemispherical Surface Burst*, US Army Armament Research and Development Center, Ballistics Research Laboratory, Aberdeen Proving Ground, MD. Technical Report ARBRL-TR-02555.

Kinney, G. F. and Graham, K. J. (1985) *Explosive Shocks in Air*, Springer-Verlag, Berlin.

Krauthammer, T. O Daniel, J. L., Guice, R. L. and Jenssen, A. (1995) 'Backfill/Backpack effects on shelter response', 7th International Symposium on Interaction of the Effects of Munitions with Structures, Mannheim, Germany, April 1995, pp. 117–24.

Kuennen, S. T. and Ross, C. A. (1991) 'Strain rate effects on compressive testing of foam', paper given at 5th International Symposium on Interaction of Conventional Munitions with Protective Structures, Mannheim, Germany, April 1991, pp. 275–83

Mainstone, R. J. (1971) *The Breakage of Glass Windows by Explosions*, BRS Current Paper CP26/71.

Mays, G. C. and Smith, P. D. (1995) *Blast Effects on Buildings*, Thomas Telford Publications, London.

Muszynski, L. C. and Rochefort, M. L. (1993) 'Materials for external shock mitigation', paper given at 6th International Symposium on Interaction of Non-Nuclear Munitions with Structures, Panama City Beach, FL, pp. 95–9.

National Building Code of Canada, National Research Council of Canada, Ottawa 1995.

Pan, Y. G. and Watson, A. J. (1996) 'Interaction between concrete cladding panels and fixings under blast loading', *Cement and Concrete Composites* **18**: 323–32.

Roberts, A. F. and Pritchard, D. K. (1982) Blast effect from unconfined vapour cloud explosions, *Journal of Occupational Accidents* **3**: 231–47.

Sadee, C., Samuels, D. E. and O'Brien, T. P. (1976) The characteristics of the explosion of cyclohexane at the Nypro (UK) Flixborough Plant on 1st June 1974, *Journal of Occupational Accidents*, **1**: 203–35.

Sierakowski, R. L. and Ross, C. A. (1993) 'Dynamic properties of novel thermoplastic honeycomb core material' paper given at 6th International Symposium on Interaction of Non-Nuclear Munitions with Structures, Panama City Beach, Florida, pp. 329–31.

Smith, P. D. and Hetherington, J. G. (1994) *Blast and Ballistic Loading of Structures*. Butterworth-Heinemann Ltd, London.

TM5-855-1, (1986) *Technical Manual, Fundamentals of Protective Design for Conventional Weapons*, US Department of the Army.

TM5-1300-1 (1991) *Structures to Resist the Effects of Accidental Explosions*, US Department of the Army.

Watson, A. J. and Chan, A. K. C. (1987) 'Some design recommendations obtained from impact experiments on model scale prestressed concrete beams', paper given at the Concrete Society International Conference, Concrete for Hazard Protection, Edinburgh, pp. 187–99.

Watson, A. J. and Ang, T. H. (1984) 'Impact response and post-impact residual strength of reinforced concrete structures', paper given at International Conference on Structural Impact and Crashworthiness, Imperial College, London.

Watson, A. J. (1988) 'Stress wave fracturing of concrete', paper given at International Conference on Decommissioning, UMIST, Manchester, March.

Watson, A. J. (1994) 'Response of Civil Engineering structures to impulsive loads', paper given at International Conference on Structures Under Shock and Impact III, Madrid, June, pp. 3–10.

Watson, A. J. and Ang, T. H. (1982) 'Reinforced microconcrete beams under impact loading', paper given at Interassociation Symposium Concrete Structures Under Impact and Impulsive Loading, Berlin, June.

Watson., A. J., Hobbs, B., Chan, A. K., Peters, J. and Westaway, R. (1991) 'A PC based program for blast load analysis and damage estimation,' paper given at 5th International Symposium on Interaction of Conventional Munitions with Protective Structures, Mannheim.

Wierzbicki, T. (1983) Crushing analysis of metal honeycombs. *Int. J. Impact Engng* **1**: 157–74.

Yang, Y-B. and Yau, J-D. (1997) 'Vehicle-Bridge interaction element for dynamic analysis', *ASCE J. Struct. Engineering* **123**(11): 1512–1518.

Yang, Y-B, Yau, J-D and Hsu, L-C. (1997) Vibration of simple beams due to trains moving at high speeds. *Eng. Struct.*, **19**(11): 936–44.

Zhao, H. and Gary, G. (1998) Crushing behaviour of aluminium honeycombs under impact loading. *Int. J. Impact Engng* **21**(10): 827–36.

Zukas, J. A., Nicholas, T., Swift, H. F., Greszczuk, L. B. and Curran, D. R. (1982) *Impact Dynamics*, Wiley Interscience Publication.

Chapter 7

Human-induced vibrations

J. W. Smith

7.1 INTRODUCTION

Human-induced forces can be critical in the design of certain types of structure. Significant dynamic loading is generated by quite normal activities such as walking, running, marching and dancing. Light or flexible structures, for example footbridges and lightweight floors, are particularly susceptible and can be made to vibrate with unacceptable intensity under the motion of a single person in some cases. The problem may be even more serious when large numbers of people jump, dance or sway in unison as at pop concerts or sports events. Designers should give particular attention to the possibilities of vibration when designing the following: footbridges, long span lightweight office floors, lightweight staircases, dance halls or gymnasiums, and grandstands or other auditoria.

There are three important elements in the design of structures for human loading. First, the overall loading and its dynamic components must be assessed. This is not easy because the behaviour of human beings is notoriously difficult to quantify. Design guidance is available for footbridges, light floors and grandstands but numerical information is generally limited to conventional structural forms. Great care should be taken with unusual structures particularly if large numbers of people are likely to be involved.

Secondly, the analytical model of the structure and loading must be considered. Relatively simple closed form solutions are generally possible for long span floors that are rectangular in plan. Simple solutions are also available for footbridges with simple structural configurations. However, in recent years there has been a trend towards footbridges with ambitious structural forms. Some have been built in busy urban environments with increased risk of dynamic crowd loading and existing design rules are inadequate in these circumstances. Simplified modelling is not satisfactory and recourse to finite element procedures will be required. An important factor is that human loading is highly mobile, and for important structures a wide range of load cases and positions should be considered. Modal analysis by finite elements will generally be required for grandstands because of their geometry. The consequences of collapse of a grandstand are very serious and every effort should be made to ensure that an extensive range of loading scenarios have been considered and accurately analysed.

Finally, the design criteria have to be considered. Detrimental consequences of human induced vibration may include over stress of the structure or perceptible motion that is unpleasant for human users. Rhythmic loading, such as marching or dancing, can result in large dynamic amplification that may result in structural damage or collapse. A famous example of this was the collapse of a cast iron bridge at Broughton in 1831 under the resonance of 60 soldiers (Tilly *et al.*, 1984). This led to the custom of troops breaking step when marching over bridges. A more recent example was the collapse of part of a temporary grandstand at Bastia, in Corsica in 1992, which was thought to have been triggered by exuberant crowd motion and resulted in tragic loss of life. On the other hand, the vibration of a structure may be unacceptable simply because of the sensitivity of humans to the perception of motion. This is an unserviceability limit state that is nevertheless very important. There are cases on record of footbridges that were found to be too lively when built and required remedial treatment in the form of additional damping to reduce the alarming intensity of motion (Brown, 1977). The London Millennium Footbridge is an even more recent example. Vibration of light floors, caused by footfall in normal usage, can be disturbing to occupants of buildings particularly if they are trying to do sensitive work.

7.2 THE NATURE OF HUMAN-INDUCED DYNAMIC LOADING

7.2.1 Vertical loads due to walking

Vertical load under a person walking was studied initially by Harper *et al.* (1961). Human locomotion is a complex phenomenon but from the point of view of vertical loading a relatively simple description suffices. It is characterized by 'heel strike' followed by a stiff legged action as the upper body passes over the foot in contact with the ground, and finally 'toe off' at the end of the stride. There is a brief period when both feet are in contact with the ground when the 'heel strike' and 'toe off' become additive resulting in a sharp impact. During this motion the centre of gravity of the upper body rises and falls by about 50 mm resulting in a vertical acceleration and corresponding periodic inertia force at the pacing frequency. Assuming a normal walking frequency of 1.6 to 2.0 Hz a simple calculation shows that the vertical force will have an amplitude of between about 150 N and 200 N.

Accurate measurements of the vertical forces during walking were determined with the aid of an orthopaedic 'gait' machine by Skorecki (1966). Force–time curves giving the vertical component of typical foot impacts are shown in Figure 7.1. Two peaks occur characteristically under 'heel strike' and 'toe off'. The sizes of the peaks increase with speed of walking. When a person runs, 'toe off'

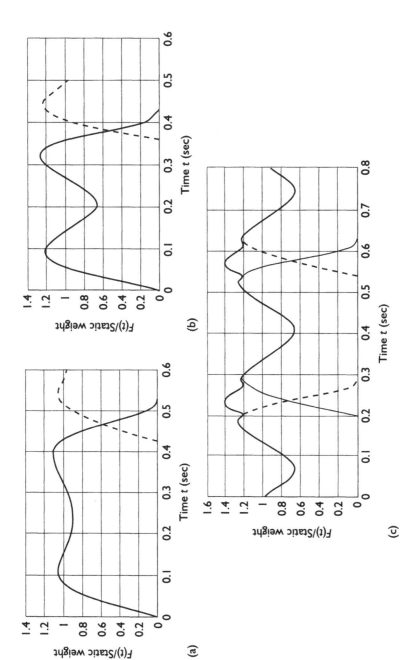

Figure 7.1 Force–time curves for walking (vertical component): (a) normal walk; (b) fast walk; (c) combined vertical force.

dominates and there is also a moment when the person is actually in flight and neither foot is in contact.

Furthermore, it should be noted that as a person walks across a structure, the point of contact changes with time. If the span is long compared with the stride length, a moving periodic force may represent the forcing function sufficiently accurately. This force may be determined by adding the vertical contributions of both feet as shown in Figure 7.1c. The result is a periodic forcing function, which may be decomposed into its Fourier series components. The first harmonic is the largest and Blanchard *et al.* (1977) recommended a magnitude of 180 N. This is particularly important in the case of footbridges and long span floors, which may be excited significantly by a single person.

Lenzen and Murray (1969) proposed the 'heel drop' test, which consists of a person of average weight rising up on his or her toes and then dropping suddenly on the heels. A typical force–time curve is shown in Figure 7.2a. The 'heel drop' test was suggested for assessing the vibration susceptibility of lightweight office floors under random walking loads. This impulse, which lasts about 1/20 of a second, not only simulates heel strike but is also considered to be representative of other miscellaneous impacts in an office environment (e.g. dropped objects). Typical vibration under 'heel drop' is shown in Figure 7.2b. In principle the locations of impacts are random but the greatest effect will occur when a person is in the vicinity of mid span of a floor. Recent studies by Ellis (2000) have indicated that individual heel strikes dominate the response of floors with high damping, but that resonance with the Fourier components of the vertical walking force is the most important factor for lightly damped floors.

Loading on staircases is similar but more intense than floor loading. This is because people often run up and down stairs resulting in very high heel strike in the latter case. This is not generally a problem for conventional reinforced concrete staircases. However, there have been instances of staircases with light or unusual supporting structures, designed for architectural effect, that have been found to vibrate excessively under dynamic loading.

7.2.2 Rhythmic excitation

Dancing, aerobics and certain gymnasium exercises are rhythmic in nature. They often involve jumping and may be co-ordinated by music or other source of regular prompting. This is usually referred to as 'dance-type' loading and because it is periodic it is particularly important from the point of view of resonance with the natural frequency of the floor structure. If the measured periodic forcing function under dance-type loading is decomposed into component frequencies it is found that the most important frequencies are between 2 and 3 Hz although significant frequencies as high as 5 Hz can be generated. Heins and Yoo (1975) investigated a dance hall in which the floor had a natural frequency of approximately 3 Hz and the vibrations during 'rock' dances were distinctly unpleasant.

The dynamic effect of crowd loading is important. Under normal circumstances,

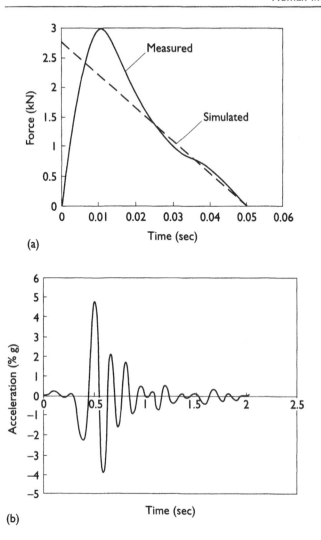

Figure 7.2 Heel drop test: (a) average force/time curve for heel impact; (b) vibration caused by impact (from Lenzen and Murray, 1969).

the combined effect of the dynamic components of large numbers of people is not significant because of randomness in their movements and a lack of co-ordination. Hence, a static load representative of the weight of closely packed people will be satisfactory. However, as with dance-type loading, the co-ordinating effect of music may give rise to large periodic excitation and the risk of resonance. Irwin (1981) reported extreme conditions at a pop concert when co-ordinated jumping of the densely packed crowd, in time with the beat of the music, generated a dynamic response factor of 1.97 at a predominant frequency of 2.5 Hz. A similar problem may occur in sports stadia or grandstands when sports fans sway to and fro rhythmically, hence generating substantial horizontal forces, but generally at a lower frequency than the vertical (Ellis *et al.* 1994). A further problem may occur due to human psychological interaction with the feedback from the motion of a structure, as outlined below.

7.2.3 Interaction between the structure and human body

Human induced forces, such as vertical loads under pedestrians, cannot be treated in isolation. This is because of interaction between the motion of the structure and the human body that is itself a complex mechanical system. This is illustrated in Figure 7.3 in which a human body is represented by a system consisting of masses, springs and damping elements, while the structure in this case is a simple beam with appropriate mass and stiffness. Ji and Ellis (1997a) showed that when a person is stationary s/he acts like a spring–mass–damper attached to the structure and affects its vibration characteristics. However, when dancing or jumping a person does not appear to affect the structure in the same way. It is as if the two systems behave independently. It is also clear that a human body never acts simply as a dead weight. For this reason the frequency analysis of floors or bridges should be based on the unloaded mass of the structure. In the case of crowd loading of stadium structures this matter is not so clear. Reid *et al.* (1997) recommend that the mass of spectators should be included when calculating the horizontal frequencies but that some judgement may be used in the case of vertical frequencies (Reid *et al.* discussion, 1998).

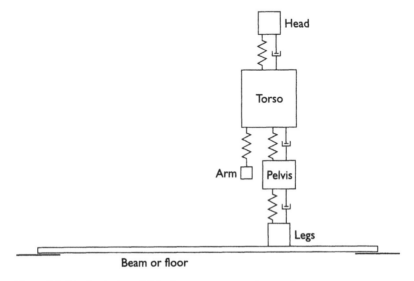

Figure 7.3 Mechanical model for human structure interaction.

A further complication arises from the psychological response of a person experiencing feedback from the motion of the structure. This has been noted in the case of excessively lively footbridges. Fujino *et al.* (1993) noted that crowds of people using a footbridge of flexible design tended to get into step with the horizontal motion of the structure. This, of course, increased the horizontal vibration. The reason for this phenomenon is thought to be the human body's subconscious desire to minimize energy usage in walking. Brown (1977) made similar observations but noted that if the motion became extreme it interfered with pedestrians' ability to walk normally, hence causing them to stop, slow down or get out of resonance.

The human body is highly sensitive to motion and it is usually the case that people on a structure will notice the vibration, and even find it unpleasant, well before there is any over stress of the structural members themselves. For this reason much research has been carried out to determine acceptable limits to vibration from the point of view of human users (Guignard and Guignard, 1970; Irwin, 1978; Irwin, 1983). Acceleration of the floor during dynamic motion, whether it be vertical or horizontal, is accepted as being the best parameter by which to measure human sensitivity to vibration, although at high frequencies velocity is a useful measure. Design criteria are provided in a British Standard document (BS 6472, 1984). This takes into account the relative sensitivity of humans in different environments. For example, people in residential properties are highly sensitive to vibration whereas manufacturing workers in a factory can tolerate higher amplitudes of vibration before becoming concerned.

A base curve representing the threshold of perception for vertical vibration, as given in BS 6472 (1984), is shown in Figure 7.4 together with curves with higher weighting factors for different tolerance criteria. The curves indicate that people are most sensitive to the perception of vibration in the frequency range from 4 to 8 Hz. Above 8 Hz sensitivity follows a constant velocity curve and it can be seen that between 8 and 15 Hz human tolerance doubles in terms of perceived acceleration.

Examples of the weighting factors are given in Table 7.1. These show that people are less tolerant of vibration if they are engaged in an activity during which mere perception of vibration is a nuisance, such as sleeping or work requiring concentration. However, people are more tolerant of infrequent or intermittent vibration. On the other hand, at pop concerts or some sporting events people are either not concerned at all by the feeling of motion of a structure or they actually enjoy it and may attempt to get in resonance with it. In these cases structures should be designed to resist collapse under resonant excitation.

7.3 METHODS FOR DETERMINING THE MAGNITUDE OF HUMAN-INDUCED LOADING

7.3.1 Footbridges

Blanchard *et al.* (1977) proposed that the worst conditions occur when a pedestrian walks in resonance with the natural frequency of a bridge with a stride length of

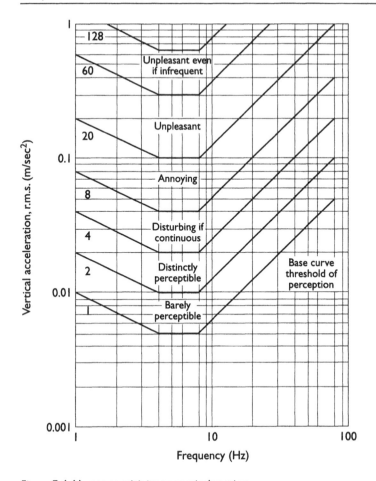

Figure 7.4 Human sensitivity to vertical motion.

Table 7.1 Weighting factors above the threshold of perception for acceptable building vibration.

Place	Time	Continuous or intermittent vibration and repeated shock	Impulsive shock with several occurrences per day
Critical working areas	Day	1	1
(e.g. hospital operating theatre)	Night	1	1
Residential	Day	2–4	60–90
	Night	1.4	20
Offices	Day	4	128
	Night	4	128
Workshops	Day	8	128
	Night	8	128

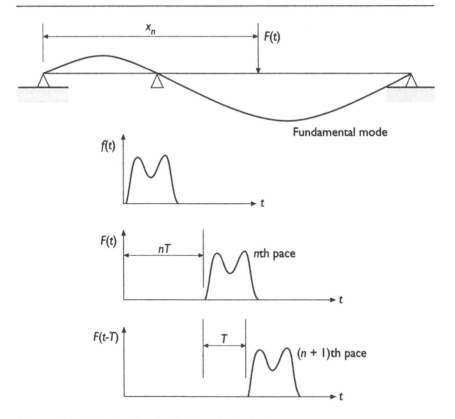

Figure 7.5 Simulated pedestrian loading of a footbridge.

0.9 m. The pacing frequency of normal walking lies between about 1.5 and 3.0 Hz, whereas frequencies above 3.0 Hz are representative of running or jogging. It is difficult to excite a footbridge with a frequency above 4 Hz.

The pedestrian forcing function may be represented by a series of point loads, each with a force–time curve of the form shown in Figure 7.1, applied at successive time intervals equal to the period of vibration T, as shown in Figure 7.5. Hence, the loading on the bridge, applied by the nth pace, is given by

$$F(t) = f[t - (n - 1)T] \tag{7.1}$$

and the position of the nth pace x_n is given by

$$x_n = 0.9n \tag{7.2}$$

Evidently there is no analytical solution to this forcing function, even for simply supported beams. Blanchard *et al.* (1977) used a numerical method to analyse foot-bridges with simple configurations under the action of the above dynamic loading. They found that the dynamic deflection could be expressed conveniently

Table 7.2 Configuration factor.

Configuration	a/L	K
	–	1.0
	1.0 0.8 <0.6	0.7 0.9 1.0
	1.0 0.8 <0.6	0.6 0.8 0.9

in the form:

$$u_{max} = u_{st} K \psi \qquad (7.3)$$

where u_{st} is the static deflection under the weight of a pedestrian at the point of greatest deflection, K is a configuration factor for the type of structure as given in Table 7.2, and ψ is a dynamic response factor. Bridge damping was included in the analysis of the standard cases and the dynamic response factor ψ was found to vary with main span length L and logarithmic decrement due to damping δ, as shown in Figure 7.6.

Blanchard *et al.* (1977) proposed a simplified loading function to permit analysis of bridges of more general configuration. As mentioned in Section 7.2.1, it consisted of a pulsating force moving across the span with a velocity of $0.9f$ m/sec and in resonance with the bridge, where f is the natural frequency of the fundamental mode of the bridge. The magnitude of the pulsating force was obtained by superimposing the individual left and right foot vertical forces as shown in Figure 7.1. It was found that the amplitude of this moving pulsating force was approximately 25 per cent of the static weight of a pedestrian to produce the same response as the rigorous method. Hence, the moving force, in Newtons, is given by

$$F(t) = 180 \sin 2\pi f t \qquad (7.4)$$

The most important criterion for dynamic design of footbridges is that they should not vibrate so much that users would be disturbed or alarmed. This is a human response criterion, as discussed in Section 7.2.3. The UK bridge code (BS 5400, 1978) recommends a maximum acceleration of footbridge decks of $\pm \frac{1}{2}\sqrt{f}$ m/sec^2 when one pedestrian walks over the main span in step with the natural frequency, f. This 'one pedestrian' test was calibrated against some real bridges that were known to be only just acceptable. It has been confirmed by Matsumoto *et al.*

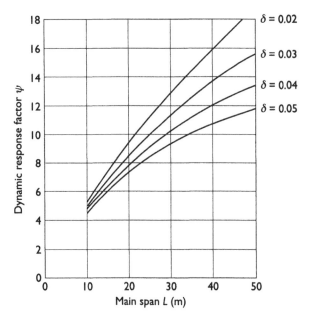

Figure 7.6 Dynamic response factor.

(1978) and Wheeler (1982) that the 'one pedestrian' test is realistic as a serviceability criterion. It is possible for two pedestrians to walk in step with the natural frequency of a footbridge with the assistance of a metronome and amplitudes are approximately twice the single pedestrian case (Tilly *et al.* 1984). However, such a condition is difficult to maintain. On the other hand, Pimentel (1997) has warned that dynamic effects of crowd loading on footbridges should be considered, especially for bridges in busy urban environments. Fujino *et al.* (1993) measured significant lateral vibration of a congested pedestrian bridge adjacent to a large sports stadium. Crowd loading will be considered in Section 7.3.5.

The horizontal component of pedestrian loading is not normally important because most bridges are stiffened laterally by their deck structures. However, the lively response of the London Millennium Footbridge, together with the observations of Fujino *et al.* (1993), showed that this component should not be ignored for bridges with flexible lateral structures.

The mechanics of horizontal load due to walking is different from the vertical component for a number of reasons. The lateral force is caused by the sway of the body from side to side at each step. However, the left and right feet produce horizontal forces in opposite directions, in contrast to the vertical forces. Hence, the frequency is approximately 1.0 Hz, being half that of the vertical forcing function. Bachmann and Ammann (1987) estimated that the amplitude of the first Fourier component of the horizontal force was 23 N. This is more than 10 per cent of the vertical amplitude given by eqn (7.4), but at half the frequency.

An important psychological factor also influences horizontal loading. Fujino *et al.* (1993) reported large horizontal vibrations of a long span footbridge that often carried as many as 2,000 pedestrians returning from sporting events at a boat race stadium. They showed that an amplitude of 23 N applied by 45 pedestrians in step ($\sqrt{n}$ to allow for random phase) was not sufficient to cause the observed vibration. They observed that when the motion had built up, pedestrians tended to synchronize their step with the frequency of the bridge, as mentioned in Section 7.2.3. This resulted in many more pedestrians being in phase than would be the case for random walking. They estimated that 10% could be fully synchronized with the bridge. Furthermore, during synchronized walking pedestrians tended to sway more and the horizontal force amplitude increased by a factor of two.

7.3.2 Foot impact on light floors for offices and dwellings

Perceptible vibrations can be induced in lightweight floors by a variety of normal activities. Polonsek (1970) considered the effects of normal walking, children playing, domestic appliances, door slam and other sources of vibration. Of these, a single person walking was the most frequently occurring and also the activity that gave rise to the greatest nuisance overall. Especially susceptible are timber floors (Whale, 1983) and long span lightweight concrete floors supported on steel joists (Pernica and Allen, 1982).

The vibration of floors under foot impact is highly dependent on their span, natural frequency and damping. Wyatt (1989) proposed that low frequency floors, which are also generally of long span, should be analysed for possible resonance with the Fourier components of footfall loading. Ellis (2000) showed that floors with frequencies as high as 12 Hz may be excited in resonance by the fourth or even fifth Fourier component. In the case of low frequency floors of long span it would be possible to apply a method similar to that used for footbridges as in the previous section. However, Wyatt (1989) suggested a simpler calculation based on the assumption that if ten or more paces were applied in the vicinity of midspan, the maximum response would be nearly as great as the steady-state response to a resonant sinusoidal force (see Chapter 2, eqn 2.44). Hence, he proposed that the displacement amplitude, u_i, under the ith Fourier component might be evaluated from:

$$u_i = \frac{F_i}{k}\frac{1}{2\xi}$$

(7.5)

where F_i = magnitude of ith Fourier component of footfall function
    k = effective stiffness of floor loaded near midspan
    ξ = critical damping factor.

In the case of floors with high damping, vibration under heel strike decays rapidly and resonance does not occur. This is particularly noticeable for high frequency floors. Structural damping in the floors of buildings is higher than in footbridges because of the forms of construction together with the added flooring and ceiling materials. Hence, in many cases it is sufficient to analyse the response of a floor to an independent impact and then check that the ensuing acceleration response is not disturbing to the building occupants (Section 7.2.3).

The 'heel drop' test of Lenzen and Murray (1969) was described in Section 7.2.1. They recommended a triangular impulse varying from 2.7 kN to zero in 1/20 second to represent foot impact loading for design purposes. This load should be applied to midspan of a simply supported floor or to the place of maximum deflection of an irregular floor. Using the theory of a Single Degree of Freedom (SDOF) system subjected to a general forcing function, it is then possible to calculate the maximum response (see Chapter 2, Section 2.2.1.3).

Allen and Rainer (1976) derived an even simpler analytical method. According to Newton's second law the rate of change of momentum of a mass is equal to the applied force. Thus

$$\frac{d}{dt}(m\dot{u}) = F(t) \tag{7.6}$$

where m is the equivalent mass of the floor, treating it as a SDOF system, and $F(t)$ is the triangular forcing function. Thus the change in momentum over a brief interval $d\tau$, brought about by an instantaneous force $F(t)$, is given by:

$$d(m\dot{u}) = F(\tau)\,d\tau = I \tag{7.7}$$

In this case the triangular forcing function can be treated as an impulse I, which would be the integral under the curve shown in Figure 7.2 (i.e. 70 N-sec). Hence:

$$d\dot{u} = \frac{I}{m} \tag{7.8}$$

This is equivalent to the velocity of the floor $\dot{u}_0$ caused by the impulse. Assuming simple harmonic motion of the ensuing vibration, the corresponding maximum acceleration is given by:

$$a_0 = 2\pi f\frac{I}{m} \tag{7.9}$$

This method has been adopted by the Canadian code for steel structures (CSA, 1984) in which a factor of 0.9 has been applied to eqn (7.9) to allow for the loss of amplitude due to damping in the first half cycle. It should also be noted that m has been referred to as the 'equivalent mass' of the floor. The reason for this is because a structure with distributed mass and stiffness, such as a beam or floor, does not oscillate with its full amplitude over its entire length or area. For example, the displacement at the supports is zero. This is illustrated in Chapter 2 with a number of examples in Figures 2.24–2.27. Hence much of the structure is participating only partially in the vibration. On the basis of tests on 42 floors the Canadian code recommends the equivalent mass to be taken as $0.4\times$ the total distributed mass of the floor.

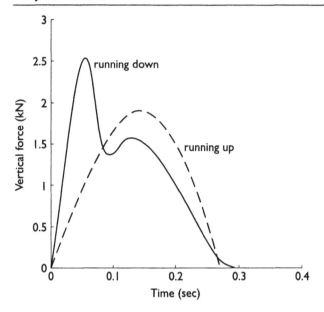

Figure 7.7 Dynamic loading on a staircase.

7.3.3 Staircases

Staircases are normally designed to carry the same static live loads as the floors to which they give access. This is nearly always satisfactory for staircases of heavy reinforced concrete construction. However, the intense loading that occurs when people run up or down stairs should be considered for light staircase structures which may be susceptible to vibration.

It was pointed out in Section 7.2.1 that the human body can generate very substantial dynamic overloads chiefly due to heel strike. Energetic walking can give rise to a peak load of up to twice the static weight of a person, G. In the heel drop test of Figure 7.2 the peak load is roughly $4 \times G$. Similarly large impacts may occur when a person runs up or down stairs. Some experiments were carried out by Smith (1988) using an orthopaedic force plate fitted into a short flight of stairs. Examples of the vertical component of foot impact are shown in Figure 7.7. When running up the peak load is generated by toe off and is about $2.5 \times G$. When running down the peak load occurs under heel strike and is generally about 3–$4 \times G$. Staircase structures that may be susceptible should be analysed under these forcing functions. In the absence of any other simpler method of analysis the 'heel drop' test is recommended.

7.3.4 Floors subjected to dance-type loads

The importance of considering the effects of dance-type loads has increased in recent years with the widespread use of light forms of construction for large span floors.

The co-ordinating effect of music results in a periodic loading that is in time with the beat. Rhythmic activities such as dancing, aerobics and military drilling are the best examples. Large dynamic magnification or resonance can occur if the forcing frequencies are close to the floor natural frequencies. The consequences may affect both serviceability and safety. Forcing functions and methods of analysis for structures subjected to dance-type loads have been proposed by Bachmann and Ammann (1987) and Pernica (1990). The analytical procedure was developed further by Ji and Ellis (1994) and is set out below.

In its most severe form, 'dance-type' loading consists of jumping in time to music. It is characterized by a high dynamic force, similar to 'toe off' when running up stairs, followed by a brief moment when the feet leave the floor and the load is zero. Finally, the person comes down and the cycle is repeated. The form of loading is similar to that produced by running and consists of a series of half-sine pulses with gaps in between when the person is airborne. This is given by

$$F(t) = \begin{cases} K_p G \sin(\pi t/t_p) & 0 \leq t \leq t_p \\ 0 & t_p \leq t \leq T_p \end{cases} \tag{7.10}$$

where G = static weight of the person
  $K_p = F_{max}/G$ = impact factor
  F_{max} = peak dynamic load
  t_p = contact duration
  T_p = period of dancing load or time between successive 'toe off'.

The contact ratio, α, depends on the nature of the dance and is defined by

$$\alpha = \frac{t_p}{T_p} \leq 1.0 \tag{7.11}$$

It has been observed that the mean value of any form of dynamic human loading is equal to the weight of the person or people engaged in the activity. Hence, integrating the force over the period of contact

$$\frac{1}{T_p} \int_0^{t_p} K_p G \sin\left(\frac{\pi t}{t_p}\right) dt = G \tag{7.12}$$

from which the impact factor can be evaluated as follows

$$K_p = \frac{\pi}{2\alpha} \tag{7.13}$$

It is first necessary to determine what values of contact ratio α, are appropriate for various activities. Ellis and Ji (1994) reviewed a number of experimental studies carried out in Canada, including those by Allen (1990) and Pernica (1990), and on the basis of these, proposed the values for contact ratio shown in Table 7.3.

Ellis and Ji (1994) demonstrated that these factors gave good agreement with the experimental observations and they have now been adopted in the UK loading code, BS 6399 (1996). It should be noted that the value of $\alpha = \frac{2}{3}$, recommended for pedestrian movements, is actually applicable to one foot only since individual

Table 7.3 Values of α for various activities.

Activity	Contact Ratio α	Impact Factor K_p
Pedestrian movements, low impact aerobics	2/3	2.4
Rhythmic exercises, high impact aerobics	1/2	3.1
Normal jumping	1/3	4.7
High jumping	1/4	6.3

footfalls overlap for pacing frequencies below about 3 Hz. For assessing the performance of floors to pedestrian movements it is probably better to use the method outlined in Section 7.3.2.

The loading model expressed by eqn (7.10) is not in the most convenient form for general design calculations. In order to obtain an analytical solution it is more useful to express the load function in terms of a Fourier series. Hence:

$$F(t) = G\left[1.0 + \sum_{n=1}^{\infty} r_n \sin\left(\frac{2n\pi}{T_p}t + \phi_n\right)\right] \tag{7.14}$$

The coefficients, r_n, and the phase lags, ϕ_n, may be evaluated and are as follows:

$$r_n = \sqrt{a_n^2 + b_n^2} \qquad \phi_n = \tan^{-1}\left(\frac{a_n}{b_n}\right) \tag{7.15}$$

When $2n\alpha = 1$; $n = 1, 2, 3 \ldots$ then $a_n = 0$ and $b_n = \pi/2$; else

$$a_n = 0.5\left[\frac{\cos(2n\alpha - 1)\pi - 1}{2n\alpha - 1} - \frac{\cos(2n\alpha + 1)\pi - 1}{2n\alpha + 1}\right] \tag{7.16}$$

and

$$b_n = 0.5\left[\frac{\sin(2n\alpha - 1)\pi}{2n\alpha - 1} - \frac{\sin(2n\alpha + 1)\pi}{2n\alpha + 1}\right] \tag{7.17}$$

This analytical model of dance-type loads is shown in Figure 7.8 together with the separate half-sine impacts of eqn (7.10). The Fourier series model of eqn (7.14) is shown taking the first three and the first six terms. It is clear that a good approximation is achieved using only the first three terms.

Ji and Ellis (1994) used the Fourier series form of the loading model to determine the response of a simply supported rectangular floor. The loading was intended to simulate a group dancing activity and therefore was assumed to be uniformly distributed over the entire area of the floor. They derived equations for the steady state response of a floor under this loading and showed that only the fundamental mode of vibration of the floor needed to be included to achieve an accurate solution. The response of the floor consists of the static deflection, under mean load, plus a dynamic component. The dynamic magnification of the fundamental mode can be obtained by summing the dynamic magnification factors of each

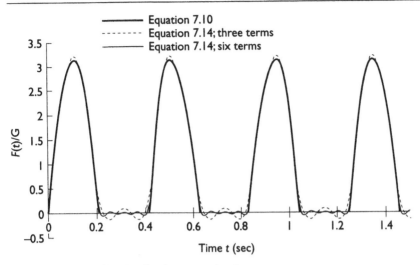

Figure 7.8 Forcing function for dance-type loading.

Fourier component used in the loading model. These are given by

$$D_n = \frac{r_n}{\sqrt{(1 - n^2\beta^2) + (2n\xi\beta)^2}} \tag{7.18}$$

where $\beta = f_p/f$ and $f_p = 1/T_p$.

Good agreement was obtained between the analytical solution and the results of laboratory tests. The load model was applied to floors with simple boundary conditions. For more general shapes it would be necessary to carry out a modal analysis of the floor.

When assessing the performance of a large span floor in a building, a number of practical points should be considered. First, the natural frequency of the floor should be calculated excluding the weight of people involved in the dancing activity (Ji and Ellis, 1997a). The value of damping should be chosen conservatively (e.g. 2 per cent of critical) since modern forms of construction are notoriously lightly damped. If the floor to be assessed was for a sports hall in a building which includes offices, it would be advisable to do a serviceability check assuming that a small number of people frequently use the floor for high impact events. If the natural frequency of the floor is in or near the range of loading frequencies then resonance may occur and it will be necessary to consider the ultimate limit state.

It should be noted that at resonance $f_p = f$ and eqn (7.18) implies very large magnification ($D_n = 25r_n$ for $\xi = 0.02$). This condition would probably occur only when dancers are spaced well apart and that therefore the static load would be small. However, it demonstrates the importance of keeping the floor frequency away from resonance.

7.3.5 Dynamic crowd loading: concert halls, grandstands and bridges

In section 7.3.4 the importance of rhythmic human loads co-ordinated by music was considered and it was shown that the periodic nature of these loads may give rise to very large dynamic response factors and possibly resonance. Specifically under consideration were loads due to dancing or aerobics that involve jumping at a set frequency. In these situations people are usually well spaced and therefore it is likely that the distributed load will be very much less than the normal floor design load. However, there are crowd events, such as pop concerts and football matches, at which the spectators may be densely packed. Ellis *et al.* (1994) suggested that six people per square metre (4.8 kN/m^2) is reasonable. Reid *et al.* (1997) suggested 2 kN/m^2 for crowds with fixed seating. The corresponding design live loads of 5 kN/m^2 and 4 kN/m^2 respectively are thought to be sufficient to include limited dynamic effects such as people rising to their feet when a goal is scored. However, co-ordinated rhythmic movement sometimes occurs and may be very intense, especially at a pop concert.

The question arises whether full co-ordination is possible for a very large crowd, say numbering in hundreds or thousands. This problem was studied by Ji and Ellis (1993). Starting with the formula for dance-type loads, eqn (7.14), they introduced a random phase lag to take account of the difference in co-ordination between one individual and another. This phase lag may lie between $-\pi$ and $+\pi$. Assuming that it was normally distributed with a mean of zero (fully co-ordinated) and a standard deviation of $\pi/3$ or 1.0, they evaluated a dynamic crowd factor of 0.68 for 100 people and 0.63 for 2,500 people. Using experimental observations Ebrahimpour and Sack (1992) obtained a value of 0.64 for 40 people. The UK code (BS 6399, 1996) recommends a factor of 0.67 to take account of the lack of co-ordination of a large crowd. The crowd factor should be applied to the dynamic component of eqn (7.14).

In section 7.3.4 the importance of avoiding resonance was pointed out since the dynamic magnification could be as much as 25 or possibly more. This would be equivalent to exceptionally high static live load even allowing for the crowd factor and the reduced partial factor of 1.0 permitted by BS6399 in this case. At the present time there is a paucity of experimental data regarding dynamic crowd loading. It is questionable whether a large crowd would be able to maintain co-ordinated jumping for as many as 30 or 40 jumps that would be required to reach the steady state amplitude at resonance. Furthermore, it requires input of energy to maintain the steady state amplitude to balance the energy lost in damping. As yet there is no firm evidence that this maximum theoretical load factor can be achieved in practice. The largest dynamic magnification actually observed is 1.97 measured by Irwin, 1981.

The design of sports stadium structures must take account of dynamic crowd loading (Scottish Office, 1997). This has arisen because of well publicized cases of crowd excitation and even failures. It is recognized that the most severe dynamic loading arises during pop concerts and since football stadiums are often used for such events it is advisable to design them accordingly. The most severe value of contact ratio in Table 7.3 ($\alpha = \frac{1}{4}$) is therefore suggested. Reid *et al.* (1997) discussed the range of frequencies over which a structure may be susceptible. BS 6399 recommends that the vertical frequency should be greater than 8.4 Hz to avoid resonance, based on three times the maximum observed coordinated jumping frequency of 2.8 Hz. This should be based on the mass of the empty structure because of the independence of the mass of the crowd and the mass of the structure during intense jumping activity.

Horizontal dynamic load due to swaying is an important component of football crowd loading. BS 6399 (1996) recommends that the horizontal frequency of susceptible structures should be greater than 4.0 Hz to avoid resonance. This may be difficult to achieve and therefore some guidance is required on the magnitude of horizontal dynamic load to consider. The CEB (Euro-international Concrete Committee) guide (1991) notes that sway loads may occur at frequencies between 0.4 to 0.7 Hz and suggests a horizontal load factor of 0.3 for sway at 0.6 Hz. Reid *et al.*, (1997) suggest a lower value and BS 6399 (1996) recommends that horizontal loads should be 10 per cent of the vertical. There is some uncertainty over whether the mass of the crowd should be included when calculating the natural frequency of horizontal vibration. This is because people will still be in contact with the structure when in swaying mode. There is a need for more data from full scale tests. The horizontal component of pedestrian crowd loading on bridges was mentioned in Section 7.3.1.

7.4 DESIGN OF STRUCTURES TO MINIMIZE HUMAN INDUCED VIBRATION

It has been found that it is often difficult to avoid the critical frequency range of human induced dynamic loading. This is particularly the case for large sports stadiums. Reid *et al.* (1997) analysed a number of stadiums including Murrayfield and Middlesbrough. They found that frequencies in the range of 1 to 3 Hz were not unusual. Sometimes the first mode may be dominated by the cantilever roof and is therefore not important. However, special consideration should be given to cantilevered decks of seating that could be excited by vertical jumping. Side to side and back to back modes should be considered for horizontal loading.

The main options available to the designer are to increase stiffness and damping. Ji and Ellis (1997b) have suggested an efficient way of arranging the bracing for steel frameworks in order to increase the stiffness. They showed that stiffness could be doubled compared with the most inefficient system, without additional steel. This is illustrated in Figure 7.9. Damping is notoriously difficult to introduce into a structure. Steelwork with composite concrete decks is generally lightly damped. The addition of cladding will help but there is little specific data available (Osborne and Ellis, 1990). Heavy reinforced concrete permanent struc-

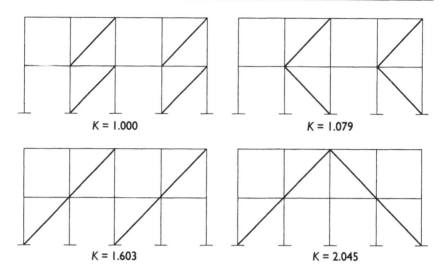

$K = 1.000$ $K = 1.079$

$K = 1.603$ $K = 2.045$

Figure 7.9 Bracing systems and normalized stiffness.

tures are likely to perform the best. On the other hand, temporary grandstands are very light and unclad and the only approach open to the designer is through increasing the stiffness with bracing. The performance of floors can be improved by ensuring that there is good transverse distribution (Whale, 1983). This will have the effect of increasing the number of longitudinal joists or stringers that contribute to the static stiffness and will also increase the proportion of the floor mass participating in the modal response (see eqn 7.9).

Footbridges with spans of over 25 m usually have natural frequencies well within the pedestrian excitation range (Pimentel, 1997). It is generally impractical to increase their natural frequencies to avoid resonance. Fortunately, footbridge loading under a single pedestrian is not true resonance because of the varying position of the load. With the addition of damping it is often possible to keep the amplitude within acceptable bounds. Brown (1977) installed a simple friction damper at one abutment of a lively bridge where the angular movement could be utilized to absorb energy. Jones *et al.* (1981) installed tuned mass–spring–damper vibration absorbers into two lively footbridges with most satisfactory results. However, crowd loading marching in step, as observed by Fujino *et al.* (1993), is capable of introducing substantial dynamic energy that may be very difficult to absorb with simple damping devices.

7.5 REFERENCES

Allen, D. E. (1990) 'Floor vibration from aerobics', *Canadian J. Civil Engng.* **17**(5): 771–9.
Allen, D. E. and Rainer, J. H. (1976) 'Vibration criteria for long span floors', *Can. J. Civil Engng.* **3**: 165–73.

Bachmann, H. and Ammann, W. (1987) *Vibration in Structures Induced by Man and Machines*, Structural Engineering Document No. 3e, International Association of Bridge and Structural Engineers, AIPC-IVBH, Zurich.

Blanchard, J., Davies, B. L. and Smith, J. W. (1977) 'Design criteria and analysis for dynamic loading of footbridges', Symposium on Dynamic Behaviour of Bridges, paper 7 given at TRRL Supplementary *Report 275*, Transport and Road Research Laboratory, Crowthorne.

Brown, C. W. (1977) 'An engineer's approach to dynamic aspects of bridge design', paper 8 given at Symposium on Dynamic Behaviour of Bridges', TRRL Supplementary Report 275, Transport and Road Research Laboratory, Crowthorne.

BS 5400 (1978) *Steel, Concrete and Composite Bridges: Part 2, Specification for Loads*, British Standards Institution, London.

BS 6399 (1996) Part 1, Code of Practice for Dead and Imposed Loads, British Standards Institution, London.

BS 6472 (1984) *Guide to Evaluation of Human Exposure to Vibration in Buildings (1 Hz to 80 Hz)*, British Standards Institution, London.

CEB (1991) *Vibration Problems in Structures*, Bulletin d'Information 209, section 1.4, August, Comité Euro-International du Béton.

CSA (1984) *Steel Structures for Buildings*, Canadian Standards Association, CAN 3–S16.1–M84.

Ebrahimpour, A. and Sack, R. L. (1992) 'Design live loads for coherent crowd harmonic movements', *J. Struct. Eng., ASCE*, **118**(4): 1121–36.

Ellis, B. R. (2000) 'On the response of long span floors to walking loads generated by individuals and crowds', *The Structural Engineer*, **78**(10), May, 17–25.

Ellis, B. R. and Ji, T. (1994) 'Floor vibration induced by dance-type loads: verification', *The Structural Engineer*, **72**(3): February, 45–50.

Ellis, Ji, and Littler (1994) 'Crowd actions and grandstands', paper given at IABSE Symposium, Places of Assembly and Long span Structures, Birmingham, 201–6.

Fujino, Y., Pacheco, B. M., Nakamura, S. and Warnitchai, P. (1993) 'Synchronization of human walking observed during lateral vibration of a congested pedestrian bridge', *Earthquake Engng Struct. Dynamics*, **22**(9): 741–58.

Guignard, J. C. and Guignard, E. (1970) *Human Response to Vibration: a Critical Survey of Published Work*, ISVR Memo. No. 373, Institute of Sound and Vibration Research, Southampton.

Harper, F. C., Warlow, W. J. and Clarke, B. L. (1961) *The Forces Applied to the Floor by the Foot in Walking. Part 1, Walking on a Level Surface*, National Building Studies Research Paper 32, HMSO, London.

Heins, C. P. and Yoo, C. H. (1975) 'Dynamic response of a building floor system', *Building Science*, **10**: 143–53.

Irwin, A. W. (1978) 'Human reponse to dynamic motion of structures', *The Structural Engineer*, **56**(A): 237–44.

Irwin, A. W. (1981) *Live Load and Dynamic Response of the Extendable Front Bays of the Playhouse Theatre during The Who Concert*, Report for Lothian Region Architecture Department.

Irwin, A. W. (1983) *Diversity of Human Response to Vibration Environments*, United Kingdom Group HRV, NIAE/NCAE, Silso, UK.

Ji, T. and Ellis, B. R. (1993) 'Evaluation of dynamic crowd effects for dance loads', paper given at IABSE Colloquium, Structural Serviceability of Buildings, Goteborg, pp. 165–72.

Ji, T. and Ellis, B. R. (1994) 'Floor vibration induced by dance-type loads: theory', *The Structural Engineer*, **72**(3): February 37–44.

Ji, T. and Ellis, B. R. (1997a) 'Floor vibration induced by human movements in buildings', in P. K. K. Lee (ed.) paper given at 4th International Kerensky Conference, Hong Kong, Structures in the New Millenium, Balkema, Rotterdam, pp. 213–19.

Ji, T. and Ellis, B. R. (1997b) Effective bracing for temporary grandstands. *The Structural Engineer*, **75**(6): March, 95–100.

Jones, R. T., Pretlove, A. J. and Eyre, R. (1981) 'Two case studies of the use of tuned vibration absorbers on footbridges', *The Structural Engineer*, **59**(B): 27–32.

Lenzen, K. H. and Murray, T. M. (1969) *Vibration of Steel Beam Concrete Slab Floor Systems*, Report No. 29, Department of Civil Engineering, University of Kansas, Lawrence, KS.

Matsumoto, Y., Nishioka, T., Shiojiri, H. and Matsuzake, K. (1978) 'Dynamic design of footbridges', *Proc Int. Assoc. Bridge Struct. Engng.* **P-17/78** (August): 1–15.

Osborne, K. P. and Ellis, B. R. (1990) 'Vibration design and testing of a long span lightweight floor', *The Structural Engineer*, **68**(10): May, 181–6.

Pernica, G. (1990) 'Dynamic load factors for pedestrian movements and rhythmic exercises', *Canadian Acoustics*, **18**(2): pp. 3–18.

Permica, G. and Allen, D. E. (1982) Floor vibration measurements in a shopping centre, *Can. J. Civ. Engng.* (CDN, **9**: 149–55.

Pimentel (1997) 'Vibration performance of pedestrian bridges due to human-induced loads', PhD dissertation, Department of Civil and Structural Engineering, University of Sheffield.

Polonsek, A. (1970) 'Human response to vibration of wood joist floors', *Wood Science*, **3**: 111–119.

Reid, W. M., Dickie, J. F. and Wright, J. (1997) 'Stadium structures: are they excited?' *The Structural Engineer*, **75**(22): November, 383–8. 'Discussion', *The Structural Engineer*, **76**(3): July 1998.

Scottish Office (1997) *Guide to Safety at Sports Grounds*, The Scottish Office and Department of National Heritage, HMSO, London.

Skorecki, J. (1966) 'The design and construction of a new apparatus for measuring the vertical forces exerted in walking: a gait machine', *J. Strain Analysis*, **1**(5).

Smith, J. W. (1988) *Vibration of Structures: Applications in Civil Engineering Design*, Chapman and Hall, London. ISBN 0–412–28020–5.

Tilly, G. P., Cullington, D. W. and Eyre, R. (1984) 'Dynamic behaviour of footbridges', *Int. Assoc. Bridge Struct. Engng.* **194**; 259–67.

Whale, L. (1983) *Vibration of Timber Floors – A Literature Review*, Research Report 2/83, Timber Research and Development Association, High Wycombe.

Wheeler, J. E. (1982) 'Prediction and control of pedestrian induced vibration in footbridges', *J. Struct. Div., ASCE*, **108**(ST9): 2045–2065.

Wyatt, T. A. (1989) *Design Guide on the Vibration of Floors*, SCI Publication 076, The Steel Construction Institute, Ascot.

Chapter 8

Traffic and moving loads on bridges

David Cooper

8.1 INTRODUCTION

In this chapter we shall consider the effects of highway traffic loading on road bridges. Because we are primarily interested in dynamic response, this limits our primary concern to the shorter bridge spans. As bridge spans increase, the static effects of vehicle convoys begin to dominate designs. However, it will be appropriate to include a description of some aspects of the assessment of long span bridge load effects.

The first UK national standard for highway bridge loading was introduced by the Ministry of Transport in 1922. This comprised a standard loading train of vehicles, with a 50 per cent allowance for impact (Henderson, 1954).

In 1932 the Ministry of Transport Loading Curve was introduced. The impact allowance was reduced in view of the improvements being made in vehicle suspension systems, and it was also reduced for longer bridge spans. The theoretical justification for these allowances for dynamic effects is unknown.

In 1954, the impact allowance was reduced to 25 per cent, which was only to be added to the effects of a single axle. This value had been obtained from some experimental observations. In the USA at the same time, an overall allowance of 30 per cent was made.

During the late 1970s (Department of Transport, 1980), the short span bridge loading provisions of the standard traffic 'HA' (highway bridge loading type A) loading model in the UK bridge design standard, BS 153 (BSI, 1972), were revised. An allowance for impact effects was derived from studies by the Transport and Road Research Laboratory (Page, 1976). This was based on vehicle suspension forces: measured by means of a system of electrical resistance strain gauges and accelerometers attached to the rear wheels of a two axled rigid heavy goods vehicle while traversing some 30 motorway over- and under-bridges. The measured impact factors varied between 1.09 to 1.47 for underbridges, and 1.16 to 1.75 for overbridges. One result, of 2.77, was described as a 'freak', and occurred over a severe step in the road surface. A value of 1.80 was selected, to be applied to the axle causing the worst load effect. This allowance was included within the static design load model. Since it only applied when a single vehicle effect governed the codified design load model, it only applied to spans below about 15 m.

Common to all of these models are:

- The difference between the effects of moving and stationary vehicles are referred to as 'impact' effects, and there is little if any reference to bridge dynamic response.
- The impact allowances were selected deterministically to represent a typically large effect that would apply to all bridges.

Design codes have become more prescriptive in recent decades, and designers have had decreasing freedom (and less need) to consider the loads on their structures from first principles. However, changes in procurement practices may begin to reverse this trend, as centralized governmental procurement is replaced by the private sector. Private service providers may need to balance safety, cost and potential liability in a different manner. It may become more common for procurement authorities to specify performance criteria rather than to specify the means of meeting those criteria and many more designers will need to consider the loads on their structures from first principles, rather as they did during the nineteenth century.

8.2 DESIGN ACTIONS

8.2.1 Probabilistic principles

Whatever means are used to produce load models for design, when a structure is faced with a complex random loading process it will not be possible to cater for all conceivable eventualities. Designers must make some rational judgement about the relationship between safety and cost of their structures. Indeed, this is recognized under the UK's Health and Safety at Work Act which requires risk to be kept 'As Low As Reasonably Practicable' (the so-called ALARP principle).

Since we cannot predict future events with precision, we cannot calculate actual costs of the risk to safety, but must content ourselves with calculating probable costs. That implies that engineers should consider probabilistic models for structural capacities and for static and dynamic load effects.

Suspicion has grown in recent years that typical allowances for dynamic (or 'impact') effects are unnecessarily onerous, and that real structures might well not respond fully to the fluctuating applied loads that have been measured. The development of new bridge design codes, including UK bridge assessment documents (Highways Agency, 1997) and Eurocodes (CEN, 1994), have led to renewed interest in bridge dynamic response, and the means of allowing for it in design.

In the UK at present, the Highways Agency memorandum BD37/88 (Highways Agency, 1988) defines design loads for bridges. A review of its derivation is provided by Flint (1990).

8.2.2 Long span bridges

Long span bridges are governed by the weight of closely spaced convoys of vehicles which, observations (Ricketts and Page, 1997) confirm, implies the presence of stationary traffic. It is conceptually not unreasonable to treat such effects as though they can be modelled as stationary random variables. Thus, it is assumed that the peak traffic load effect is potentially the same day after day, since it is caused by the random association of a large number of events (provided that non-random factors such as deliberate sabotage are neglected).

Therefore, a long span bridge load model may be based on statistical analysis of the effects of convoys of traffic. Usually, these effects will be simulated, using convoy models based either on automatically measured records from large numbers of vehicles, or on models regenerated from statistical models of vehicles and traffic composition.

The current UK bridge design code (Flint and Neill Partnership, 1986), as well as Eurocode 1 (CEN, 1994) defines most actions (loads) and resistances (capacities) in relation to 'characteristic' values, where the characteristic value is considered to be the upper 5 percentile for loads, and the lower 5 percentile for capacities. The design rules for long span bridges are intended to provide load effects that have approximately a 5 per cent probability of exceedence in a nominal structural life of 120 years. This is calculated more rationally by taking a one in 2,400 probability of exceedence per year. For bridge assessment, this might be derived from information about current traffic, whereas for design purposes it might be necessary to consider foreseeable future changes in traffic legislation or growth in volume.

The BS5400 traffic load model approximates to $1/1.2$ times the characteristic. Thus, the partial factor of 1.5 on traffic loading effectively provides a factor of 1.25 on the characteristic load effect.

TRL (Transport Research Laboratory) Contractor Report CR16 (Flint and Neill Partnership, 1986) describes the derivation of the present long span bridge loading rules. Much of the document refers to the manner in which a model of future traffic was developed, since at the time of collection of the background data there was a 32 tonne weight restriction, and 38 tonne vehicles were about to be legalized.

Since the publication of that report, advances in computer speed have allowed much more extensive traffic load effect simulations to be undertaken. Wherever possible, this author believes that it is preferable to use real traffic records (obtained by weigh-in motion sensors) in such simulations, rather than to attempt to mathematically model traffic and then to regenerate data (as was performed for the 38 tonne vehicles in the CR16 models).

There are still relatively poor data to describe the spaces between vehicles in long traffic convoys. TRL describe the results of analysis of relatively recent video tape records of traffic behaviour, including lane selection and the behaviour of traffic convoys, in Ricketts and Page's (1997) report, and it is recommended that these observations (or actual site observations) should be used rather than the models which are described in CR16.

8.2.3 Short span bridges

Short span bridge load effects present more difficulties. Peak load effects are caused by the joint extreme of the combined static and dynamic effects from all individual vehicles, moving much more quickly than jammed traffic. On the shorter span bridges on any particular route, there will be a very much larger number of load events of potential concern than on the longer bridges. Furthermore, the highest load effects on the shortest bridges are likely to be caused by unusual and possibly illegally configured vehicles, whose existence might not be predictable by statistical analysis of measured traffic data. The bridge owner must decide whether such vehicles need be considered.

The present UK design rules for the effects of normal traffic loading on short span bridges are derived from a deterministic assessment of the envelope of load effects that would be produced by all vehicles that conform to the current UK Construction and Use Regulations. Deterministic allowances are included for impact and for overload (Department of Transport, 1980).

The rules used for assessment of short span bridges (Highways Agency, 1997; Cooper and Flint, 1997), unlike the design rules, are based on probabilistic principles. However, they were 'calibrated' against the current design rules. They are intended to provide adjustments to cater for different types of traffic and road surface roughness, whilst retaining reliabilities that are consistent with those of current designs of similar dimensions and construction, used in onerous situations.

8.2.4 Determination of design action

Whether or not a probabilistic method is used to determine the relationship between potential loads and design capacities, it will be necessary to derive a model of the effects of traffic loads that will cater for static and dynamic effects.

The static design model may be based on deterministic or probabilistic assessment of extreme load effects, determined from the traffic weight and classification data: as described for 'long span bridges' in Section 8.2.2. Then, any dynamic amplification to such static effects is usually considered separately, to be combined later.

8.2.5 Dynamic amplification factor

The dynamic amplification factor is usually defined to be the ratio between the effects of moving traffic on bridges to the effect of stationary traffic. Thus: if the maximum response to stationary traffic (or slowly moving traffic traversing the entire length of the bridge) $= R_s$ and the maximum response to moving traffic crossing the bridge $= R_d$ Then:

$$\mathrm{DAF} = R_d/R_s \tag{8.1}$$

where DAF = Dynamic Amplification Factor.

Values of R_r may be derived from analysis of the types and weights of vehicle which use the bridge, as indicated above. However, derivation of the appropriate DAF is then necessary.

8.3 DETERMINATION OF STRUCTURAL RESPONSE

In theory, bridge response cannot be separated from the vehicle loading (the action), since the movement of a real bridge affects the wheel loads that initiate the original response. Therefore an iterative analysis is required at each time step to ensure compatibility between the suspension and bridge deflections and interacting forces. A number of theoretical studies have been performed for road vehicles and rail vehicles (AEA Technology (Bailey, 1996; Green *et al.*, 1995)) in which multi degree of freedom vehicle models have been used in conjunction with theoretical road surface profiles and elastic bridge models in order to model the interactions between bridges and vehicles, and thus to obtain the bridge responses. These methods appear to be most useful in very specific applications, for example:

● when refining the design of vehicle suspensions (where the vehicle models are under the direct control of the analyst);
● in military bridging design (where vertical deflections can easily be three or four times larger than the suspension travel of a typical vehicle).

However, they possess drawbacks when used in more conventional bridge assessment. In particular:

● they require the user to have access to realistic models of many details of vehicle suspension design: knowledge which is likely to be commercially confidential to vehicle manufacturers;
● road surface profiles at bridge sites are not stationary random variables, and they cannot reliably be recreated from frequency domain (spectral analysis) models;
● analysis is slow, and requires relatively complex input. It is difficult to analyse sufficient cases to build up a large enough set of results from which generalized conclusions can confidently be drawn.

In civil applications the feedback effect from bridge response to suspension response is normally small, since highway bridges are usually so much stiffer than vehicle suspension systems. In recent studies sponsored by the UK Highways Agency and undertaken by TRL (Ricketts and Page, 1997), axle weights were recorded during heavy goods vehicle transits over a small number of bridges which were equipped with strain and deflection measuring equipment. It was found that the biggest vertical deflection during the transit of a 38 tonne articulated truck over a relatively flexible 10 metre span bridge was just over 1 mm, which would have negligible effect on suspension forces.

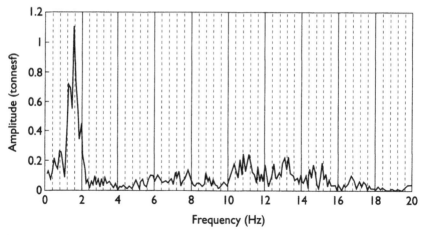

Figure 8.1 Typical relationship between frequency and the amplitude of variation of effective total weight.

8.3.1 Vehicle dynamic forces

Various workers have investigated the interaction between vehicles and road surfaces. In particular, the UK Transport Research Laboratory (Ricketts and Page, 1997) has instrumented individual vehicles and bridges in order to measure the variations in loading imposed by vehicle wheels onto road surfaces.

A frequency domain approach might appear to provide a useful means of characterizing the load model, and TRL used the Fast Fourier Transform procedure to obtain the relationships between wheel load magnitudes and frequencies. They observed that vehicle dynamic behaviour can be separated for practical purposes into two distinct parts. There is the oscillation of the mass of the whole vehicle on its suspension: the so-called 'heave' or 'bounce' response; and there are the oscillations of individual axles, responding to road roughness and discontinuities: the 'wheel hop' response. Typically, the heave mode has a frequency between about 2 and 3 Hz, whereas wheel hop frequency is between about 12 and 16 Hz.

Figure 8.1 shows the relationship between the frequency and amplitude of oscillation of the total weight of a modern five-axled articulated air suspension heavy goods vehicle. The 'bounce' mode has a frequency of 1.6 Hz.

Figure 8.2 shows a typical plot of the variation of the sum of all wheel forces of a 38 tonne articulated vehicle against time, for transit speeds of 40 mph (17.9 m s^{-1}) and 10 mph (4.5 m s^{-1}). The peak dynamic increment in the vehicle weight is very nearly 8 tonnes force for the 40 mph transit, but less than 2 tonnes force for the 10 mph transit.

Correlation coefficients may be calculated between the various wheel loads. Table 8.1 shows a set obtained by analysis of the wheel load record obtained at 40 mph from the same five-axled vehicle as referred to above.

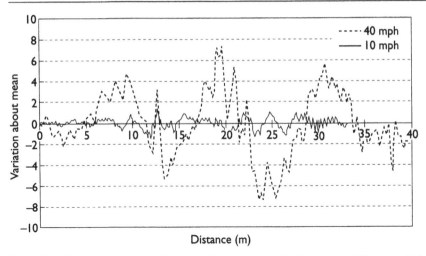

Figure 8.2 Typical variation of effective vehicle weight with distance for different vehicle speeds.

Table 8.1 Wheel load correlation coefficients.

	a	b	c	d	e	f	g	h	i	j
a	1	0.82	0.34	0.34	0.10	0.18	0.22	0.18	−0.21	−0.18
b		1	0.32	0.41	0.14	0.21	0.23	0.20	−0.22	−0.16
c			1	0.81	0.39	0.31	0.24	0.11	−0.03	−0.04
d				1	0.35	0.46	0.22	0.16	−0.07	−0.03
e					1	0.77	0.29	0.15	0.14	0.11
f						1	0.21	0.18	0.02	0.07
g							1	0.72	0.06	0.07
h								1	0.00	0.02
i									1	0.79
j										1

Wheel pairs at each axle are a–b, c–d, e–f, g–h and i–j. The steering wheels are i-j, driving wheels are g–h, and the remainder are the trailer wheels.

The relatively high correlations between the pairs of wheel loads on each axle (enclosed in boxes in the table) contrast with the low correlations elsewhere.

A mathematical model based on such frequency analysis would appear to be attractive as the basis from which statistical load models might be generated. However, bridge specific road profiles such as those which occur at movement joints, or due to approach road settlement, would still need to be included when assessing response.

This complicates the process of creating load models, and it would appear to be preferable to use real measurements obtained at real sites as much as possible. Then there will be no need to transform measured data into a mathematical model simply in order to use that model to reproduce the original values.

8.3.2 Vehicle and structure interaction

Dynamic effects due to moving loads on bridges are of most concern at shorter spans. They are essentially transient effects. The magnitude of the forcing function will be changing with time and will have a definite beginning and end. Therefore, it is more convenient to analyse bridge dynamic response in the time domain by performing a 'time history' analysis rather than by using a spectral analysis approach in the frequency domain. Furthermore, it is preferable to use recorded wheel data rather to mathematically characterize it and regenerate it using a Monte Carlo simulation approach. Regeneration of continuous records from frequency domain spectral analysis data has been criticized because it 'tends to produce too many peaks' (Elnashai, 1995).

Various commercial Finite Element Method (FEM) programs are available with the ability to perform time history calculations. It is not always easy to model multiple loads which are changing in space and time, and it is useful to consider more economical and simpler alternatives. These may also provide means of obtaining results for a variety of structures relatively quickly and economically.

It is possible to analyse the structural response to a particular loading history independently in each of a number of independent modes of vibration, and use the principle of mode superposition to combine them. This would require prior analysis (using FEM or classical theory) to obtain the elastic properties which define each mode of vibration (mode shapes, frequencies, masses). Chapter 2 (Section 2.3) describes modal analysis methods.

8.3.3 Flexural response

The dynamic response characteristic of a simple beam bridge that is likely to be of most concern is that in bending. The frequencies of the modes of vibration of a simple beam are given by (see also Section 2.4.1):

$$\omega = n^2 \pi^2 \sqrt{\frac{EI}{\bar{m}L^4}} \tag{8.2}$$

where: n = Mode of vibration (1, 2, 3 . . .)
    L = Span length
    $\bar{m}$ = Mass per unit length
    EI = Flexural rigidity
    ω = Circular frequency (rad/sec).

For most bridge construction types, it has been established that a crude mathematical model of the frequency of the first mode of vibration is given by the form: $f = 82L^{-0.9}$ Hz (Paultre *et al.* 1992). Thus a 15 m bridge beam will typically have a natural frequency in its first mode of vibration equal to approximately 7 Hz, whereas its second mode frequency will be 28 Hz.

However, it should be remembered that real bridges decks are primarily two-dimensional surfaces that may be excited in many different modes in their third (out of plane) dimension by highway traffic. The first torsional mode may have a very similar frequency to the first bending mode, since the strained shape of the principal elements will be similar. However, there will be many other vibration modes, most of which will have much higher natural frequencies.

The contribution to response from the lowest modes will be much greater than for higher modes, since road vehicle excitation frequencies are not much in excess of 16 Hz.

Each mode's contribution to stress effects is proportional to the square of the response frequency, so the contribution of each higher mode to moments will be more significant than its contribution to deflections.

8.3.4 Time intervals

A step by step time history analysis based on linear relationships between displacement, velocity and acceleration within each time step is only stable when the time step is sufficiently small. Typically, the vibration periods must be in the order of 5 to 10 times the integration period (see Chapter 2). There are methods of stabilizing the analysis, but the highest mode responses may have little physical meaning. When the ratio of excitation to response frequency falls towards zero, the dynamic magnification approaches unity and static analysis will suffice.

Since a 15 m span bridge will have a first mode period in the order of 7 Hz, the time step must be approximately 1/100 sec or less.

8.3.5 Shear response

Stresses that are dominated by shear loading seldom if ever appear to be discussed. They are not referred to in any of the summaries of findings that appear in the 1992 paper by Paultre, Chaallal and Proulx. Shear deflections are normally so small that the high stiffness leads to very high natural frequencies of vibration. The result is that shear sensitive elements will tend to respond in direct proportion to rapid changes in applied force. Therefore, dynamic analysis of structural response is not needed, and analysis of the possible variation in the applied force due to the response of the vehicle suspension system to road irregularities will suffice.

This discussion concentrates on the effects of road vehicles. These have pneumatic tyres, which prevent very high transient loads from occurring. The effects of railway rolling stock are very different, and there is anecdotal evidence that damaged wheels may cause load spikes that are as much as six times greater than the average rolling load. Such very high frequency spikes are quickly attenuated in most structures, although they do cause serious local problems such as premature fatigue damage and fractures in railway lines.

Dynamic magnification of shear effects due to wheel loads running on or off bridges will be small. However, there might well be significant dynamic amplification of end reactions due to bending responses. If the shear vibration mode shapes are to be considered in a multi-modal vibration analysis, a very large number of modes of vibration will need to be considered before they can sum (even moderately accurately) to the correct shape, and the dynamic amplification in these higher modes will be negligible.

8.3.6 Effect of influence line shapes

When conducting a time history analysis of response in any one mode, the forcing function will be obtained by taking the sum of the products, at each interval of time, of the instantaneous value of wheel load and a modal influence coefficient. This influence coefficient is equal to the local magnitude of the normalized mode response shape, which is obtained from the structure's eigenvectors in the usual manner (see Chapter 2).

If it a static analysis solution is to be compared with a dynamic analysis, it is important to notice that the static influence line for midspan bending of a simply supported beam is triangular, whereas the first flexural mode shape is approximately sinusoidal. Therefore, even if there is no dynamic amplification of the mode 1 response, the time history analysis will appear to give a larger response. The precision of the results can theoretically be improved by increasing the number of modes considered in the analysis, but this leads to practical analytical problems.

A pragmatic approach is to arbitrarily assume the static and dynamic influence line shapes to be identical. The absolute value of response will not be obtained exactly, but it will allow the difference between dynamic and static response to be found.

8.3.7 Use of bridge strain measurements

A number of workers have reported analyses of recorded values of bridge strains, in which the higher frequency oscillations are attributed to dynamic response, and lower frequency oscillations to static responses.

If a bridge span is, say, 30 m, a typical transit time will be in the order of 2 sec. The mode 1 frequency will be in the order of 3.8 Hz (eqn 8.2), so there will be in the order of eight full oscillations in mode 1. Since the static effect of the vehicle will only cause approximately one half oscillation, it is conceptually reasonable to consider separating the two effects.

However, even when there appears to be no significant vibration, the load effects may well be strongly affected by overall road profile effects caused by the bridge's being in a dip or on a hump. Such effects are likely to be at least as important as any dynamic vibration response.

Figure 8.3 illustrates strain gauge readings obtained at 1/100 sec intervals from the lower flange centre of a steel beam from a composite beam plus reinforced concrete slab bridge with a 10 m span. The peak 40 mph strains are actually less than the 10 mph strains, but there is little sign of periodic oscillation on either trace (the unevenness seems largely to be due to signal noise). The differences

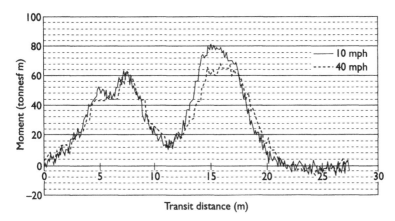

Figure 8.3 Comparison between midspan bending strains in 10 m span bridge due to 10 mph and 40 mph transits.

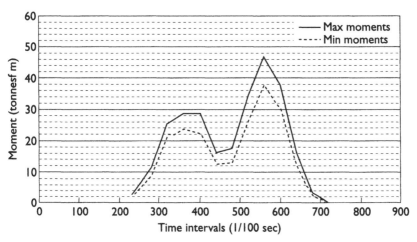

Figure 8.4 Envelope of moments calculated from a series of theoretical transits of a 10 m span bridge at 10 mph. Maximum = 47; minimum = 38; mean = 42; standard deviation = 0.84; peak potential dynamic amplification factor = 1.12.

appear to be caused by the uneven road profile, and not by vibration response of the bridge. The DAF is here actually less than unity, although bridge vibrations were very small.

Figure 8.4 illustrates an envelope of the 'pseudo-static' effects caused by variations in loading caused by vertical acceleration of vehicle mass due to uneven road surfaces, excluding dynamic response of the supporting surface. The plotted values are potential load effects derived for all possible locations of a 10 m midspan beam bending influence line relative to the entire set of all wheel load measurements for a 15 sec period during a 10 mph passage of the same 5-axled air

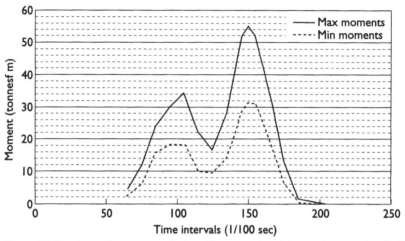

Figure 8.5 Envelope of moments calculated from a series of theoretical transits of a 10 m span bridge at 40 mph. Maximum = 55; minimum = 31; mean = 42; standard deviation = 3.2; peak potential dynamic amplification = $\frac{55}{42}$ = 1.31.

suspension articulated 38 tonne heavy goods vehicle. The maximum possible amplification of the static moment due to the measured variations in wheel loads for this period was equivalent to a DAF of 1.12, but the 'characteristic' (upper 5 percentile) value was approximately a DAF of 1.03.

Figure 8.5 shows the same plot for approximately 8 seconds at 40 mph. The peak DAF was potentially 1.31, and the characteristic was approximately 1.12.

8.3.8 Trial time history analysis

Some trials were performed in order to establish the validity or otherwise of possible analysis methods, and values of damping parameters. The approach which was chosen was to use the Duhamel integral method (Clough and Penzien, 1993). This may be conveniently implemented in a computer spreadsheet and is described in the form of a hand analysis spreadsheet in the early (1975) edition of Clough and Penzien.

The response equation that is used is given by the following convolution integral. The response is obtained by integrating a series of harmonic vibration responses due to a series of short duration impulses. Thus (see eqn 2.38):

$$v(t) = \frac{1}{m\omega_D} \int_0^t p(\tau) \sin \omega_D(t - \tau) \exp[-\xi\omega(t - \tau)] \, d\tau \qquad (8.3)$$

where:

$v(t)$ = Displacement at time t
τ = Time at each impulse
ξ = Damping ratio

$p(\tau) = $ Impulsive force at time τ
$\omega_D = $ Modal natural frequency

The applied load function $p(\tau)$ is obtained by summing the products of each of the instantaneous values of the vehicle axle loads and a modal influence coefficient at each time step. Thus:

$$p(\tau) = \int_0^L f(x, t)\phi(x)\, dx \qquad (8.4)$$

where:

$p(\tau) = $ Impulsive force at time τ
$L = $ Span of bridge
$x = $ Position along span
$f(x, \tau) = $ Forces applied at locations x at time τ
$\phi(x) = $ Value of mode 1 vibration shape at locations x

As explained above, since dynamic response is only calculated in Mode 1, and the mode shape is not identical to the static influence line shape, it is convenient to assume that the modal influence coefficient and static influence lines are both sinusoidal, and have equal maxima at midspan. (The maximum value $= 0.25L$, since that is the influence line magnitude for midspan bending due to a central load on a beam.)

At a particular time T, therefore, the static load effect is merely given by $p(T)$. The DAF is then given by:

$$\text{DAF} = \frac{\max\left\{ \dfrac{1}{m\omega_D} \displaystyle\int_0^T p(\tau)\sin \omega_D(t - \tau)\exp[-\xi\omega(t - \tau)]\, d\tau \right\}}{\max\{\bar{p}(x)\}} \qquad (8.5)$$

where:

$$\bar{p}(x) = \int_0^L \bar{f}(x)\phi(x)\, dx$$

where:

$\bar{p}(x) = $ Impulsive force when vehicle is at location x
$L = $ Span of bridge
$x = $ Position along span
$\bar{f}(x) = $ Average (static) axle loads at positions x
$\phi(x) = $ Value of mode 1 vibration shape at locations x

Figure 8.6 compares the static 'Input' and dynamic response 'Response' from a time-history analysis considering mode 1 response on a 10 m bridge span during a 10 mph transit, superimposed on a plot of the lower flange strains 'Measured Strain' during the period of the vehicle transit.

The first flexural mode natural frequency for a 10 m span for a simple beam from eqn (8.2) is (approximately) 10 Hz. Mode 2 was omitted since it has no curvature

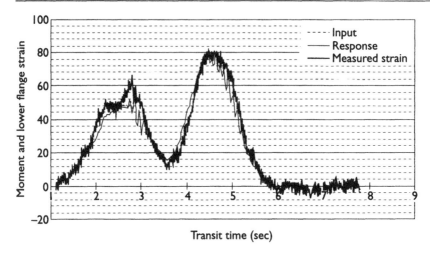

Figure 8.6 Mode I response due to 10 mph transit: Damping ratio = 0.10.

or deflection at mid span. The third mode s natural frequency would be 90 Hz, which is so far in excess of the vehicle's excitation frequency that it, too, was to be ignored.

8.3.9 Effect of damping

Structural damping must be obtained by observation. In dynamic analysis, it is usually expressed as the ratio, ξ to the critical damping value. Structural engineers often find it convenient to observe the ratio between two successive peak values of an oscillation as it decays following some test excitation. The logarithm of the ratio between successive peaks is known as the logarithmic decrement ('log dec' see also Section 2.2.2) δ, where:

$$\delta = \frac{2\pi\xi}{\sqrt{1 - \xi^2}} \tag{8.6}$$

which, for small damping becomes:

$$\delta \approx 2\pi\xi \tag{8.7}$$

In the trial analyses, the best match between predicted and measured response was found when using a damping ratio, ξ, of 0.10. Green *et al.* (1995) report ξ values in tests of 0.045. The peak moment calculated for the 40 mph transit was somewhat less than that for the 10 mph transit, owing to the form of local road profile. (This effect also appears in the strain gauge readings plotted in Figure 8.3.) The 'Input' line represents the changing static midspan moment taking account of the variation in effective vehicle weight that appears in Figure 8.2. The 'Response' line includes bridge dynamic response in the first flexural mode. Both lines fit the measured strain gauge changes almost equally well, which implies that

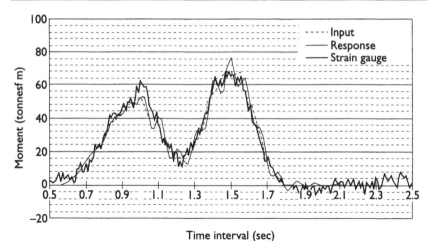

Figure 8.7 Mode I response due to 40 mph transit: Damping ratio = 0.10.

(provided that measured wheel loads were available) dynamic response analysis was not necessary for this structure under this load.

Figure 8.7 suggests that (at least for this type of composite steel beam plus concrete slab structure) a pseudo-static analysis which takes account of the change in effective vehicle weight but which does not concern itself with dynamic response of the bridge will be adequate for all practical purposes.

8.3.10 Interpretation and implementation of dynamic analysis

Practical bridge design codes usually provide load models which will provide 'nominal' load effects which have some pre-determined probability of exceedence.

If the load model has been derived separately for static and dynamic effects, there remains the problem of combining the two analysis results into a single design model, which is related in some pre-determined manner to the statistically determined extreme of the joint effects of static and dynamic loading.

It does appear that, for most practical structures, dynamic magnification or reduction of static load effects is caused mainly by the effects of uneven road profile. To a first approximation, therefore, the DAF is a unique (although uncertain) property of each bridge (or, at least, of the transit of each individual type of vehicle).

Thus, the extreme static load effect will be a function of the lifetime exposure of the bridge to traffic, but the extreme dynamic load effect will be a property of the bridge. When the Highways Agency's (1997) assessment rules were developed, it had to be assumed that there were generally no site specific strain records, and the uncertainty in DAF was treated as a structural property. After much consideration, the rules were finally based on reviewing variations in static load effects derived from a large number of continuous wheel load measurements from a set of vehicles which was broadly representative of the types of vehicle in common use in the UK.

8.4 REFERENCES

AEA Technology *Program VAMPIRE*, Internal report, Derby.

Bailey, S. F. (1996) 'Basic principles and load models for the structural safety evaluation of existing road bridges', Thesis No. 1467, Ecole Polytechnique Fédérale de Lausanne.

BSI (1972) BS 153: Part 3A, *Specification for Steel Girder Bridges*, British Standards Institution, London.

BSI (1980) BS 5400, *Code of Practice for Steel, Composite and Concrete Bridges*, British Standards Institution, London.

CEN Technical Committee 250 (1994) Eurocode 1, *Basis of Design and Actions on Structures – Part 3, Traffic Loads on Bridges (ENV 1991-3)*, CEN, Brussels.

Clough, R. W. and Penzien, J. (1975, 1993) *Dynamics of Structures*, McGraw-Hill, New York.

Cooper, D. I and Flint, A. R. (1997) 'Development of short span bridge-specific assessment live loading', in P. C. Das (ed.) *Safety of Bridges*, Institution of Civil Engineers, London.

Das, P. C. (ed.) Safety of Bridges, T. Telford, London 1997.

Department of Transport, BES Division. (1980) *Revision of Short Span Loading*. Unpublished. London.

Elnashai, A. S. (1995) Institution of Civil Engineers Lecture, February, London.

Flint, A. R. (1990) 'Current UK bridge assessment rules and traffic loading criteria', in P. C. Das (ed.) *Safety of Bridges*, Institution of Civil Engineers, London.

Flint and Neill Partnership. (1986) *Interim Design Standard: Long Span Bridge Loading.* TRL Contractor Report CR16, TRL, Crowthorne.

Green, M. F., Cebon, D. and Cole, D. J. (1995) 'Effects of vehicle suspension design on dynamics of highway bridges', *J. Struct. Engng. ASCE.* **121**(2).

Henderson, W. (1954) 'British highway bridge loading', paper given at ICE Proceedings, Road Engineering Division Meeting: Road Paper No. 34, 2 March.

Highways Agency (1988) *Design Manual for Roads and Bridges*. Vol. 1 Section 3 Part 6 BD37/88 Loads for Highway Bridges, HA, London.

Highways Agency (1997) *Design Manual for Roads and Bridges*. Vol. 3 Section 4 Part 3 BD21/97 Chapter 5 Loading, HA, London.

Page, J. (1976) *Dynamic Wheel Load Measurements on Motorway Bridges*, TRRL Laboratory Report LR722, TRL, Crowthorne.

Paultre, P., Chaallal, O. and Proulx, J. (1992) 'Bridge dynamics and amplification factors: a review of anaytical and experimental findings', *Canadian J. Civil Engng.* **19**.

Ricketts, N. J, and Page, J. (1997) *Traffic Data for Highway Bridge Loading*, TRL Report 251, TRL, Crowthorne.

Acknowledgement

I wish to acknowledge the assistance of the UK Highways Agency and Transport Research Laboratory who respectively sponsored and collected the bridge and vehicle response data upon which the analysis of UK bridge responses is based.

Chapter 9

Machine-induced vibrations

J. W. Smith

9.1 INTRODUCTION

Many industrial processes give rise to large dynamic forces. The unbalanced masses of large rotating machines generate oscillating forces, while forge hammers and rock crushers apply transient shocks and impacts. Rotating or reciprocating machines generally operate at fixed frequencies. It is essential to design the foundations or supports to avoid resonance. Periodic forcing functions will always induce dynamic responses and these should be evaluated to ensure that they do not damage the fabric of the enclosing buildings, the machines themselves or other sensitive processes nearby. The vibration due to transient shocks and impacts should also be evaluated for the same reason. Furthermore, human beings are very sensitive to vibrations and the amplitudes should not cause discomfort to personnel working nearby or to other occupants of buildings that contain machines.

There are four important steps in the successful design of machine foundations. First, the dynamic forces have to be assessed accurately. This is the task of the mechanical designer of the machinery itself, and largely consists of forces exerted by the inertia of the moving parts. These generally occur at harmonics of machine speed. Other forces arise from cylinder ignition, rock crushing, impact from hammers, and certain fault conditions such as short circuits in electrical machinery. Secondly, the ground conditions have to be assessed. This requires a geotechnical investigation with the aim of determining reliable values of the effective elastic resistance provided by the foundation material. A balance has to be achieved between the cost of a detailed survey and the value and importance of the project. There will inevitably be considerable uncertainty about the numerical values eventually adopted for the design calculations and the designer needs to be confident about the upper and lower bounds. Thirdly, the numerical model of the system should be suitable for the purpose. Many machine foundations consist of large mass concrete blocks. These are effectively rigid, and reliable design calculations can be done using quite simple methods of analysis by treating the soil foundation as an elastic supporting medium. However, in principle the ground is a non-linear solid with infinite boundaries. Vibration of a machine block results in wave propagation within the soil radiating away from the centre of disturbance. This should be reflected in the analytical modelling for major projects. Advanced analysis using finite element and layered half space methods may be required. Finally, the predicted vibration should be compared with criteria chosen to ensure that personnel are not discomfited and that equipment performance is not impaired. Some design rules exist but only for limited types of industrial machinery. Information often has to be obtained from experience within the relevant industry or from research.

9.2 DYNAMIC LOADING BY MACHINERY

The dynamic loading from industrial machinery derives principally from the inertia effects of moving parts. Every machine behaves differently and it is usually the responsibility of the manufacturer to calculate the forces that will be imposed on the supporting structure. The rotation speeds or frequencies at which machines operate are also important and should be specified.

9.2.1 Reciprocating engines

Large multi-cylinder diesel engines are often used to provide the primary motive power for electrical generating plant in remote regions lacking indigenous fossil fuels or where there is no access to other forms of energy production. The reciprocation together with the cylinder ignition sequence give rise to periodic forces.

A typical arrangement of diesel engine and alternator mounted on a foundation block is shown in Figure 9.1. A crank–piston linkage is shown in Figure 9.2 where the masses of the crank, connecting rod and piston are m_1, m_2, and m_3 respectively. Engine speeds are normally quoted in revolutions per minute (r.p.m.) in which case the crankshaft rotation frequency Ω is given by:

$$\Omega = 2\pi N/60 \text{ (rad/sec)} \tag{9.1}$$

where N is the engine speed. It is evident that there will be oscillatory inertia forces due to the moving masses m_1, m_2, and m_3. The first two will have vertical and horizontal components while m_3 will have a vertical component only. The magnitude of these forces may be evaluated if the various masses and lengths are known. For example, it can be shown that the inertia force due to the piston will be:

$$P(t) = m_3\Omega^2 R[\cos \Omega t + (R/L) \cos(R\Omega t/L)] \tag{9.2}$$

This is not a simple harmonic excitation because of the second term. Therefore, in addition to the primary engine forces applied at engine speed, there will be higher harmonics applied at integral multiples of the engine speed. The mechanical designer of the engine should evaluate these forces.

In multi-cylinder engines it is possible to balance most of the inertia forces, depending on the number and arrangement of the cylinders. However, there will

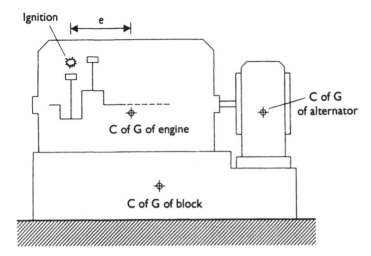

Figure 9.1 Diesel engine and alternator.

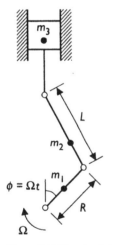

Figure 9.2 Crank–piston linkage.

always be some residual unbalanced forces due to tolerances on weights and geometry. Furthermore, ignition of a cylinder will create a dynamic moment about the centre of gravity of the engine, as may be seen in Figure 9.1. In a four-stroke engine this will result in a pitching moment whose frequency will be:

$$\Omega = \frac{2\pi N}{60}\frac{n}{4} \tag{9.3}$$

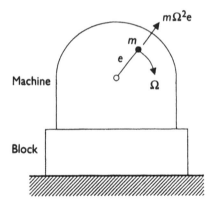

Figure 9.3 Rotating unbalanced mass.

where n is the number of cylinders. The firing order is selected to minimize the effect of this moment.

9.2.2 High speed rotating machines

Turbines, centrifugal pumps, fans and electrical generators are examples of high speed rotating machinery. Even though considerable care is taken to balance the rotating parts, residual imbalances will always exist. An inertia force is generated by the eccentricity e of an unbalanced mass m about the centre of rotation as indicated in Figure 9.3. Strictly speaking, this force acts radially and rotates with the shaft. But it will have oscillatory vertical and horizontal components that may excite the corresponding modes of vibration. This happens in everyday experience with a domestic spin dryer. It should also be noted that the speeds of turbines are many times greater than those of reciprocating engines of similar power and therefore the out of balance force will be significantly amplified because of the frequency being squared ($m\Omega^2 e$).

9.2.3 Transient torques in electrical machines

There are two important cases of transient dynamic loading that occur with driven electrical generators. The first is known as *short circuit torque*. Consider an electrical alternator being driven by an engine as in Figure 9.1. If a fault occurred, which had the effect of creating a short circuit in the output of the alternator, a very large current would be demanded (for a fraction of a second, perhaps). This would be experienced as a suddenly applied load or brake on the system. As a result the engine would apply a torque about the axis of the drive shaft. The second case is known as *faulty synchronizing torque*. This occurs when an engine generator system starts up and feeds power into the national grid. If the output of the generator is not synchronized with the a.c. waveform of the national grid a braking effect, or torque, is experienced by the generator until such time as it is in phase. This will have a similar effect to short circuit torque, in that a sudden torque is applied to the alternator which has to be reacted by the machine supports. The dynamic effect of a suddenly applied load is twice the static effect. The same applies to a suddenly applied torque. The machine manufacturer should provide the magnitude and direction of transient torques that may occur during operation of a machine.

9.2.4 Gyratory rock crushers

In the quarrying industry there is a need to crush excavated rock into stones of varying size for different end uses. The raw excavated rock is often in very large pieces (perhaps in excess of one metre across and weighing a couple of tons), whereas the end product may be required for highway chippings of 20 mm size or less. Gyratory crushers are usually used in modern quarries to process the rock.

There are several designs in existence but one of the most common is the base supported cone crusher. The principle of operation is illustrated in Figure 9.4. The fixed parts of the machine consist of a drum, hopper and bowl (inverted). Uncrushed rock is loaded into the hopper and onto the feed plate of the crushing head, which ensures that the rock falls into the space between the bowl and the mantle of the crushing head. The crushing head is driven in a gyratory motion by

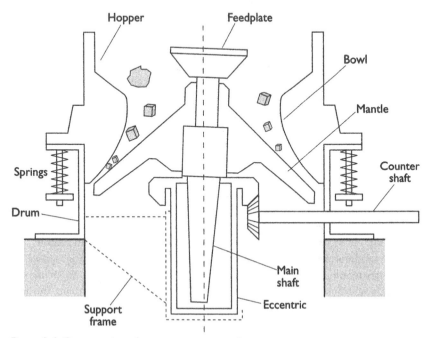

Figure 9.4 Cross section of a gyratory cone crusher.

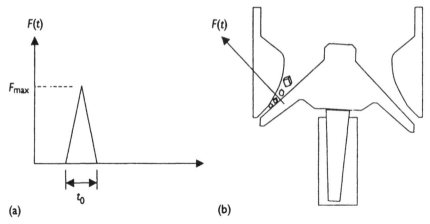

Figure 9.5 Schematic diagram of transient force due to stone crushing: (a) forcing function; (b) direction of dynamic force.

the main shaft which is seated in the eccentric shaft at an angle designed to achieve the desired size of stone after crushing. Note that the gap between the crushing mantle and the bowl varies around the perimeter of the bowl. The eccentric shaft is driven through a gear mechanism by an electrically powered counter shaft. Crushing is achieved when lumps of rock are trapped between the bowl and mantle and then crushed when the gap diminishes during a revolution of the eccentric shaft.

It will be appreciated that a number of dynamic forces will occur due to rotation of the moving parts (Szczepanik *et al.* 1990). The largest will be caused by the gyratory motion of the mantle and main shaft due to their eccentricity. There will also be the unbalanced counter shaft force, which is usually at twice the frequency of the main shaft, and vibration of the springs holding down the hopper and bowl.

However, the most important dynamic forces are transient shocks arising from the crushing action itself. Smith (1993) showed that the dynamic load when a stone is crushed consists of an impulse as the power of the drive motor compresses the rock between the bowl and mantle, followed by a sudden release as the stone fractures. The peak value of the force could be as much as two or three tons, according to the size of the crusher, and is shown schematically in Figure 9.5a. There is a need for experimental data on the magnitude of the typical crushing force, but in its absence an estimate can be made from knowledge of the motor torque and the mechanical advantage available between motor and mantle. The duration of the impulse is based on the assumption that typically four stones are crushed per revolution. The force is applied in a direction normal to the face of the mantle, which is usually about $45°$, but could occur in any direction in the horizontal plane because of the rotation of the pinch point of the crusher (see Figure 9.5b). These forces are transient and random, as rocks are fed into the machine, but may occur several times per revolution of the eccentric shaft.

It has been found (Szczepanik *et al.* 1990) that vibration of the crushing machine on its foundation leads to dynamic stresses in the arms of the supporting frame that holds the eccentric shaft and head in position. Ultimately, fatigue cracks may occur in the arms, causing breakdown of the machine. It is important that the supporting foundation is sufficiently stiff to minimize this vibration.

9.2.5 Hammers

There are many industrial processes, typically impact forging, which require single or repeated blows with a hammer. Kinetic energy is given to the hammer head either by some external source of power such as steam, compressed air, or more usually by gravity. Velocity is imparted to the anvil and work piece by transfer of momentum. A schematic arrangement for a drop hammer is shown in Figure 9.6. The mass of the hammer head, or tup as it is called, is denoted by m_0, the anvil by m_2 and the foundation by m_1. Some kind of elastic layer, often hardwood, is interposed between the anvil and the foundation block. The block is either supported elastically by the foundation material or by specially designed springs to minimize the transmission of vibration to nearby buildings. Thus the system has two degrees of freedom.

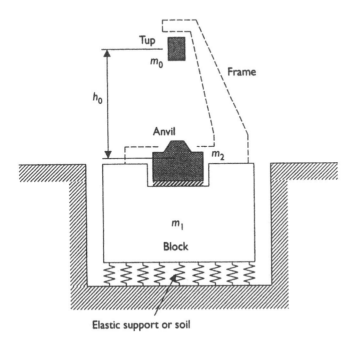

Figure 9.6 Schematic arrangement for a drop hammer.

The velocity of the tup before impact is given by:

$$v = \sqrt{2gh_0} \tag{9.4}$$

In the case of pneumatic or steam powered drop hammers Barkan (1962) found that, in practice, this should be reduced by an empirical factor of 0.65 to allow for friction and the resistance of exhaust air or steam.

Following the method of Barkan (1962), conservation of momentum can be expressed by:

$$m_0 v = m_0 v' + m_2 v_0 \tag{9.5}$$

where $(-v')$ is the rebound velocity of the hammer head and v_0 is the velocity imparted to the anvil. The relative velocity after impact depends on the elastic characteristics of the colliding bodies and is obtained from the expression:

$$C_r = (v_0 - v')/v \tag{9.6}$$

where C_r is the coefficient of restitution. This constant varies between 0 (fully plastic) to 1 (perfectly elastic). Thus the velocity of the anvil after impact may be obtained from eqns (9.5) and (9.6) and is given by:

$$v_0 = \frac{1 - C_r}{1 + \mu_0} v \tag{9.7}$$

where $\mu_0 = m_2/m_0$. This velocity may be used as an input to the equations of motion of a two degree of freedom system (see Section 2.3). It is possible that the hammer might strike the anvil eccentrically, thus imparting a rotational component. This condition should also be considered.

Novak and El Hifnawy (1983) have verified that the above procedure is satisfactory provided that the duration of the impact is much shorter than the natural period of the foundation. This may not be so if the foundation support is very stiff (e.g. piles). It would then be necessary to take account of the force–time function of the impulse.

9.2.6 Vibrating screens

These are used extensively in the mining and quarrying industries for washing, separating and grading processes. Rock is passed over a horizontal screen that is supported at its ends, as shown in Figure 9.7. The screen is vibrated vertically and horizontally by motor driven eccentric cranks or cams. Stones that are small enough pass through the screen and are collected, while larger stones pass over the sloping upper surface and continue to the next size of screen. Sinusoidal inertia forces are thus applied to the supporting frame structure and to the foundations. In the quarrying industry rock is graded into different sizes by passing over a sequence of vibrating screens. These are usually housed in a single large building, which is therefore subjected to continuous dynamic loading while in

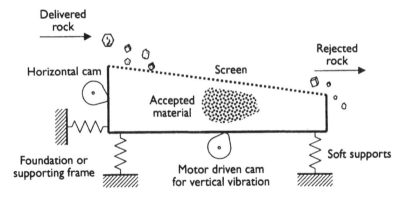

Figure 9.7 Schematic diagram of a vibrating screen.

operation. The frame of the building should be stiff enough to prevent unacceptable vibration amplitude.

9.2.7 Rolling mills

In processes such as rolling of steel sections, the shaping is achieved by passing a hot bloom or billet through the shaping rollers which are driven by a d.c. motor. As a billet enters the rollers, the initial resistance acts like a suddenly applied torque to the shaft of the driving motor. The resulting dynamic couple applied to the foundation block is analogous to the 'short circuit' torque in electrical generators. The torque can be estimated by knowing the speed and power of the driving motor.

9.3 DESIGN OF STRUCTURES TO MINIMIZE MACHINE-INDUCED VIBRATION

9.3.1 Dynamic response of supporting structure and foundation

Most types of heavy industrial machines are provided with one of the five types of supporting foundation shown in Figure 9.8. These are mass concrete blocks, box-type foundations, wall foundations, reinforced concrete frames and table top foundations. The choice of foundation is influenced by the type of machine, the magnitude of the dynamic forces and the access required around the machine.

Power generating sets, comprising large diesel engines and alternators, are often mounted on mass concrete foundation blocks which distribute the load over the base area to the supporting soil or rock. Piled foundations would be preferred where the ground conditions are poor. The block may be anything from twice to

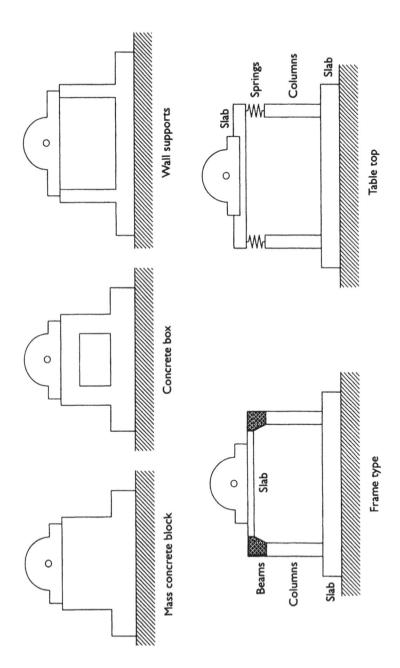

Figure 9.8 Foundation systems for industrial machinery.

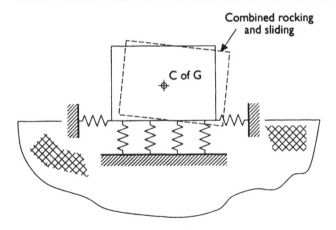

Combined rocking
and sliding

C of G

Figure 9.9 Simplified model of rigid foundation block resting on elastic supports.

five times the mass of the machinery. The performance of block foundations was discussed fully by Smith (1989). The idealized structural system consists of a large rigid mass resting on a semi-infinite elastic medium with dynamic forces and moments applied to the mass. This is an example of ground structure interaction for which the mathematical analysis is relatively advanced and inconvenient for general design practice (Arnold *et al.* 1955). Various attempts have been made to derive simplified formulae including Hsieh (1962) who proposed an equivalent mass restrained by elastic elements, the system having an appropriate amount of damping derived from semi-infinite elastic theory. The theory of ground structure interaction is discussed briefly in Section 9.3.2.

The method of analysis in most widespread use at the present time originates from the work of Barkan (1962). His simplified system is shown in Figure 9.9 where the soil stiffness is represented by vertical and horizontal springs. Provided that the centre of gravity of the mass coincides with the centre of stiffness of the soil, vertical vibration can be treated as a single degree of freedom mass–spring system. In the case of horizontal vibration, sliding motion will be accompanied by rocking, resulting in a two-degree of freedom system. Other motions that should be considered include pitching and horizontal motion in a plane perpendicular to the one shown, and also yawing motion about a vertical axis. Barkan (1962) believed that the participating soil mass does not make a sufficient difference to the calculation of the natural frequency and could be safely neglected. Furthermore, provided that the machine frequency is sufficiently different from the natural frequency of the system, amplitudes of forced vibration may be calculated with reasonable accuracy by ignoring damping. Full details of the analytical procedure are provided in the Code of Practice for Foundations for Machinery (BSI, 1974). Information and methods for calculating the stiffness of the foundation material whether it be soil, rock or piles is provided by Skipp (1966). On the basis of experience of the performance of a very large machine, Smith (1989) has commented that the BSI (1974) method of calculation (which is based on Barkan, 1962) is generally satisfactory for machines with very stiff bases. However, industrial practice is moving towards lighter and more flexible bases for which the flexing of the base should be considered.

In an effort to reduce the cost, and also to provide access to equipment, box-type foundations may be employed. The BSI (1974) method may be adopted for analysis but consideration should be given to the possible flexing of the structure, especially when designing for large machines. Wall foundations are particularly useful for machines which require equipment such as conveyors beneath them. The machine may be fixed directly to the tops of the walls or supported on steel bearers that span between the walls. This kind of foundation is potentially very flexible and particular care should be taken in the design. For example, it may not be wise to assume that the foundation slab is rigid since transverse vibration of the walls may be accompanied by flexing of the slab.

Turbines and other high speed rotating machines are often mounted on reinforced concrete frames and slabs (Srinavasulu and Vaidyanathan, 1976). In this case the supporting structure cannot be treated as a rigid block and analysis of its dynamic characteristics may require numerical computation. The 'table top' arrangement is similar to the frame type of supporting structure, as shown in Figure 9.10 in which a turbine and alternator are mounted on a slab, which in turn is supported by columns. In order to confine the vibration caused by the machinery to the slab, the slab itself is mounted on isolating springs and dampers at the tops of the columns. The stiffness and damping of these spring–damper units must be chosen to minimize transmission of the vibration to the columns. The slab/machine system has to be checked for all possible modes of vibration including flexural motion of the slab.

9.3.2 Ground–structure interaction

The above methods are highly simplified theoretically, and take no account of the infinite or semi-infinite extent of the ground that is supporting a machine foundation. These simple methods have been used extensively for most types of foundations for general industrial machinery, including quite large generator sets. However, in the case of very large, important, expensive or safety critical projects it may be necessary to carry out more rigorous analysis using advanced theory. For example, nuclear installations in seismic zones should be analysed taking full account of ground–structure interaction. The principles are discussed briefly below.

An idealized model of a foundation block supported on soil or rock is shown in Figure 9.11. The mass consists of the concrete block together with the machinery. The ground is a semi-infinite region of layered elastic media. The layers represent different soil or rock strata and would have different density and elastic characteristics. Dynamic forces are applied to the block as a result of the operation of the

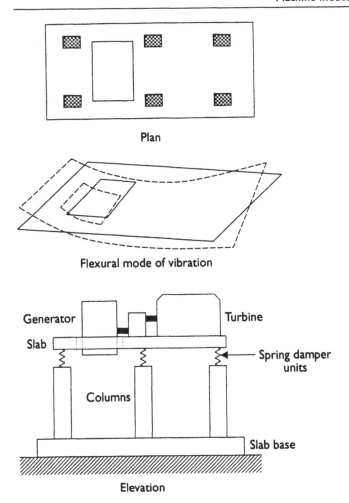

Plan

Flexural mode of vibration

Generator

Turbine

Slab

Spring damper
units

Columns

Slab base

Elevation

Figure 9.10 Table top supporting structure for turbine and alternator system.

machine. Generally these would be periodic forces arising from the out of balance masses.

Although the ground provides elastic resistance, which may be evaluated by the theory of elasticity, the lack of boundaries means that vibration in a steady state mode shape does not occur. Instead, vibration of the block results in displacement of the ground in the form of waves that propagate away from the source of disturbance. The most important waves occur at the free surface and are analogous to ripples on a pond radiating in the form of concentric rings increasing in diameter with time. The energy of motion at the block is confined to a small volume of soil whereas it is spread over a much larger volume after some time when the ripple has widened. Thus, energy of vibration is radiated away from the source of

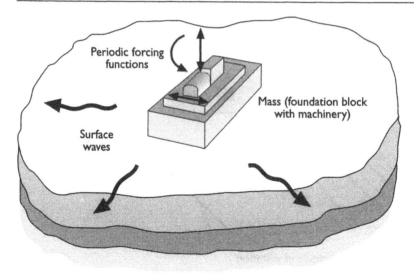

Semi-infinite layered elastic material

Figure 9.11 Idealized model of a foundation block supported on soil or rock.

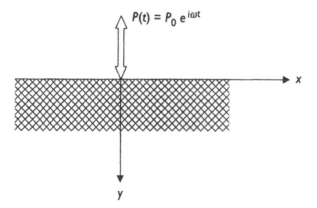

Figure 9.12 Dynamic point load on surface of an elastic half space.

disturbance. Even if frictional damping in the soil is ignored, this loss of energy due to the propagating waves is always present and is called *radiation damping*. This surface wave effect is of great importance when evaluating the vibration of structures supported by unbounded media.

A preliminary step in the analysis is to consider a dynamic point load on the surface of an elastic half space as shown in Figure 9.12. The most basic loading function is a steady state periodic force given by:

$$P(t) = P\,e^{i\omega t} \tag{9.8}$$

noting that $e^{i\omega t} = \cos \omega t + i \sin \omega t$.

The displacement response of the surface $y(t)$ is of a similar form

$$y(t) = y\,e^{i\omega t} \tag{9.9}$$

Even if there was no material damping in the soil, the displacement amplitude could never go to infinity at resonance, as in the case of structures with boundaries (see eqn 2.42 and Figure 2.14). This is because the surface waves travelling away from the source of disturbance absorb energy and give rise to apparent damping or radiation damping. A further feature of the behaviour of wave motion in elastic half-spaces is that there is a characteristic frequency below which wave motion will not occur.

If eqns (9.8) and (9.9) are substituted into an equation of motion such as eqn (2.27) given in Chapter 2, it can be shown that:

$$[k(1 + 2\zeta i) - \omega^2 M]y = P \tag{9.10}$$

where y is the displacement and ζ is the effective critical damping ratio (see eqns 2.33, 2.34 and 2.35). The term $[k(1 + 2\zeta i) - \omega^2 M]$ is referred to as the dynamic stiffness and is a function of frequency. This relationship also holds for multi-degree of freedom systems for which the equation of motion is given by:

$$[S]\{u\} = \{P\} \tag{9.11}$$

where $\{u\}$ is the vector of displacements of all the degrees of freedom and $\{P\}$ is the load vector. $[S]$ is the dynamic stiffness matrix and is given by:

$$[S] = [K](1 + 2\zeta i) - \omega^2[M] \tag{9.12}$$

It is generally easier to obtain solutions for periodic loading using analysis in the frequency domain as above. Note that the dynamic stiffness matrix is frequency dependent. Analysis of general forcing functions can also by synthesized from frequency domain solutions using Fourier transforms.

Wolf (1985) has provided a detailed treatment of ground–structure interaction in the frequency domain for earthquake analysis of large structures. He proposed a system using substructures, suitable for finite element analysis. This is illustrated in Figure 9.13 where the structure–soil system is reduced to two substructures, one being the main structure and the other being the surrounding soil of infinite extent. The dynamic stiffness matrix of the main structure, $[S]$, has an order equal to the number of degrees of freedom in the finite element model. It may be evaluated by conventional finite element methods. The displacement vector may be decomposed into sub vectors $\{u_s\}$ and $\{u_b\}$. The subscript b denotes all the nodes at the interface while s denotes the remaining nodes of the structure, as shown in Figure 9.13. Similarly, the stiffness matrix may be decomposed into the submatrices $[S_{ss}]$, $[S_{sb}]$ and $[S_{bb}]$. Hence, the equation of motion of the structure may be expressed as:

$$\begin{bmatrix} [S_{ss}][S_{sb}] \\ [S_{bs}][S_{bb}] \end{bmatrix} \begin{Bmatrix} u_s \\ u_b \end{Bmatrix} = \begin{Bmatrix} P_s \\ 0 \end{Bmatrix} \tag{9.13}$$

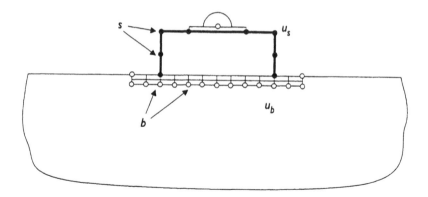

Figure 9.13 Finite element model of structure embedded in an unbounded region of soil.

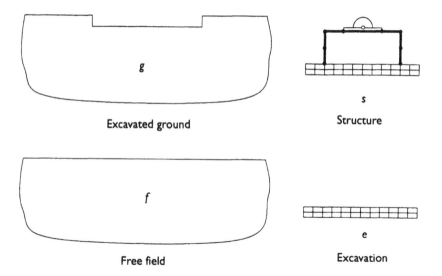

Excavated ground

Structure

Free field

Excavation

Figure 9.14 Reference subsystems for the structure and excavated ground.

where P_s are the loads applied to nodes of the structure other than at the interface where it is assumed there are no external loads.

In order to obtain the dynamic stiffness matrix of the soil structure system, it is necessary to add the dynamic stiffness matrix of the excavated ground. This is shown in Figure 9.14 and the equation of motion of the system will become:

$$\begin{bmatrix} [S_{ss}][S_{sb}] \\ [S_{bs}][S_{bb}^s + S_{bb}^g] \end{bmatrix} \left\{ \begin{matrix} u_s \\ u_b \end{matrix} \right\} = \left\{ \begin{matrix} P_s \\ 0 \end{matrix} \right\} \tag{9.14}$$

where the superscripts s and g denote the stiffness sub matrices belonging to the structure and the ground, respectively. It should be noted that the stiffness matrix of the excavated ground $[S_{bb}^g]$ in principle represents the dynamic stiffness of generalized springs and dash pots joining the nodes b to adjacent fixed virtual nodes. They take account of the density and stiffness of the ground material together with radiation damping due to the wave motion.

This stiffness matrix is not easy to obtain because it is a semi-infinite region of irregular shape. However, it can be obtained by subtracting the stiffness matrix of the excavated soil $[S_{bb}^e]$ from that of the interface nodes in the free field $[S_{bb}^f]$ as follows:

$$[S_{bb}^g] = [S_{bb}^f] - [S_{bb}^e] \tag{9.15}$$

Hence, the equation of motion becomes:

$$\begin{bmatrix} [S_{ss}][S_{sb}] \\ [S_{bs}][S_{bb}^s - S_{bb}^e + S_{bb}^f] \end{bmatrix} \begin{Bmatrix} u_s \\ u_b \end{Bmatrix} = \begin{Bmatrix} P_s \\ 0 \end{Bmatrix} \tag{9.16}$$

$[S_{bb}^s] - [S_{bb}^e]$ may be interpreted as the dynamic stiffness matrix of the structure, with the stiffness matrix of the excavated soil subtracted. The latter may be evaluated by conventional finite element methods.

The free field dynamic stiffness matrix $[S_{bb}^f]$ must be evaluated using the theory of semi-infinite layered media. This requires advanced mathematical treatment that is beyond the scope of this book. Wolf (1985) has derived a number of useful special cases for two-dimensional and axi-symmetric foundations using Green's functions. Moreover, the method has been developed further for analysis in the time domain (Wolf, 1988). Wolf and Paramesso (1992) solved a practical example of a hammer foundation with uplift of the anvil. This was a non-linear problem in which they used a lumped parameter model for a rigid cylindrical foundation embedded in a soil layer.

9.3.3 Design criteria

The primary requirement of the machine/foundation system is that resonance is avoided at the operating machine frequencies. It is sometimes recommended that for important installations any natural frequency of the system should differ from significant operating frequencies by a factor of 2.0 (BSI, 1974). The factor is reduced to 1.5 for installations of lesser importance. If this degree of separation of natural frequencies and driving frequencies can be achieved, no further analysis is generally required. In practice this usually requires a compromise because the dynamic system will have numerous natural frequencies of vibration while the machinery may generate harmonics in addition to the principal operating frequency. More recent specifications allow the frequencies to be even closer (DIN 4024, 1988).

In the case of an installation where the forcing and natural frequencies are close, the maximum amplitudes of the foundation/structure system should be evaluated under the principal forcing functions. These would include: periodic loading from unbalanced rotating masses; transient loads such as short circuit torque; and shock or impulse loads from industrial processes.

There are three principal concerns when considering whether the predicted vibrations are acceptable or not. First, the vibrations in the vicinity of the machinery should not be large enough to disturb personnel working nearby (e.g. maintenance or control room staff). Human beings are surprisingly sensitive to vibration. A vibration of small amplitude can be disturbing or annoying especially if it is continuous. It can impair concentration, cause fatigue and other physical symptoms, including headaches and sickness in extreme cases. Human response to vibration was discussed in Chapter 7. Secondly, excessive vibration of the supports of a machine may result in over stress of components of the machine itself. An example of this is the supporting framework for the eccentric shaft of a gyratory cone crusher (see Figure 9.4). If this steel framework is subjected to continuous vibration induced stresses it may be susceptible to fatigue damage (Szczepanik et al. 1990). The foundation should be stiff enough to prevent the machine vibrating excessively on its mountings. Bearings of turbines and engines may be adversely affected by excessive vibration. Thirdly, transmission of machine-induced vibration to the structure of the building in which the machinery is housed may be undesirable. The vibration may be disturbing to personnel working in the same or adjacent buildings. Electronic control equipment is often housed in boxes or panels and fixed to the floors, columns or walls of enclosing buildings. Possible damage to such equipment by continuous vibration should be checked. A good general principle is to keep the foundations of the machines and the building separate.

There is a scarcity of code provisions dealing with machine-induced vibrations. Plant manufacturers often work to their own standards and by default set the standards for the relevant industry. Suppliers of sensitive equipment may specify limits to the acceptable environment in which their equipment operates satisfactorily.

Information on limits to the vibration environment of rotating machines may be found in Moore (1985). The acceptable operating amplitude decreases with frequency and a limit is often specified in terms of velocity of vibration at the bearings of the machine. A vibration velocity of less than 2.0 mm/sec would be expected to provide smooth running conditions whereas over 16 mm/sec the operation of the machine would probably be very rough. The British Standard for rotating electrical machines (BSI, 1987) recommends a limit of about 2.5 mm/sec r.m.s., although this is intended for relatively small machines. In the quarrying industry in the UK a limit of 0.36 mm/sec is generally recommended for the foundations of crushers.

The vibration limits recommended in the British code of practice for foundations for machinery (BSI, 1974) are actually human tolerance criteria. They are widely used for applications other than foundations for reciprocating engines. Further information on human tolerance criteria is provided in Chapter 7. In the absence of other information, the application of human tolerance criteria will often help to minimize other adverse effects of vibrations.

9.4 REFERENCES

Arnold, R. N., Bycroft, G. N. and Warburton, G. B. (1955) 'Forced vibrations of a body on an infinite elastic solid', *J. Applied Mech.* **22**: *Trans. ASME, series E.* **77**: 391–400.

Barkan, D. D. (1962) *Dynamics of Bases and Foundations*, McGraw-Hill, New York.

BSI (1974) *Code of Practice for Foundations for Machinery: CP 2012: Part 1, Foundations for Reciprocating Machines*, British Standards Institution, London.

BSI (1987) BS 4999, *General Requirements for Rotating Electrical Machines: Part 142, Specification for Mechanical Performance Vibration*, British Standards Institution, London.

DIN 4024 (1988) *Machine Foundations: Part 1, Elastic Foundations for Rotary Machines*, Deutsches Institut für Normung, Berlin.

Hsieh, T. K. (1962) 'Foundation vibrations', *Proc. ICE*, **22**: 211–26.

Moore, P. J. (ed.) (1985) *Analysis and Design of Foundations for Vibration*, Balkema, Rotterdam/Boston.

Novak, M. and El Hifnawy, L. (1983) 'Vibration of hammer foundations', *Int. J. Soil Dyn. Earthqu. Engng* **2**: 45–53.

Skipp, B. O. (ed.) (1966) *Vibration in Civil Engineering*, Butterworths, London.

Smith, D. G. E. (1989) 'Foundation for diesel generating set in Naval Dockyard, Gibraltar', *Proc. ICE*, **86**: Part 1, February, 109–37.

Smith, J. W. (1993) *Vibration of Cliffe Hill Quarry Crusher Foundations*, Report by the University of Bristol to M. J. Crowson and Associates, No. UBCE/JWS/93/03.

Srinavasulu, P. and Vaidyanathan, C. V. (1976) *Handbook of Machine Foundations*, Tata McGraw-Hill, New Delhi.

Szczepanik, A., Roy, I. and Kuhnell, B. T. (1990) 'Vibration and stress analysis for condition monitoring of Symon cone crushers, *Trans ASME, J. Vibn. Acoustics*, **112**: 268–73.

Wolf, J. P. (1985) *Dynamic Soil-Structure Interaction*, Prentice-Hall, Inc., Englewood Cliffs, NJ.

Wolf, J. P. (1988) *Soil-Structure-Interaction Analysis in Time Domain*, Prentice-Hall, Inc., Englewood Cliffs, NJ.

Wolf, J. P. and Paramesso, A. (1992) 'Lumped-parameter model for a rigid cylindrical foundation embedded in a soil layer or rigid rock', *Earthquake Engng Struct. Dynamics*, **21**(12): 1021–38.

Chapter 10

Random vibration analysis

George D. Manolis

10.1 INTRODUCTION

This chapter serves as an introduction to the field of random vibrations, which in recent years has found extensive applications in structural dynamics, machine vibrations, earthquake engineering, as well as in non-destructive testing and identification. Essentially, it is an extension of Chapter 2, which focused on deterministic structural dynamics. We note that the concepts of random variables and random (or stochastic) processes, the latter being functions of both space and time in their most general form, appear in most of the intervening chapters. For instance, wind, water wave and earthquake-induced ground motions are loadings of random nature. Specifically, the former two types of loads can be viewed as comprising a rapidly fluctuating part superimposed on a slowly varying mean value. They can be classified as stationary random loads in the sense that there is a certain periodicity (and hence some predictability) in the fluctuating part. Earthquake loads are fully random and classified as non-stationary, a term that will be explained later on. Finally, there is some mild stochasticity inherent in traffic induced loads, simply because the movement of vehicles cannot be fully controlled.

The presentation of such a vast subject within the confines of a single chapter is by necessity brief. Thus, it is assumed that the reader is familiar with the basic ideas and concepts underlying probability theory and elementary statistics. This way, the present chapter serves a dual purpose, namely to refresh the reader's memory on the subject of stochastic processes and then to move on to an elementary, yet basic review of random vibrations. The chapter is structured as follows: First, we look at random functions of time and of frequency. In the interest of brevity, a list of references at the end includes several excellent textbooks on probabilistic methods, random vibrations and numerical methods for stochastic problems (Augusti *et al.*, 1984; Crandall and Mark, 1963; Ghanem and Spanos, 1991; Klieber and Hien, 1992; Nigam, 1983), which the reader may want to consult. The second part of this work examines the response of both single and multiple Degree Of Freedom (DOF) structural systems to stochastic input. Both time domain and frequency domain techniques are covered, as is the case of non-linear systems. In analysing multiple DOF systems, the Finite Element Method (FEM) is often used in the numerical modelling of a complicated structural system, although alternative methods such as the Boundary Element Method (BEM) are becoming increasingly popular. Next, a simple example serves to illustrate the concepts and methodologies presented herein. Finally, some material is presented on structures with uncertain properties, so as to introduce this very important source of stochasticity that stems from randomness in the material properties and in the geometry, as opposed to randomness in the applied loads only.

10.2 RANDOM PROCESSES

10.2.1 General remarks

If the outcome of a (conceptual) experiment is to assign a real value to variable x, then x is known as a random variable. Furthermore, if x assumes only a finite number of values, it is called a discrete random variable. Finally, if x assumes a continuous range of values, it is called a continuous random variable.

Probabilities associated with a random variable are conveniently described by a distribution function such that the probability of x assuming a value less than X is $P(x \leq X)$. Note that $P(x < -\infty) = 0$ and $P(x < +\infty) = 1$ where zero denotes impossibility and unity denotes certainty. The probability that x lies in the interval (a, b) is simply:

$$P(x \leq b) - P(x \leq a) = P(a < x \leq b) \geq 0 \tag{10.1}$$

From the above equation it is seen that if $b \geq a$, $P(x \leq b) > P(x \leq a)$ and hence the distribution function is a monotonically non-decreasing function of X. Figure 10.1 shows the distribution function for both discrete and continuous random variables.

By differentiating the distribution function $P(x \leq X)$ in the regions where the derivative exists, we obtain the Probability Density Function (PDF) as:

$$p(x) = \lim_{\Delta X \to 0} \frac{P(X \to \Delta X) - P(x)}{\Delta X} \tag{10.2}$$

In the case of a discrete random variable, the PDF can be represented by a series of impulses or Dirac delta functions at the location of each jump, as shown in Figure 10.2(a). Each impulse is of area equal to the magnitude of the corresponding jump in P. The probability that $X - \Delta X < x \leq X$ is then approximated as $p(X)\Delta X$. Figure 10.2(b) finally shows the PDF corresponding to a continuous random variable.

10.2.2 Random time functions

Consider a random process that generates an infinite ensemble (or collection) of sample functions (or records) $x(t)$. An example of this would be all possible

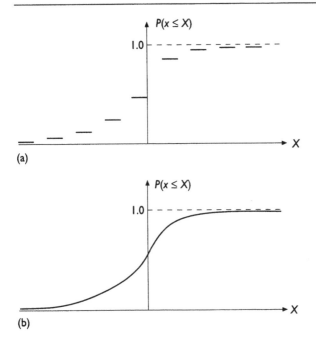

(a)

(b)

Figure 10.1 Distribution function for (a) discrete and (b) continuous random variables.

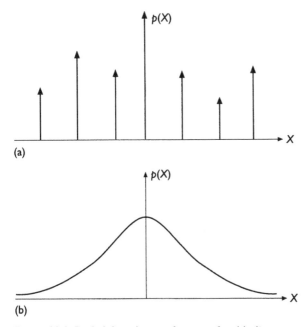

(a)

(b)

Figure 10.2 Probability density function for (a) discrete and (b) continuous random variables.

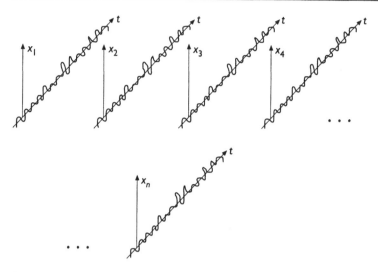

Figure 10.3 Ensemble of records $x(t)$.

acceleration records at a given locality, or wind pressure readings in tall buildings in a city. We then proceed to define probabilities for such an ensemble. For example, at any time t, a first order distribution function and a first order PDF may be defined across the ensemble (i.e. in the horizontal direction of Figure 10.3) as a limiting process in the form:

$$p(X, t)\, dX = P(X - dX < x(t) \leq X) \tag{10.3}$$

Similarly, a joint PDF can be defined as:

$$p(X_1, t_1, \ldots, X_n, t_n)dX_1 \ldots dX_n = P(X_1 - dX_1 < x(t_1)$$
$$\leq X_1, \ldots, X_n - dX_n < x(t_n) \leq X_n) \tag{10.4}$$

with the following properties:

$$p(x_1, t_1, \ldots, x_n, t_n) \geq 0$$

$$\int_{-\infty}^{\infty} \ldots \int_{-\infty}^{\infty} p(x_1, t_1, \ldots, x_n, t_n)\, dx_1 \ldots dx_n = 1 \tag{10.5}$$

where X_1 can be replaced by x_1 and so forth, if there is no danger of confusion. Lower order joint PDF (index m) can be found from higher order ones (index n), where $m < n$, by integrating across $x_{m+1}, \ldots, x_n$ as in eqn (10.5). Also, a joint PDF which is invariant to shifts in the time axis is said to be stationary, that is:

$$p(x_1, t_1, \ldots, x_n, t_n) = p(x_1, t_1 + T, \ldots, x_n, t_n + T) \tag{10.6}$$

Since it is not possible in practice to determine the joint probabilities necessary for completely defining a random process, one has to settle for a few easily obtainable averages which partially specify the random process. At first, the mean (or statistical average or ensemble average) of $x(t)$ is:

$$E[x(t)] = \int_{-\infty}^{\infty} Xp(X, t)\,dX \tag{10.7}$$

where E (or $<>$) is known as the expected value (or expectation) of $x(t)$. Of particular interest is a set of averages called central moments:

$$E[x_t - E(x_t)]^n = \int_{-\infty}^{\infty} (x_t - E(x_t))^n p(x_t)\,dx_t \tag{10.8}$$

where n is an integer. In the above, subscript t serves to emphasize that the averages refer to a particular instant of time. In the case of a zero mean random process, the central moments are simply referred to as moments. The second central moment is very important in many applications and is known as the variance σ_x^2, that is:

$$\sigma_x^2 = E[x_t - E(x_t)]^2 = E[x_t^2] - [E(x_t)]^2 \tag{10.9}$$

In the stationary case, the above averages do not vary with time.

The operation of finding an expected value was shown to involve an averaging across the ensemble of sample functions $x(t)$ of a random process. We may also form time averages along a particular member of the ensemble. We therefore have that:

$$\bar{x}(t) = \lim_{T \to \infty} \frac{1}{2T} \int_{-T}^{T} x(t)\,dt \tag{10.10}$$

where the overbar indicates a time averaged value. If the time averages and the ensemble averages are identical, the random process is ergodic. Obviously, this property holds for stationary processes only, because in a non-stationary process the ensemble average will vary in time. Ergodicity is a very desirable property and a stationary process in random vibrations is assumed to be ergodic unless there are strong reasons to the contrary.

The correlation coefficient ρ_{xy} between two random variables x and y with joint PDF $p(x, y)$ is defined as:

$$\rho_{xy} = E[(x - E[x])(y - E[y])]/\sigma_x \sigma_y \tag{10.11}$$

It is common practice to normalize both x and y such that their means are zero and their variances are equal to unity. In such cases, $\rho_{xy} = \rho_{yx} = \rho = E[xy]$, where ρ is the slope of a straight line that best fits (by minimizing the mean square error) the data of a normalized (x, y) scatter plot, as shown in Figure 10.4. Also, $\rho \leq 1$ and intermediate values measure the degree of linear statistical dependence between x and y.

For a random process x, we may express the correlation between $x(t_1)$ and $x(t_2)$ through the autocorrelation function:

$$R_x(t_1, t_2) = E[x(t_1)x(t_2)] \tag{10.12}$$

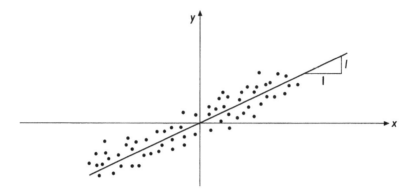

Figure 10.4 Correlation coefficient ρ for a scatter of sample values.

In the stationary case, only the time difference τ between t_1 and t_2 is important, that is:

$$R_x(\tau) = E[x(t)x(t + \tau)] = R_x(-\tau) \tag{10.13}$$

where t is arbitrary. Also note that:

$$R_x(0) = E[x^2] = \sigma_x^2 + (E[x])^2 \tag{10.14}$$

In the ergodic case, R_x can be found by averaging any sample function of the ensemble across time as:

$$R_x(\tau) = \lim_{T \to \infty} \frac{1}{2T} \int_{-T}^{T} x(t)x(t + \tau)\, dt \tag{10.15}$$

The correlation between two samples from random processes $x(t)$ and $y(t)$ is described by the cross-correlation functions

$$R_{xy}(t_1, t_2) = E[x(t_1)y(t_2)] \tag{10.16}$$

$$R_{yx}(t_1, t_2) = E[y(t_1)x(t_2)] \tag{10.17}$$

As before, in the stationary case, $R_{xy}(t_1, t_2) = R_{xy}(\tau)$, $R_{yx}(t_1, t_2) = R_{yx}(\tau)$, and $R_{xy}(\tau) = R_{yx}(-\tau)$. It is observed that although the autocorrelation is an even function of τ, the cross-correlation functions are not.

Using the fact that differentiation and expectation are linear operators and as such commute, the time derivatives of the autocorrelation function in the stationary case are:

$$\frac{d}{d\tau} R_x(\tau) = E[x(t)\dot{x}(t + \tau)] = E[\dot{x}(t - \tau)x(t)] \tag{10.18}$$

and

$$\frac{d^2}{d\tau^2} R_x(\tau) = -E[\dot{x}(t-\tau)\dot{x}(t)] = -E[\dot{x}(t)\dot{x}(t+\tau)] \tag{10.19}$$

where dots indicate derivatives with respect to t. The above two equations indicate that the second derivative of the autocorrelation function of x is the negative of the autocorrelation function of $\dot{x}$. Finally, both $R_x(\tau)$ and $R_{xy}(\tau)$ tend to zero as $\tau \to \infty$, provided x does not have any periodic components.

10.2.3 Spectral analysis

The Fourier Transform (FT) (Zayed, 1996) of the autocorrelation function for a stationary process is the Power Spectral Density Function (PSDF) $S_x(\omega)$, that is:

$$S_x(\omega) = \frac{1}{2\pi} \int_{-\infty}^{\infty} R_x(\tau) \exp(-i\omega\tau) d\tau \tag{10.20}$$

Also, the inverse Fourier transformation gives:

$$R_x(\tau) = \int_{-\infty}^{\infty} S_x(\omega) \exp(+i\omega\tau) d\omega \tag{10.21}$$

In the above, ω is the frequency, $i^2 = -1$ and the factor $1/2\pi$ may be associated with either member of the above pair or may be evenly split between them. Since $R_x(\tau)$ is a real and even function, Fourier cosine transforms may be used in lieu of the exponential transform shown above. The PSDF is also known as the mean square spectral density because:

$$R_x(0) = \int_{-\infty}^{\infty} S_x(\omega) d\omega = E[x^2] \tag{10.22}$$

This implies that $S_x(\omega) d\omega$ can be interpreted as the power or mean square density contained in an infinitesimal band of complex exponentials (sinusoids and co-sinusoids) into which the random function is resolved. The PSDF is a positive, real valued function and is even in ω. Since physical meaning can only be assigned to positive frequencies, an experimentally obtained spectrum is plotted by halving the measured $S_x(\omega)$ at each frequency and plotting the result for both positive and negative ω. A spectrum $S_{xy}(\omega)$ for the cross-correlation function $R_{xy}(\tau)$ can also be defined for the stationary case as in eqns (10.20) and (10.21).

As expected, the PSDF of an ergodic process and the FT of a sample function $x(t)$ of the random process are related. When $x(t)$ is a non-periodic function, its FT $X_T(\omega)$ is given as:

$$X_T(\omega) = \int_0^T x(t) \exp(-i\omega t) dt \tag{10.23}$$

where $x(t)$ is assumed to be zero before $t = 0$ and after $t = T$. An energy density spectrum for $x(t)$ is:

$$\Phi_x(\omega, T) = |X_T(\omega)|^2/2\pi \tag{10.24}$$

and the power density spectrum is:

$$S_x(\omega, T) = \Phi_x(\omega, T)T = |X_T(\omega)|^2/(2\pi T) \tag{10.25}$$

The power density spectrum is now a random variable dependent on both $x(t)$ and T. Although it can be shown that:

$$\lim_{T \to \infty} E[S_x(\omega, T)] = S_x(\omega) \tag{10.26}$$

the manner in which the power density spectrum approaches the PSDF needs to be investigated in each case. For a normal (or Gaussian) process, it is known that the variance of $S_x(\omega, T)$ does not approach zero as $T \to \infty$, and hence measurements of $S_x(\omega, T)$ provide questionable estimates for the PSDF.

10.3 SYSTEM RESPONSE TO RANDOM INPUT

10.3.1 Single Degree-Of-Freedom systems (SDOF)

Consider an SDOF linear system (Hurty and Rubinstein, 1964) described by:

$$\ddot{x}(t) + 2\omega_0 \zeta \dot{x}(t) + \omega_0^2 x(t) = \tilde{f}(t)/m = f(t), \quad t > t_0 \tag{10.27}$$

where the natural frequency is $\omega_0 = (k/m)^{1/2}$, the damping ratio is $\zeta = c/2m\omega_0$, and $\tilde{f}(t)$ (or $f(t)$) is the forcing function. Note that t_0 is taken as equal to or greater than zero to avoid having an SDOF system operating at negative times. Also, m, c and k are the usual mass, damping and stiffness constants, while $x(t)$ is the displacement response of the system to a Gaussian stochastic input $f(t)$, which is a member function (or sample) of a stochastic process $\{f(t)\}$. Eqn (10.27) is accompanied by initial conditions of the form:

$$x(t_0) = x_0, \qquad \dot{x}(t_0) = \dot{x}_0 \tag{10.28}$$

In general, the probability law for a random process cannot be fully determined solely from knowledge of the mean and covariance of that process (Augusti et al., 1984). The exception to this comes when the functional form of the probability law is known and utilizes parameters which are simply related to the mean and covariance, as in the case of a normal (or Gaussian) distribution. In what follows, it is assumed that the input process in eqn (10.27) is Gaussian, and so is the output process. Typical representations of SDOF systems are shown in Figure 10.5.

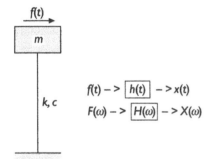

Figure 10.5 SDOF representations.

The mean square (as well as the deterministic) response of the SDOF system is given by:

$$x(t) = x_0 g(t - t_0) + \dot{x}_0 h(t - t_0) + \int_0^{t-t_0} h(\xi) f(t - \xi) \, d\xi \tag{10.29}$$

where

$$g(t) = \exp(-\zeta \omega_0 t)(\cos \bar{\omega} t + \zeta \omega_0 \sin \bar{\omega} t / \bar{\omega})$$

$$h(t) = \exp(-\zeta \omega_0 t)(\sin \bar{\omega} t / \bar{\omega}) \tag{10.30}$$

$$\bar{\omega} = \omega_0 \sqrt{1 - \zeta^2}, \quad 0 < \zeta < 1$$

Function $h(t)$ is referred to as the unit impulse response of a linear SDOF system. The mean $m_x(t)$ of the output process is obtained by averaging across the ensemble as:

$$m_x(t) = E[x(t)] = x_0 g(t - t_0) + \dot{x}_0 h(t - t_0) + \int_0^{t-t_0} h(\xi) m_f(t - \xi) \, d\xi \tag{10.31}$$

For a system with infinite operating time, $t = \infty$ and:

$$m_x(t) = \int_0^\infty h(\xi) m_f(t - \xi) \, d\xi \tag{10.32}$$

The covariance $K_{xx}(t_1, t_2)$ of the output process, which is the autocorrelation function of eqn (10.12) taken about the mean, is given as:

$$K_{xx}(t_1, t_2) = E[(x(t_1) - m_x(t_1))(x(t_2) - m_x(t_2))]$$

$$= \int_{t_0}^{t_1} \int_{t_0}^{t_2} h(t_1 - \tau_1) h(t_2 - \tau_2) E[(f(\tau_1) - m_f(\tau_1))(f(\tau_2)$$

$$- m_f(\tau_2))] \, d\tau_1 \, d\tau_2 \tag{10.33}$$

$$= \int_0^{t_1-t_0} \int_0^{t_2-t_0} h(\xi_1) h(\xi_2) K_{ff}(t_1 - \xi_1, t_2 - \xi_2) \, d\xi_1 \, d\xi_2$$

where a change of variables takes place in the form $\xi_1 = t_1 - \tau_1$ and $\xi_2 = t_2 - \tau_2$. As before, the upper limits are replaced by $+\infty$ for a system with infinite operating time. Furthermore, the variance $\sigma_x^2(t)$ of the output process is obtained by setting $t_1 = t_2 = t$ in the expression for the covariance (i.e. $\sigma_x^2(t) = K_{xx}(t, t)$). Finally, given a normal or Gaussian input, the PDF of $x(t)$ is given as:

$$p(x) = \exp\{-(x - m_x(t))^2/(2\sigma_x^2(t))\}/(\sqrt{2\pi}\sigma_x(t)) \qquad (10.34)$$

so that the probability of x lying in the interval $(x, x + dx)$ at time t is given by $p(x)\,dx$. Note that the above development was for non-stationary processes. For stationary processes eqns (10.32) and (10.33) still hold true, but the averages employed no longer vary with time.

10.3.2 Multiple Degree-Of-Freedom systems (MDOF)

Consider now the response of an MDOF system to non-stationary random input. The development follows along the lines of the SDOF system, expect for the introduction of matrix notation. At first, the governing equation of motion of an MDOF system is:

$$\text{with} \qquad \begin{aligned} [M]\{\ddot{x}\} + [C]\{\dot{x}\} + [K]\{x\} &= \{f(t)\}, \quad t > t_0 \\ \{x(t_0)\} = \{x_0\}, \{\dot{x}(t_0)\} &= \{\dot{x}_0\} \end{aligned} \qquad (10.35)$$

In the above equation, $[M]$, $[C]$ and $[K]$ are symmetric, $N \times N$ matrices, while $\{x\}$ and $\{f\}$ are $N \times 1$ column vectors denoting the input and output processes, respectively. In particular, the mass matrix $[M]$ is positive definite, while the damping $[C]$ and stiffness $[K]$ matrices are non-negative definite. Also, $\{f(t)\}$ is a vector of Gaussian random variables whose mean and covariance matrix are given by:

$$\text{and} \qquad \begin{aligned} \{m_f(t)\} &= E[\{f(t)\}] \\ [K_{ff}(t_1, t_2)] &= E[\{f(t_1) - m_f(t_1)\}\{f(t_2) - m_f(t_2)\}^T] \end{aligned} \qquad (10.36)$$

respectively, with superscript T denoting transposition.

We first focus on the case where the system of eqn (10.35) possesses real eigenvectors, else known as classical (or normal) modes, which was the case presented in Chapter 2. Thus, the matrix of normalized eigenvectors $[A]$ defines a set of modal co-ordinates:

$$\{y\} = [A]^{-1}\{x\} = [A]^T[M]\{x\} \qquad (10.37)$$

the employment of which results in an uncoupled system of governing equations of motion given by:

$$[I]\{\ddot{y}\} + [\bar{C}]\{\dot{y}\} + [\bar{K}]\{y\} = [A]^T\{f(t)\} = \{q(t)\} \qquad (10.38)$$

In the above, $[I]$ is the identity matrix and $[\bar{C}]$ and $[\bar{M}]$ are diagonal matrices. Taking the ith row ($i = 1, 2, \ldots, N$) of the above equation gives the SDOF-like equation:

$$\ddot{y}_i + 2\omega_i\zeta_i\dot{y}_i + \omega_i^2 y_i = q_i(t) = \sum_{j=1}^{N} A_{ij}f_j \tag{10.39}$$

for the ith mode. As before, Gaussian input results in Gaussian output that is a linear combination of Gaussian variables. Finally, the transformation of co-ordinates defined by eqn (10.37) is known as a congruent transformation.

The mean square (and deterministic) response of the MDOF system is given in modal co-ordinates as:

$$\{y\} = [U(t - t_0)]\{y(t_0)\} + [H(t - t_0)]\{\dot{y}(t_0)\} + \int_{t_0}^{t} [H(t - \tau)]\{q(\tau)\}\, d\tau$$

$$\tag{10.40}$$

where $[U]$ and $[H]$ are diagonal matrices with elements:

$$U_{ii}(t) = \exp(-\zeta_i\omega_i t)(\cos\bar{\omega}_i t + \zeta_i\omega_i \sin\bar{\omega}_i t/\bar{\omega}_i)$$

$$H_{ii}(t) = \exp(-\zeta_i\omega_i t)\sin\bar{\omega}_i t/\bar{\omega}_i \tag{10.41}$$

$$\bar{\omega}_i = \omega_i\sqrt{1 - \zeta_i^2}, \quad 0 < \zeta_1 < 1$$

Reverting to the physical co-ordinates via the transformation defined by eqn (10.37) gives:

$$\{x\} = [A][U(t - t_0)][A]^T[M]\{x_0\} + [A][H(t - t_0)][A]^T[M]\{\dot{x}_0\}$$

$$+ \int_{t_0}^{t} [A][H(t - \tau)][A]^T\{f(\tau)\}\, d\tau \tag{10.42}$$

Given the above solution for the mean square (deterministic) response, the stochastic means are given by:

$$\{m_x(t)\} = [A][U(t - t_0)][A]^T[M]\{x_0\} + [A][H(t - t_0)][A]^T[M]\{\dot{x}_0\}$$

$$+ \int_{t_0}^{t} [A][H(t - \tau)][A]^T\{m_f(\tau)\}\, d\tau \tag{10.43}$$

while the covariance matrix is given by the stochastic average of the outer product of the zero mean response vector evaluated at two different times, that is:

$$[K_{xx}(t_1, t_2)] = E[\{x(t_1) - m_x(t_1)\}\{x(t_2) - m_x(t_2)\}^T]$$

$$= \int_0^{t_1 - t_0} \int_0^{t_2 - t_0} [A][H(\xi_1)][A]^T[K_{ff}(t_1 - \xi_1, t_2 - \xi_2)][A]$$

$$\times [H(\xi_2)][A]^T\, d\xi_1\, d\xi_2 \tag{10.44}$$

Finally, the PDF for the ith component of the response $\{x(t)\}$ is given by:

$$p(x_i) = \exp(-(x_i(t) - m_{xi}(t))^2/(2\sigma_{xi}^2(t)))/(\sqrt{2\pi}\sigma_{xi}(t)) \qquad (10.45)$$

where variance σ_{xi}^2 is the ith diagonal component of the covariance matrix evaluated at $t_1 = t_2 = t$ (i.e. $\sigma_{xi}^2(t) = K_{ii}(t, t)$).

If the components of the input $\{f\}$ are jointly normally distributed, so are the components of the output $\{x\}$ with a joint PDF given by:

$$p(x_1, x_2, \ldots, x_N) = \exp(-0.5\{x(t) - m_x(t)\}^T[K_{xx}(t)]^{-1}\{x(t)$$

$$- m_x(t)\})/((2\pi)^{N/2}(\det[K_{xx}(t)])^{1/2}) \qquad (10.46)$$

As before, for stationary processes all statistical averages are time independent.

If the damping and stiffness matrices are non-symmetric, then the classical normal mode approach fails and a more general approach must be sought. This occurs when damping is no longer of the proportional kind. The key idea here is to convert the second order matrix differential equation of eqn (10.35) into a first order matrix differential equation by defining

$$\{z(t)\}^T = \lfloor\{\dot{x}(t)\}^T, \{x(t)\}^T\rfloor \qquad (10.47)$$

By combining the above equation with the matrix equation of motion which has been premultiplied by $[M]^{-1}$, the following $2N \times 2N$ matrix differential equation is obtained:

$$\{\dot{z}\} = [B]\{z\} + \{b(t)\}, \quad t > t_0 \qquad (10.48)$$

where

and
$$[B] = \begin{bmatrix} -[M]^{-1}[C] & -[M]^{-1}[K] \\ [I] & [0] \end{bmatrix} \qquad (10.49)$$

$$\{b(t)\}^T = \lfloor([M]^{-1}\{f(t)\}^T), \{0\}^T\rfloor$$

with $\{z(t - t_0)\} = \{z_0\}$ as initial condition. It is assumed that the input vector $\{b(t)\}$ is Gaussian with mean $\{m_b(t)\}$ and covariance $[K_{bb}(t_1, t_2)]$.

It is well known (Coddington and Levinson, 1955) that for any real valued matrix $[B]$ there exists a similarity transformation $[T]$ that will reduce it to an upper diagonal (Jordan canonical) form $[J]$. By letting:

$$\{z\} = [T]\{y\} \qquad (10.50)$$

and substituting in eqn (10.48), we obtain:

$$\{\dot{y}\} = [T]^{-1}[B][T]\{y\} + [T]^{-1}\{b(t)\} = [J]\{y\} + [T]^{-1}\{b(t)\} \qquad (10.51)$$

along with $\{y_0\} = [T]^{-1}\{z_0\}$ as initial condition. The mean square (deterministic) solution of the above equation is given by:

$$\{y(t)\} = \exp((t - t_0)[J])\{y_0\} + \int_{t_0}^{t} \exp((t - \tau)[J])[T]^{-1}\{b(\tau)\}\, d\tau \qquad (10.52)$$

In the case where $[J]$ is strictly diagonal:

$$\exp(t[J]) = \begin{bmatrix} \exp(\lambda_1 t) & & & 0 \\ & \exp(\lambda_2 t) & & \\ & & \ddots & \\ 0 & & & \exp(\lambda_{2N} t) \end{bmatrix} \tag{10.53}$$

where λ_i, $i = 1, 2, \ldots, 2N$, are the eigenvalues of $[B]$. By making the substitution $\xi = t - \tau$ and reverting to the physical co-ordinates $\{z\}$, we obtain:

$$\{z(t)\} = [T]\exp((t - t_0)[J])[T]^{-1}\{z_0\} + \int_0^{t-t_0} [T]\exp(\xi[J])[T]^{-1}b(t - \xi)\,d\xi \tag{10.54}$$

Given the above solution to eqn (10.48), the vector of stochastic means of $\{z(t)\}$ is:

$$\{m_z(t)\} = [T]\exp((t - t_0)[J])[T]^{-1}\{z_0\}$$

$$+ \int_0^{t-t_0} [T]\exp(\xi[J])[T]^{-1}\{m_b(t - \xi)\}\,d\xi \tag{10.55}$$

and the matrix of covariances is (ignoring the initial conditions):

$$[K_{zz}(t_1, t_2)] = \int_0^{t_1-t_0} \int_0^{t_2-t_0} [T]\exp(\xi_1[J])[T]^{-1}[K_{bb}(t_1 - \xi_1, t_2 - \xi_2)]([T]^{-1})^T$$

$$\times \exp(\xi_2[J])^T[T]^T\,d\xi_1\,d\xi_2 \tag{10.56}$$

For Gaussian input, the output process is completely specified in terms of the above means and covariances. The individual and joint PDF may be obtained by using eqns (10.45) and (10.46), provided $\{z(t)\}$ is decomposed according to eqn (10.47).

10.3.3 Application of Fourier transforms

As was shown in Section 10.2.3, FTs play a central role in the analysis of stationary random variables by relating the autocorrelation (or autocovariance) to the PSDF and vice versa. These relations can be extended to non-stationary processes following (Lampard, 1954).

Consider $f_T(t)$ to be a member of a real valued, non-stationary process. First, define:

$$f_T = \begin{cases} f(t) & \text{for } |t| < T \\ 0 & \text{for } |t| \geq T \end{cases} \tag{10.57}$$

Next, assume for simplicity that $f(t)$ is a zero mean process and subsequently define the FT of $f_T(t)$ (see eqn (10.20)) as:

$$F_T(\omega) = \frac{1}{2\pi} \int_{-\infty}^{\infty} f_T(t) \exp(-i\omega t) \, dt \tag{10.58}$$

Using the definition of eqn (10.57) for $f_T(t)$, the above equation can be recast as:

$$F_T(\omega) = \frac{1}{2\pi} \int_{-T}^{T} f(t) \exp(-i\omega t) \, dt \tag{10.59}$$

Finally, use of the inverse FT (see eqn (10.21)) gives:

$$f_T(t) = \int_{-\infty}^{\infty} F_T(\omega) \exp(i\omega t) \, d\omega \tag{10.60}$$

Since f_T is a real function, it is equal to its complex conjugate f_T^* so that:

$$f_T(t) = \int_{-\infty}^{\infty} F_T^*(\omega) \exp(-i\omega t) \, d\omega \tag{10.61}$$

Equations (10.60) and (10.61) can now be used in conjunction with definition of the covariance of f_T, that is:

$$K_{ff}(t_1, t_2, T) = E[f_T(t_1) f_T(t_2)]$$

$$= \int_{-\infty}^{\infty} \int_{-\infty}^{\infty} E[F_T^*(\omega_1) F_T(\omega_2)] \exp(-i(\omega_1 t_1 - \omega_2 t_2)) \, d\omega_1 \, d\omega_2 \tag{10.62}$$

The covariance of $f(t)$ is given by:

$$K_{ff}(t_1, t_2) = \lim_{T \to \infty} K_{ff}(t_1, t_2, T)$$

$$= \int_{-\infty}^{\infty} \int_{-\infty}^{\infty} S_{ff}(\omega_1, \omega_2) \exp(-i(\omega_1 t_1 - \omega_2 t_2)) \, d\omega_1 \, d\omega_2 \tag{10.63}$$

where

$$S_{ff}(\omega_1, \omega_2) = \lim_{T \to \infty} E[F_T^*(\omega_1) F_T(\omega_2)] \tag{10.64}$$

is the generalized PSDF for the random process $f(t)$. Applying the inverse FT to eqn (10.62) yields

$$S_{ff}(\omega_1, \omega_2) = \frac{1}{4\pi^2} \int_{-\infty}^{\infty} \int_{-\infty}^{\infty} K_{ff}(t_1, t_2) \exp(i(\omega_1 t_1 - \omega_2 t_2)) \, dt_1 \, dt_2 \tag{10.65}$$

For a linear system with infinite operating time, the response $x_T(t)$ can be determined as:

$$x_T(t) = \int_0^{\infty} h(\xi) f_T(t - \xi) \, d\xi \tag{10.66}$$

where $h(\xi)$ is the unit impulse response of eqn (10.30). Taking the FT of the above equation yields:

$$X_T(\omega) = \frac{1}{2\pi} \int_{-\infty}^{\infty} x_T(t) \exp(-i\omega t) \, dt = H(\omega) F_T(\omega) \tag{10.67}$$

where

$$H(\omega) = \int_{-\infty}^{\infty} h(t) \exp(-i\omega t) \, dt = \frac{1}{(\omega_0^2 - \omega^2) - i(2\omega\omega_0\zeta)} \tag{10.68}$$

is known as the complex frequency response of an SDOF system. If eqn (10.67) is multiplied by its complex conjugate, that is:

$$X_T^*(\omega_1) X_T(\omega_2) = H^*(\omega_1) H(\omega_2) F_T^*(\omega_1) F_T(\omega_2) \tag{10.69}$$

then the generalized power spectrum of $x(t)$ is:

$$S_{xx}(\omega_1, \omega_2) = \lim_{T \to \infty} E[X_T^* X_T] = H^* H \lim_{T \to \infty} E[F_T^* F_T]$$

$$= H^*(\omega_1) H(\omega_2) S_{ff}(\omega_1, \omega_2) \tag{10.70}$$

The above equation may be regarded as the generalization of the equation given below, namely:

$$S_{xx}(\omega) = |H(\omega)|^2 S_{ff}(\omega) \tag{10.71}$$

that holds true for stationary processes. As before, the inverse FT gives the co-variance of the output process as:

$$K_{xx}(t_1, t_2) = \int_{-\infty}^{\infty} \int_{-\infty}^{\infty} H^*(\omega_1) H(\omega_2) S_{ff}(\omega_1, \omega_2) \exp(-i(\omega_1 t_1 - \omega_2 t_2)) \, d\omega_1 \, d\omega_2 \tag{10.72}$$

Finally, the variance of the response is:

$$\sigma_x^2(t) = E[x^2(t)] = K_{xx}(t, t)$$

$$= \int_{-\infty}^{\infty} \int_{-\infty}^{\infty} H^*(\omega_1) H(\omega_2) S_{ff}(\omega_1, \omega_2) \exp(-it(\omega_1 - \omega_2)) \, d\omega_1 \, d\omega_2 \tag{10.73}$$

10.3.4 Non-linear systems

Non-linearities in dynamic systems are usually exhibited by the stiffness terms and, to a lesser extent, by the damping terms. In this section we focus on a SDOF system that is governed by eqn (10.27) and has the initial conditions of eqn (10.28), except that the stiffness term $\omega_0^2 x(t)$ is replaced by the general restoring force $g(x)$, that is:

$$\ddot{x}(t) + 2\omega_0 \zeta \dot{x}(t) + g(x) = f(t) \tag{10.74}$$

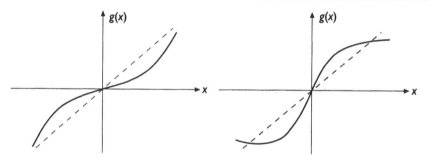

Figure 10.6 Restoring forces: (a) linear plus cubic (Duffing oscillator) and (b) sinusoidal (pendulum type).

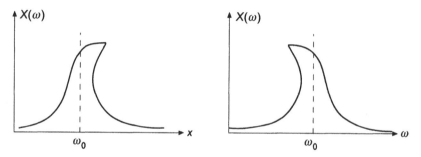

Figure 10.7 Non-linear resonance plot for (a) hardening spring and (b) softening spring.

Invariably, $g(x)$ is a single valued, odd function of the response and represents either a hardening spring or a softening spring, as shown in Figure 10.6. The presence of a non-linear spring in a SDOF system results in a period T that is amplitude-dependent. Also, under steady state vibrations, the peak response amplitude versus frequency of the excitation plot exhibits a backbone at the resonant peak, which is manifested at the natural frequency ω_0 of the linear case. This backbone points backwards in the case of a hardening spring and forwards in the case of a softening spring, as shown in Figure 10.7.

A particular case for which there is considerable information regarding the response $x(t)$ of a non-linear SDOF system is when the excitation is ideal white noise (i.e. when the PSDF of the input $f(t)$ is $S_{ff}(\omega_1, \omega_2) = S_0$, a constant). In that case, the joint distribution of x and $\dot{x}$ at time t is described by the joint conditional PDF $p = p(x_0, \dot{x}_0, x, \dot{x}, t)$ that diffuses in time from a Dirac delta function at $t = t_0$ towards a steady state condition at large times. This diffusion process is governed by the well known Fokker–Planck equation (Caughey, 1963):

$$\frac{\partial p}{\partial t} = -\dot{x}\frac{\partial p}{\partial x} + \frac{\partial}{\partial \dot{x}}(2\omega_0\zeta\dot{x}p) + g(x)\frac{\partial p}{\partial \dot{x}} + \pi S_0 \frac{\partial^2 p}{\partial \dot{x}^2} \tag{10.75}$$

which is linear in p and has variable coefficients. Although closed form solutions to the above equation do not exist at present, the stationary case which is obtained as $t \rightarrow \infty$ (and $\partial p/\partial t = 0$) has a unique solution in the form:

$$p(x, \dot{x}) = C\exp\left[-2\omega_0\zeta\left(\frac{\dot{x}^2}{2} + G(x)\right)\Big/\pi S_0\right] \tag{10.76}$$

where constant C is determined from the normalization requirement:

$$\int_{-\infty}^{\infty}\int_{-\infty}^{\infty} p(x, \dot{x})\, dx\, d\dot{x} = 1 \tag{10.77}$$

and $G(x)$ depends on the type of non-linearity exhibited by the SDOF system. This result implies that $\dot{x}$ and x are statistically independent and that $\dot{x}$ has a Gaussian distribution with variance $\sigma_{\dot{x}}^2 = \pi S_0/(2\omega\zeta)$. Also, x does not have a Gaussian distribution unless $g(x)$ is linear.

The two most prevalent techniques for an approximate solution of a non-linear SDOF system are the perturbation method and the equivalent linearization method. The key idea behind the former approach is expansion in terms of a small parameter ε. In particular, the stiffness is decomposed into a predominant linear part and a small non-linear part $g_0(x)$ as:

$$g(x) = \omega_0^2 x + \varepsilon g_0(x) \tag{10.78}$$

and the response is expanded in powers of ε as:

$$x(t) = x_0(t) + \varepsilon x_1(t) + \varepsilon^2 x_2(t) + \cdots \tag{10.79}$$

Such a solution is assumed to satisfy the equation of motion (10.74) identically in ε so that the coefficients corresponding to each power of ε vanish separately. Therefore, a re-arrangement in terms of powers of ε gives the following sequence of linear equations:

$$\ddot{x}_0 + 2\omega_0\zeta\dot{x}_0 + \omega_0^2 x_0 = f(t)$$

$$\ddot{x}_1 + 2\omega_0\zeta\dot{x}_1 + \omega_0^2 x_1 = g_0(x_0(t)) \tag{10.80}$$

$$\ddot{x}_2 + 2\omega_0\zeta\dot{x}_2 + \omega_0^2 x_2 = x_1(t)\,\partial g_0(x_0(t))/\partial x_0$$

where a Taylor series expansion of $g_0(x)$ about x_0 has been used. Note that in the above system of equations, the nonlinearity has been shifted to the right-hand side in a sequence of equations involving the same linear operator. Therefore, the excitation for the ith solution involves a non-linear combination of all previous $i-1, i-2, \ldots, 0$ solutions.

Equations (10.80) are solved through use of the unit impulse response $h(t)$ given by eqn (10.30). In particular, for the case of infinite operating time:

$$x_0(t) = \int_{-\infty}^{\infty} h(\tau) f(t - \tau) \, dt$$

$$x_1(t) = \int_{-\infty}^{\infty} h(\tau) g_0(x_0(t - \tau)) \, dt \tag{10.81}$$

$$x_2(t) = \int_{-\infty}^{\infty} h(\tau) x_1(t - \tau) \, \partial g_0(x_0(t - \tau))/\partial x_0 \, dt$$

The above solution applies irrespective of $f(t)$ being a deterministic excitation or a random process. In the latter case, eqn (10.81) gives the components of the mean square solution which is synthesized according to eqn (10.79). For a zero mean process, the next statistical average of interest is the variance of the response given by:

$$\sigma_x^2(t) = E[x^2(t)]$$

$$= E[x_0^2(t)] + 2\varepsilon E[x_0(t)x_1(t)] + \varepsilon^2 (E[x_1^2(t)] + 2E[x_0(t)x_1(t)])$$

$$\tag{10.82}$$

For a first order perturbation only, the first two terms of eqn (10.82) need to be retained, that is:

$$E[x_0^2(t)] = \int_0^{\infty} \int_0^{\infty} h(\xi_1) h(\xi_2) E[f(t - \xi_1) f(t - \xi_2)] \, d\xi_1 \, d\xi_2$$

and

$$E[x_0(t)x_1(t)] = \int_0^{\infty} \int_0^{\infty} h(\xi_1) h(\xi_2) E[f(t - \xi_1) g_0(x_0(t - \xi_2))] \, d\xi_1 \, d\xi_2$$

$$\tag{10.83}$$

In principle, the above expressions apply to both stationary and non-stationary processes. In practice, it may not be possible to evaluate the expectations on the right-hand side of eqn (10.83) unless the excitation process has special properties and the non-linear function $g_0(x)$ is of a simple form. When the excitation is a stationary Gaussian process and $g_0(x)$ is an odd polynomial in x, then $x_0(t)$ is also a Gaussian process with autocorrelation

$$E[x_0(t - \xi_1)x_0(t - \xi_2)] = R_{x_0}(\tau)$$

$$= \int_{-\infty}^{\infty} \int_{-\infty}^{\infty} h(\xi_1) h(\xi_2) R_f(\tau - \xi_1 + \xi_2) \, d\xi_1 \, d\xi_2$$

$$\tag{10.84}$$

Also, the expectation between x_0 and g_0 consists of even order moments of $x_0(t)$ (Crandall, 1963). For example, if $g_0(x) = x^3$ (Duffing oscillator), then:

$$E[x_0(t - \xi_1)x_0^3(t - \xi_2)] = 3R_{x_0}(0)R_{x_0}(\xi_1 - \xi_2) \tag{10.85}$$

For this special case, the first order perturbation approximation of eqn (10.83) can be completely evaluated.

A second technique that has been extensively used for non-linear systems is equivalent linearizaton (Caughey, 1971). We begin by introducing a linear term λx to both sides of eqn (10.74), that is:

$$\ddot{x}(t) + 2\omega_0 \zeta \dot{x}(t) + \lambda x(t) = f(t) + \lambda x(t) - g(x(t)) \tag{10.86}$$

where parameter λ is unknown but will be chosen so as to optimize the linearization process. Note that the above equation now describes a linear system that is subjected to a non-linear forcing function $\Phi = \lambda x - g(x)$. Therefore, the variance of the response of this system to stationary random excitation with spectral density $S_{ff}(\omega)$ is given (see eqns (10.67) and (10.71)) as:

$$E[x^2] = \sigma_x^2 = \int_{-\infty}^{\infty} S_{xx}(\omega)\, d\omega = \int_{-\infty}^{\infty} |H(\omega)|^2 S_{ff}(\omega)\, d\omega$$

$$= \int_{-\infty}^{\infty} (\lambda - \omega^2 + i2\omega_0\zeta\omega)^{-2} S_{ff}(\omega)\, d\omega \tag{10.87}$$

In general, it is impossible to choose a parameter λ so that Φ will be identically zero. Since the simplest statistical measure of the magnitude of Φ is its variance, a natural optimization is achieved by choosing a λ that minimizes $E[\Phi^2]$. This requires that:

$$\partial E[\Phi^2]/\partial\lambda = 2\lambda E[x^2] - 2E[xg(x)] = 0 \tag{10.88}$$

where the term $2E[g^2]$ goes to zero since $g(x)$ is an odd valued function. The above equation gives:

$$\lambda = E[xg(x)]/E[x^2] \tag{10.89}$$

and all that remains is to eliminate λ between eqns (10.87) and (10.89). The result is invariably too complicated to permit an exact algebraic solution, but it provides a starting point for a perturbation expansion. For the simple case of ideal white noise, where $S_{ff}(\omega) = S_0$, the integral in eqn (10.87) yields:

$$E[x^2] = \pi S_0/(2\omega_0\zeta\lambda) \tag{10.90}$$

so that elimination of λ between eqn (10.89) and eqn (10.90) gives:

$$E[xg(x)] = \pi S_0/(2\omega_0\zeta) = E[\dot{x}^2] = \sigma_{\dot{x}}^2 \tag{10.91}$$

This result was encountered in the earlier part of this section in conjunction with the Fokker–Planck equation. If the restoring force $g(x)$ is split into a linear and a non-linear part according to eqn (10.78), then:

$$E[x^2] = (\sigma_{\dot{x}}^2 - \varepsilon E[xg_0(x)])/\omega_0^2 \tag{10.92}$$

and a perturbation technique needs to be employed.

10.3.5 Example: non-stationary case

As an example, consider the simple case of a SDOF system with a finite operating time $t_0 = 0$ subjected to a stationary random process. Although the input is stationary, the output is not, by virtue of the fact that the system has a finite operating time. Consider therefore eqn (10.27) under zero initial conditions and where input $f(t)$ is a member function of a zero mean stochastic process which is stationary, ergodic and described by a PSDF equal to $S_{ff}(\omega)$. First we have that the output process $x(t)$ also has a zero mean, as can be seen by recourse to eqn (10.32). Next, the variance of $x(t)$ is (see eqn (10.33)):

$$\sigma_x^2(t) = K_{xx}(t = t_1 = t_2) = \int_0^t \int_0^t h(\xi_1)h(\xi_2)K_{ff}(t - \xi_1, t - \xi_2)\, d\xi_1\, d\xi_2$$

(10.93)

Since $f(t)$ is stationary $K_{ff}(t - \xi_1, t - \xi_2) = R_{ff}(\xi_1 - \xi_2)$ and this autocorrelation function is related to the PSDF via the Wiener–Khinchine relation (Caughey, 1963, 1971) as:

$$R_{ff}(\xi_1 - \xi_2) = \int_0^\infty S_{ff}(\omega)\cos(\omega(\xi_1 - \xi_2))\, d\omega$$

(10.94)

where it is assumed that $\int_0^\infty S_{ff}(\omega)\, d\omega < 0$. Using eqn (10.94) in eqn (10.93) gives:

$$\sigma_x^2(t) = \int_0^t \int_0^t \int_0^\infty S_{ff}(\omega)\cos(\omega(\xi_1 - \xi_2))h(\xi_1)h(\xi_2)\, d\xi_1\, d\xi_2\, d\omega$$

(10.95)

Since the integrals involved in the above equation are convergent, the order of integration may be reversed. Using the definition of $h(t)$ in eqn (10.30) and carrying out the integrations gives:

$$\sigma_x^2(t) = \int_0^\infty \frac{S_{ff}(\omega)\, d\omega}{|H(\omega)|^{-2}} \left[1 + e^{-2\omega_0\zeta} \left(1 + \frac{2\omega_0}{\bar{\omega}}\zeta \sin\bar{\omega}t\cos\bar{\omega}t \right.\right.$$
$$- \exp(\omega_0\zeta t)\left(2\cos\bar{\omega}t + \frac{2\omega_0\zeta}{\bar{\omega}}\sin\bar{\omega}t \right)\cos\omega t$$
$$\left.\left. - \exp(\omega_0\zeta t)\frac{2\omega}{\bar{\omega}}\sin\bar{\omega}t\sin\omega t + \frac{(\omega_0\zeta)^2 - \bar{\omega}^2 + \omega^2}{\bar{\omega}^2}\sin^2\bar{\omega}t \right) \right]$$

(10.96)

where $|H(\omega)|^{-2}$ can be found by recourse to eqn (10.68) as:

$$|H(\omega)|^{-2} = (\omega_0^2 - \omega^2)^2 + (2\omega\omega_0\zeta)^2$$

(10.97)

As $t \to 0$ in eqn (10.96), $\sigma_x^2(0) \to 0$ as expected. Furthermore, as $t \to \infty$, $\sigma_x^2(\infty) \to \int_0^\infty S_{ff}(\omega)\, d\omega/|H(\omega)|^{-2}$, a result in agreement with harmonic (i.e. steady state) analysis of the SDOF system that was also recovered in conjunction with eqn (10.87). Finally, a common approximation for a lightly damped SDOF system is to set $S_{ff}(\omega) = 2S_0/\pi$, as shown in Figure 10.8. In that case,

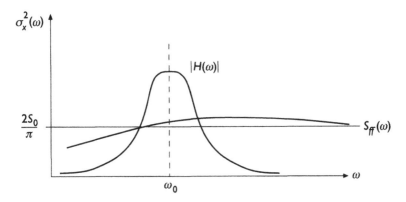

Figure 10.8 Random SDOF system response.

$\sigma_x^2(\infty) = S_0/(2\zeta\omega_0^3)$, a result for stationary conditions that can be found in many references (Hinch, 1991).

10.4 STRUCTURES WITH UNCERTAIN PROPERTIES

10.4.1 Static analysis

So far, we have examined the case where a structure is deterministic and its excitation is random. We will now look at a FEM formulation for stochastic cases where randomness can be expressed in the general form $\chi = \chi^0(1 + \gamma)$, with χ^0 being the deterministic value of a material property (such as the elastic modulus) or a structural component (such as the moment of inertia of a member) and γ being a random, zero mean small fluctuation about χ^0. Following (Vanmarke *et al.*, 1986), we will utilize the FEM stiffness approach which gives the following system of algebraic equations for the static case:

$$[K]\{x\} = \{f\} \tag{10.98}$$

As before, $[K]$ is the $N \times N$ stiffness matrix and $\{x\}$ and $\{f\}$ are $N \times 1$ column vectors containing the nodal displacements and forces, respectively. The important distinction to be made here is that the uncertainty in the structure is reflected in the stiffness matrix and, upon solution, on the nodal displacements. Also, since the case of random input was examined in the previous sections, $\{f\}$ is assumed to be deterministic here.

The stiffness matrix can now be expanded about the uncertainty using Taylor series as:

$$[K] = [K^0] + \sum_{i=1}^{n} [K]_{,i}\gamma_i + 0.5 \sum_{i=1}^{n} \sum_{j=1}^{n} [K]_{,ij}\gamma_i\gamma_j \tag{10.99}$$

where n denotes the total number of random parameters γ_i. As before, superscript denotes a deterministic quantity, while the next two terms in the expansion respectively denote first and second rates of change which are evaluated by differentiating $[K]$ with respect to the random parameters γ_i. Note that the use of commas indicates partial differentiation with respect to the subscript that follows.

The same type of expansion can also be used for the displacements, that is:

$$\{x\} = \{x^0\} + \sum_i \{x\}_{,i} + 0.5 \sum_i \sum_j \{x\}_{,ij}\gamma_i\gamma_j \qquad (10.100)$$

where the range of the summation indices is omitted for reasons of notational convenience. Substitution of the above two expansions in eqn (10.98) and a subsequent perturbation-type ordering of the terms gives the following system of equations:

$$\left.\begin{array}{l} [K^0]\{x^0\} = \{f\} \\[2mm] [K^0]\{x\}_{,i} = -[K]_{,i}\{x^0\} \\[2mm] [K^0]\{x\}_{,ij} = -[K]_{,i}\{x\}_{,j} - [K]_{,j}\{x\}_{,i} - [K]_{,ij}\{x^0\} \end{array}\right\} \qquad (10.101)$$

The structure of the above system of equations is similar to that of eqn (10.80) which was obtained for non-linear systems in Section 10.3.4 using perturbations. Thus, all unknown displacement terms can be obtained sequentially, starting from the deterministic solution $\{x^0\}$ and substituting the newly found terms in the right-hand side of the next equation. As a result, the deterministic stiffness matrix needs to be inverted only once, resulting in an efficient solution scheme. Also, the non-zero terms in $[K]_{,i}$ and $[K]_{,ij}$ are relatively few so that the right-hand sides can be quickly formed. The same approach can be used for problems involving lack of fit in structural members by introducing the concept of initial strains, as well as for structures on an elastic foundation with uncertain foundation modulus by introducing the foundation reaction matrix. Finally, uncertainty in the boundary conditions can be accounted for by inserting virtual springs at the boundaries and taking the spring constants as uncertain.

Following the displacement solution, the unknown stress tensor on any point within a finite element can be found after the stress terms $\{\sigma^0\}$, $\{\sigma\}_{,i}$ and $\{\sigma\}_{,ij}$ have been evaluated in the usual way from their corresponding displacement terms $\{x^0\}$, $\{x\}_{,i}$ and $\{x\}_{,ij}$. Thus, the final expression for the stress tensor is:

$$\{\sigma\} = \{\sigma^0\} + \sum_i \{\sigma\}_{,i} + 0.5 \sum_i \sum_j \{\sigma\}_{,ij} \qquad (10.102)$$

Based on the above equation, the expectation and variance of the stresses are:

$$E[\{\sigma\}] = \{\sigma^0\} + 0.5 \sum_i \sum_j \{\sigma\}_{,ij} E[\gamma_i\gamma_j] \qquad (10.103)$$

and

$$E[\{\sigma\}\{\sigma\}^T] = \sum_i \sum_j \{\sigma\}_{,i}\{\sigma\}_{,j} E[\gamma_i\gamma_j] + \sum_i \sum_j \sum_k \{\sigma\}_{,i}\{\sigma\}_{,jk}^T E[\gamma_i\gamma_j\gamma_k]$$

$$+ 0.25 \sum_i \sum_j \sum_k \sum_l \{\sigma\}_{,ij}\{\sigma\}_{,kl}^T (E[\gamma_i\gamma_j\gamma_k\gamma_l]$$

$$- E[\gamma_i\gamma_j]E[\gamma_k\gamma_l]) \tag{10.104}$$

respectively. The second moments of the random variables γ_i are related to the power spectrum $S_{\gamma\gamma}(k)$ via the Wiener–Khinchine relation (Lampard, 1954) as:

$$E[\gamma_i\gamma_j] = R_{\gamma\gamma}(r) = \int_{-\infty}^{\infty} S_{\gamma\gamma}(k)\exp(i2\pi kr)\,dk \tag{10.105}$$

where r is the distance between nodal co-ordinates and k denotes the wave number.

Since local changes in a structural parameter cause non-linear changes in the structural response, a second order Taylor series expansion such as the one used here is necessary to cover such non-linearities. Third order expansions are preferable, but computation becomes prohibitively expensive since sixth moments of the random variables γ_i are necessary for compatibility in the computation of stress variances.

10.4.2 Dynamic analysis

As a first step, we consider the eigenvalue problem:

$$([K] - \lambda[M])\{\varphi\} = \{0\} \tag{10.106}$$

where $[M]$ is the mass matrix, λ are the eigenvalues and $\{\varphi\}$ are the eigenvectors. As before, uncertainty in the stiffness and mass matrices filters, upon solution, to the eigenproperties of the structure. We begin (Liu *et al.*, 1986) by expanding both eigenvalues and eigenvectors in a Taylor series about the randomness γ as:

$$\lambda = \lambda^0 + \sum_i \lambda_{,i}\gamma_i + 0.5\sum_i \sum_j \lambda_{,ij}\gamma_i\gamma_j \tag{10.107}$$

and

$$\{\varphi\} = \{\varphi^0\} + \sum_i \{\varphi\}_{,i}\gamma_i + 0.5\sum_i \sum_j \{\varphi\}_{,ij}\gamma_i\gamma_j \tag{10.108}$$

respectively. Substitution of the above two expressions in eqn (10.106) along with eqn (10.99) and a similar expansion for the mass gives, after the usual perturbation-type ordering, the following system of equations:

$$([K^0] - \lambda^0[M^0])\{\varphi^0\} = [H^0]\{\varphi^0\} = \{0\}$$

$$[H^0]\{\varphi\}_{,i} = -([K]_{,i} - \lambda^0[M]_{,i} - \lambda_{,i}[M^0])\{\varphi^0\}$$

$$[H^0]\{\varphi\}_{,ij} = -([K]_{,i} - \lambda^0[M]_{,i} - \lambda_{,i}[M^0])\{\varphi\}_{,j}$$
$$- ([K]_{,j} - \lambda^0[M]_{,j} - \lambda_{,j}[M^0])\{\varphi\}_{,i}$$
$$- ([K]_{,ij} - \lambda_{,i}[M]_{,j} - \lambda_{,j}[M]_{,i}$$
$$- \lambda_{,ij}[M^0] - \lambda^0[M]_{,ij})\{\varphi^0\}$$

$$\tag{10.109}$$

By taking advantage of symmetry in $[H^0]$, $\lambda_{,i}$ can be computed from the second of eqns (10.109) as:

$$\lambda_j = \{\varphi^0\}^T([K]_{,i} - \lambda^0[M]_{,i})\{\varphi^0\}/(\{\varphi^0\}^T[M]\{\varphi^0\}) \tag{10.110}$$

Determination of $\{\varphi\}_{,i}$ from the second of eqns (10.109) is not, however, feasible because of the singularity of $[H^0]$. To overcome this drawback, a reduction in the rank of $[H^0]$ is necessary. The same situation holds for the evaluation of $\{\varphi\}_{,ij}$ since only the right-hand side of eqns (10.109) changes. As with the static case, each new eigenvalue solution depends on the previously obtained eigenproperties.

As far as time history analyses are concerned, the most rational approach is to go to a modal co-ordinate environment and assume that properties such as the modal damping ratios ζ are uncertain. Although this ignores the fact that uncertainty is first manifested at the physical co-ordinate level in terms of uncertain stiffness and mass, the convenience of decoupled modal equations is too tempting to ignore.

The analysis at the modal co-ordinate level is essentially the same as the perturbation approach used in Section 10.3.4 for a non-linear SDOF system under random input. In particular, the modal damping ratio is written as:

$$\zeta_i = \zeta_i^0(1 + \gamma_i), \quad i = 1, 2, \ldots, N \tag{10.111}$$

where i is a modal DOF, while the modal co-ordinate y_i is expanded as

$$y_i(t) = y_i^0(t) + y_i^1(t)\gamma_i + y_i^2(t)\gamma_i^2 \tag{10.112}$$

where superscripts (1), (2) on y respectively denote first and second order perturbation terms which are random processes. Substitution of the above two expansions in the ith uncoupled equation of motion (see eqn (10.39)) gives the following system of equations:

$$\ddot{y}_i^0 + 2\zeta_i^0\omega_i\dot{y}_i^0 + \omega_i^2 y_i^0 = -2\{\varphi_i^0\}^T[M^0]\{\ddot{x}_g\}$$

$$\ddot{y}_i^1 + 2\zeta_i^0\omega_i\dot{y}_i^1 + \omega_i^2 y_i^1 = -2\zeta_i^0\omega_i\dot{y}_i^0$$

$$\ddot{y}_i^2 + 2\zeta_i^0\omega_i\dot{y}_i^2 + \omega_i^2 y_i^2 = -2\zeta_i^0\omega_i\dot{y}_i^1$$

$$\tag{10.113}$$

Numerical integration of the above system can proceed without difficulties. Following solution for all expansion terms of $y_i(t)$, one may return to physical co-ordinates (see eqn (10.37)) and apply the statistical averaging using the expectation to find the response statistics.

10.5 REFERENCES

Augusti, G., Barrata, A. and Casciati, F. (1984) *Probabilistic Methods in Structural Engineering*, Chapman and Hall, London.

Caughey, T. K. (1963) 'Derivation and application of the Fokker–Planck equation to discrete nonlinear dynamic systems subjected to white noise excitation', *Journal of the Acoustical Society of America* **35**(11): 1683–92.

Caughey, T. K. (1971) 'Nonlinear theory of random vibrations', *Advances in Applied Mechanics* **11**: 209–53.

Coddington, E. A. and Levinson, N. (1955) *Theory of Ordinary Differential Equations*, McGraw-Hill, New York.

Crandall, S. H. (1963) 'Zero crossings, peaks and other statistical measures of random response', *Journal of the Acoustical Society of America* **35**(11): 1693–9.

Crandall, S. H. and Mark, W. D. (1963) *Random Vibration in Mechanical Systems*, Academic Press, New York.

Ghanem, R. and Spanos, P. D. (1991) *Stochastic Finite Elements: A Spectral Approach*, Springer-Verlag, New York.

Hinch, E. J. (1991) *Perturbation Methods*, Cambridge University Press, Cambridge.

Hurty, W. C. and Rubinstein, M. F. (1964) *Dynamics of Structures*, Prentice-Hall, Englewood Cliffs, NJ.

Klieber, M. and Hien, T. D. (1992) *The Stochastic Finite Element Method*, John Wiley, New York.

Lampard, D. G. (1954) 'Generalization of the Wiener–Khintchine theorem to nonstationary processes', *Journal of Applied Physics* **25**: 802–3.

Liu, W. K., Belytscko, T. and Mani, A. (1986) 'Random field finite elements', *International Journal for Numerical Methods in Engineering* **23**: 1831–45.

Nigam, N. C. (1983) *Introduction to Random Vibrations*, MIT Press, Cambridge, MA.

Vanmarke, E., Shinozuka, M., Nakagiri, S., Schueller, G. I. and Grigoriou, M. (1986) 'Random fields and stochastic finite elements', *Structural Safety* **3**: 143–66.

Zayed, A. I. (1996) *Handbook of Function and Generalized Function Transforms*, CRC Press, Boca Raton, FL.

Index

elastic spectra (of seismic motion) 125,
128–130
inelastic spectra 140–3
site specific spectra 137–40
uniform hazard spectra 125, 133, 135
Restitution 330
Return period 24–5, 136, 161
Reynolds number 86–7, 92, 187
Rhythmic motion 285, 299
Rice's formula 15
Rigid foundation 47
Rigid impactor 251
Rise and decay functions 234–5, 238, 255–7,
259
Robustness 233, 275–8
Rock crushers 323, 327
Rolling mills 331
Ronan Point 232–3, 253, 275
Rotating machines 323, 326, 340
Running 285

Safety margin 5, 13, 18
Scabbing 247, 252
Scanlan's formulation (aeroelasticity) 104
Scruton number 83–4, 89–91, 99
Sea state 183
Seismic hazard assessment 122–5, 127
Seismic zones 125, 334
Seismic waves 110, 112
Semi-infinite (media) 333–4, 339
Sensitivity factors 9
Shear response (bridges) 315–6
Ship impact 279–80
Shock mitigating (absorbing) systems
271–2, 274–5
Shock wave 234–6, 238–9, 241–6
Short circuit torque 323, 326
Single degree-of-freedom system 31, 350
Site effects 120–121
Size number 86, 90, 93
Slurry infiltrated concrete 275
Soft impact 249
Software for dynamic analysis 64
Soil-structure interaction 334
Spatial variability (of seismic motion) 122,
154
Spectral analysis 349
Spring-damper units 334
Spectral sequence (Davenport's algorithm)
74–5
Standard normal variable 6, 8, 12
Staircases 285, 298

Stationary process 348, 357
Statistical average 347
Stiffness, stiffness matrix 33, 264, 303, 337,
350, 352
Stochastic system 343
Strain rate (high) 234, 247, 261, 272–4, 276,
278
Stress waves 247–8, 271
Structural integrity 275–6, 278
Sub-matrices 337
Swaying 303
Synchronized motion 296

Table-top foundations 331
Tectonic plates 110
Tension-leg platform 212
Terrain roughness 67–8, 71
Theodorsen function 102, 105
Threshold of perception 291, 292
Time-history (time domain) analysis 52,
155–9, 169, 204, 314–5, 318
TNT equivalence 238–9
Toe-off 286, 298, 299
Topography effects 122
Torsion effects (seismic) 150, 154
Transient load 36
Transient shocks 323, 328
Transient torques 326
Transmission 334, 340
Tsunamis 109
Tup 329
Turbines 326

UBC provisions 136–7, 139–40, 148, 150–2,
154–5, 159, 169
Unbalanced masses 323, 325
Uncertain properties 363
Unit impulse response 351, 353
Unit observation time 23, 25

Variability 254, 278–9
Variance 357
Vehicle suspension 311–2
Velocity potential 180
Venting 246, 280
Vibrating screens 330
Viscous forces 186
Vortex shedding
countermeasures 85, 94
deterministic models 83–4
stochastic models 88–9